Soil Biology

Series Editor: Ajit Varma

2

Springer

Berlin
Heidelberg
New York
Hong Kong
London
Milan
Paris
Tokyo

Ajay Singh • Owen P. Ward (Eds.)

Biodegradation and Bioremediation

With 40 Figures and 12 Tables

Dr. AJAY SINGH
Director R&D
Petrozyme Technologies Inc.
7496 Wellington Road 34, R.R.#
Guelph, Ontario N1H 6H9
Canada

e-mail:
asingh@petrozyme.com

Adjunct Faculty Member
Department of Biology
University of Waterloo
Waterloo, Ontario N2L 3G1
Canada

e-mail:
asingh@sciborg.uwaterloo.ca

Dr. OWEN P. WARD
Professor of Microbial
Biotechnology
Department of Biology
University of Waterloo
Waterloo, Ontario N2L 3G1
Canada

e-mail:
opward@sciborg.uwaterloo.ca

ISSN 1613-3382
ISBN 3-540-21101-2 Springer-Verlag Berlin Heidelberg New York

Library of Congress Cataloging-in-Publication Data

Biodegradation and bioremediation / Ajay Singh, Owen P. Ward, eds.
p. cm. – (Soil biology ; v. 2)
Includes bibliographical references and index.
ISBN 3-540-21101-2 (alk. paper)
1. Bioremediation. 2. Hazardous wastes – Biodegradation. I. Singh, Ajay, 1963– II. Ward, Owen P., 1947– III. Series

TD192.5.B544 2004
628.5′5 – dc22

Springer-Verlag is a part of Springer Science + Business Media
springeronline.com

Springer-Verlag Berlin Heidelberg 2004
Printed in Germany

Cover design: Design & Production, Heidelberg
Typesetting: SNP Best-set Typesetter Ltd., Hong Kong
31/3150-WI – 5 4 3 2 1 0 – Printed on acid-free paper

Preface

The diverse metabolic capabilities of microorganisms and their interactions with complex organic and inorganic substrates have long been recognized and are now being exploited for the treatment of hazardous soil contaminants. Microbial reactions range from single-step transformations of pollutants into more or less toxic forms (biodegradation) to complete mineralization, yielding water and either carbon dioxide or methane. Microbial processes are environmentally compatible, operate under mild conditions and can be integrated with non-biological processes with treatment goals including detoxification, destruction and immobilization of the contaminants. The application of biological methods, through bioremediation, has been used successfully for remediation of soils with a variety of pollutants.

Biodegradation and Bioremediation, Volume 2 of the series *Soil Biology*, presents a selection of contributions related to microbiological and biochemical processes with an emphasis on their use in bioremediation. Topics include bioavailability and biodegradation, anaerobic biodegradation of environmental pollutants, microbial community dynamics during bioremediation of hydrocarbons, biodegradation of halogenated compounds, polycyclic aromatic hydrocarbons and nitrogen-containing xenobiotics. The roles of aromatic hydrocarbon dioxygenases and bacterial reductive dehalogenases and engineering of improved biocatalysts in bioremediation are discussed. Innovative methods for monitoring bioremediation processes and approaches for combined biological and abiological degradation of xenobiotic compounds are also included. Volume 1 of the series *Soil Biology*, *Applied Bioremediation and Phytoremediation*, included topics related to the applied aspects of bioremediation and phytoremediation technologies.

This book contains contributions from experts in the area of environmental microbiology, biotechnology and bioremediation. The authors are from diverse institutions, universities, government laboratories and industry, with basic, applied and industrial research background. This book should prove to be useful to under- and post-graduate students of biotechnology, microbiology, biochemistry, and soil and environmental sciences and engineering. We hope that teachers, scientists and engineers, whether in academia, industry or government, will find the contents, including its basic and practical aspects, helpful.

We gratefully acknowledge the continuous support of all the contributing authors, help and encouragement provided by our colleagues and valuable guidance provided by Dr. Ajit Varma, Editor-in-Chief of the series *Soil Biology*, and Dr. Jutta Lindenborn, Springer-Verlag, during the preparation of this volume and at various stages of this editorial work.

Guelph, Waterloo, Ontario, April 2004
Ajay Singh and Owen Ward

Contents

List of Contributors

Michael D. Aitken
Department of Environmental Sciences and Engineering School of Public Health, The University of North Carolina, Chapel Hill, North Carolina 27599-7400, USA

Wilfred Chen
Department of Chemical and Environmental Engineering University of California, Riverside, CA 92521, USA

Roland L. Crawford
Environmental Biotechnology Institute, University of Idaho, Moscow, 83844-1052, Idaho, USA

John D. Coates
Department of Plant and Microbial Biology 271 Koshland Hall, University of California, Berkeley Berkeley, CA 94720, USA

E. Ann Greene
Department of Biological Sciences, University of Calgary, 2500 University Drive NW, Calgary, Alberta, T2N 1N4, Canada

Marc B. Habash
Depatment of Environmental Biology, University of Guelph, Guelph, Ontario, N1G 2W1, Canada

Thomas F. Hess
Biological and Agricultural Engineering Department, 401 Engineering & Physics, University of Idaho, Moscow, Idaho 83844-0904, USA

Ramesh C. Kuhad
Department of Microbiology, University of Delhi South Campus, New Delhi-110 020, India

Hung Lee
Department of Environmental Biology, University of Guelph, Room 3218, Bovey Building, Guelph, Ontario Canada N1G 2W1

Thomas C. Long
Department of Environmental Sciences and Engineering, School of Public Health, University of North Carolina, Chapel Hill, NC 27599-7431, USA

William W. Mohn
Department of Microbiology & Immunology, University of British Columbia, 300-6174 University Boulevard, Vancouver, British Columbia V6T 1Z3, Canada

Ashok Mulchandani
Department of Chemical and Environmental Engineering, University of California, Riverside, CA 92521, USA

Rebecca E. Parales
Section of Microbiology 266 Briggs Hall, University of California Davis, CA 95616, USA

Andrzej Paszczynski
Environmental Biotechnology Institute, Food Research Center 103, University of Idaho, Moscow, Idaho 83844-1052, USA

Sol M. Resnick
Biotechnology R&D, The Dow Chemical Company, San Diego, CA 92121, USA

Ajay Singh
Petrozyme Technologies Inc., 7496 Wellington Road 34, R.R. #, Guelph, Ontario N1H 6H9, Canada

Jack T. Trevors
Depatment of Environmental Biology, University of Guelph, Guelph, Ontario, N1G 2W1, Canada

Jonathan D. Van Hamme
Department of Biology, The University College of the Cariboo, 900 McGill Road, Kamloops B.C., V2C 5N3, Canada

Gerrit Voordouw
Department of Biological Sciences, University of Calgary, 2500 University Drive NW, Calgary, Alberta T2N 1N4, Canada

Owen P. Ward
Department of Biology, University of Waterloo, Waterloo, Ontario N2L 3G1, Canada

Jing Ye
Department of Biology, University of Waterloo, Waterloo, Ontario N2L 3G1, Canada

1 Biotechnology and Bioremediation – An Overview

Ajay Singh[1] and Owen P. Ward[2]

1 Introduction

The large-scale manufacturing, processing and handling of chemicals have led to serious surface and subsurface soil contamination with a wide variety of hazardous and toxic hydrocarbons. Many of the chemicals, which have been synthesized in great volume, including polychlorinated biphenyls (PCBs), trichloroethylene (TCE) and others, differ substantially in chemical structure from natural organic compounds and are designated as xenobiotics because of their relative recalcitrance to biodegradation. Other compounds, for example the polycyclic aromatic hydrocarbons (PAHs), are also toxic and the high molecular weight PAHs (having four or more fused rings) are typically recalcitrant to biodegradation. The latter compounds, the products of incomplete combustion of natural organic materials and hydrocarbons, occur in the soil environment as a result of naturally ignited forest fires. However, intensification of energy-related and other industrial processes with associated production of wastes and by-products, rich in PAHs, has led to serious soil contamination of many industrial sites. The resultant accumulations of the various organic chemicals in the environment, particularly in soil, are of significant concern because of their toxicity, including their carcinogenicity, and also because of their potential to bioaccumulate in living systems. A wide variety of nitrogen-containing industrial chemicals are produced for use in petroleum products, dyes, polymers, pesticides, explosives and pharmaceuticals. Major chemical groups involved include different nitroaromatics, nitrate esters and nitrogen-containing heterocycles. Many of these chemicals are toxic and threaten human health and are classified as hazardous by the United States Environmental Protection Agency.

[1] Petrozyme Technologies Inc., 7496 Wellington Road 34, R.R. #3, Guelph, Ontario N1H 6H9, Canada, e-mail: asingh@petrozyme.com, Tel.: 519-767-2299, Fax: 519-767-9435
[2] Department of Biology, University of Waterloo, Waterloo, Ontario N2L 3G1, Canada

Soil Biology, Volume 2
Biodegradation and Bioremediation
(ed. by. A. Singh and O. P. Ward)

Interest in bioremediation of polluted soil and water has increased in the last two decades primarily because it was recognized that microbes were able to degrade toxic xenobiotic compounds which were earlier believed to be resistant to the natural biological processes occurring in the soil. Although microbial activity in soil accounts for most of the degradation of organic contaminants, chemical and physical mechanisms can also provide significant transformation pathways for these compounds (Rogers 1998). Bioremediation is generally considered a safe and less expensive method for the removal of hazardous contaminants and production of non-toxic by-products (Providenti et al. 1993; Ward et al. 2003).

There have been many experimental successes with the more difficult to degrade contaminants, but there have also been many notable failures. However, it has been suggested that, although microorganisms have the primary catalytic role in bioremediation, our knowledge of the alterations occurring in the microbial communities remains limited and the microbial community is still treated as a "black box" (Iwamoto and Nasu 2001; Dua et al. 2002). Put in a more positive light, bioremediation remains a developing field, largely because it has traditionally been carried out in a natural environment where many of the organisms are uncharacterized and because no two environmental projects are identical (Watanabe 2001; Verstraete 2002).

Biotechnology has the potential to play an immense role in the development of treatment processes for contaminated soil. As with any microbial process, optimizing the environmental conditions in bioremediation processes is a central goal in order that the microbial, physiological and biochemical activities are directed towards biodegradation of the target contaminants. Environmental factors influencing microbial growth and bioactivity will include moisture content, temperature, pH, soil type, contaminant concentrations and oxygen for aerobic degradation and redox potential for anaerobic degradation. Deviations of these parameters away from optimal conditions will reduce rates of microbial growth and transformation of target substrates and perhaps cause premature cessation and failure of the bioremediation process. Biodegradation potential may also be limited by the toxicity of the pollutants to the degrading microbes. Some species have developed cellular defenses enabling them to tolerate high concentrations of toxic contaminants.

Understanding the biochemical and physiological aspects of bioremediation processes will provide us with the requisite knowledge and tools to optimize these processes, to control key parameters and to make the processes more reliable. Since the majority of bioremediation processes rely on the activities of complex microbial communities, we have much to learn about the interactive and interdependent roles played by individual species in these communities. We need to develop strategies for improving the bioavailability of the many hydrophobic contaminants which have an extremely low water solubility and tend to be adsorbed by soil particles and persist there. We need to continue to elucidate the complex aerobic and anaerobic metabolic pathways which microbes have evolved to degrade organic contaminants and to understand the nature of rate-limiting steps, bottlenecks and underlying genetic and biochemical regulatory mechanisms. We need to continue to characterize many of the key enzymatic reactions that participate in contaminant transformation and to relate contrasting reaction rates, substrate specificities and enzyme mechanisms to differences in protein structures. Such new knowledge can provide us with the requisite information to test, design and engineer biocatalysts with improved substrate specificities, reaction rates or other desired catabolic properties and ultimately to engineer improved catabolic pathways for bioremediation. We must recognize that some chemical species are inherently intractable to enzyme transformation and we should be open to the possibility of combining chemical or physical strategies with biological systems to achieve overall effective remediation. We must also continue to devise better methods for monitoring and assessing the progress and effectiveness of microbial biodegradation processes at both the research and process implementation level. Clearly, the availability of advanced molecular techniques provides a new impetus and enhances our abilities to address many of these issues.

These topics, introduced below, represent the main focus of this book.

2 Microbial Communities and Bioremediation

Since microbial communities play a significant role in biogeochemical cycles, it is important to analyze the community structure and its changes during bioremediation processes. The challenge of characterizing the roles of a range of hydrocarbon-metabolizing organisms in degrading the myriad of petroleum substrates present in hydrocarbon-contaminated soils is clearly substantial. However, such studies can provide major insights into important biochemical and physiological aspects of bioremediation and microbial catabolism. Culture-dependent and -independent methods are being applied to microbial community characterization. The temporal and spatial changes in bacterial populations and the diversity of the microbial community during bioremediation can be determined using sophisticated molecular methods (Stapleton et al. 1998; van Elsas et al. 1998; Widada et al. 2002). Recent advances in molecular techniques, combined with genomic information, are greatly assisting microbiologists in unraveling some of the mysteries related to the diverse roles of microbes in these communities. One of the exciting outcomes of these studies is an advance in our understanding of the degree and importance of lateral gene transfer in complex microbial communities. Catabolic genes have the ability to spread through a microbial community at high frequencies (Top and Springael 2003; Van der Meer and Senchilo 2003). An assessment of microbial community dynamics during hydrocarbon bioremediation is presented in Chapter 2.

3 Contaminant Bioavailability

For efficient microbial degradation of chemical contaminants to occur, the contaminants must be bioavailable to the degrading cells. The biodegradation rate of a contaminant depends on the rate of contaminant uptake and mass transfer. Bioavailability of a contaminant in soil is influenced by a number of factors such as desorption, diffusion and dissolution. The decrease in bioavailability due to long-term contamination of soil, often referred to as aging or weathering, is a result of chemical oxidation reactions and slow chemical diffusion of the contaminant into small pores incorporating contaminants into the organic matter. Use of chemical or bio-surfactants during the biodegradation process helps overcome bioavailability problems (Van Hamme et al. 2003). The molecular structure of the contaminant and hydrophobicity may also affect the pollutant uptake by the microorganisms. Indeed, the cells may also have active or selective systems for transporting the contaminants into the cell. Given that many of these contaminants have low solubility in aqueous media, understanding mechanisms of their uptake by the degrading microbes and developing strategies to promote or accelerate their accession represent important aspects of effective bioremediation processes. Chapter 3 provides a critical analysis of the bioavailability of organic pollutants. The hydrophobicity/low water solubility properties of PAHs cause them to associate with hydrophobic components in soil, thereby limiting their accession to microorganisms. This topic is discussed in detail in Chapter 5.

4
Microbial Catabolism of Organic Pollutants

Biodegradation involves the breakdown of organic compounds either through biotransformation into less complex metabolites or through mineralization into inorganic minerals, H_2O, CO_2 (aerobic) or CH_4 (anaerobic). Both bacteria and fungi have been extensively studied for their ability to degrade a range of environmental pollutants including recalcitrant polycyclic aromatic hydrocarbons, halogenated hydrocarbons and nitroaromatic compounds. The biochemical pathways/enzymes required for the initial transformation stages are often specific for particular target environmental contaminants, converting them to metabolites which can be assimilated into more ubiquitous central bacterial pathways. An overview of some of the biodegradation systems used by microorganisms in the catabolism of key organic contaminants in soil is presented in Table 1. The extent and rate of biodegradation depend on many factors including pH, temperature, oxygen, microbial population, degree of acclimation, accessibility of nutrients, chemical structure of the compound, cellular transport properties, and chemical partitioning in growth medium. Figure 1 provides a schematic of aspects of the biodegradation system involved in bioremediation.

Some recalcitrant chemicals contain novel structural elements that seldom occur in nature and which may be incompletely transformed as microbes lack the degradative pathway for complete degradation of these xenobiotics. While microbes may not have the metabolic pathways for mineralization of certain newly introduced synthetic chemicals, there is evidence that microorganisms have the capacity to evolve such catabolic systems over time. In bioremediation processes, it is generally an objective to exploit microbial technology to accelerate the rate of pollutant removal.

Many contaminants in soil exist in anaerobic environments. A couple of decades ago, by observing the anaerobic dechlorination of PCBs over time in Hudson River sediments, it became clear that microbes could transform contaminants under anaerobic conditions. By the late 1980s, there was conclusive evidence that hydrocarbons could be degraded in the absence of oxygen. These anaerobic degradation systems required terminal electron acceptors such as iron (III), manganese oxide or nitrate to replace that function of oxygen in aerobic systems. We have now entered a period of intensive research and discovery focused on the catalytic mechanisms which facilitate the anaerobic catabolism of pollutants. Anaerobic aspects of hydrocarbon bioremediation are discussed in detail in Chapter 4. As is mentioned above, anaerobic processes are

Table 1. Overview of some of the biodegradation systems used by microorganisms in catabolism of key organic contaminants in soil

Contaminant	Catabolism	References
Petroleum hydrocarbons	*Alkanes (aerobic)* – Monooxygenase/hydroxylase-mediated conversion of alkanes to alcohols with subsequent aldehyde and acid formation prior to b-oxidation (example: OCT plasmid of *Pseudomonas putida* Gpo1) – Also a dioxygenase-mediated conversion of alkanes to aldehydes (e.g. *Acinetobacter* sp. M1) – Other mechanisms initiated by desaturation reactions	van Beilen et al. (1994); Atlas and Cerniglia (1995); Ward et al. (2003)
	Monoaromatics (aerobic) Oxygenase-mediated attack of the aromatic ring typically producing a dihyrodiol, with subsequent conversion to catechol, oxygenase-mediated catechol ring cleavage with production of muconic acid-type metabolites which can be degraded via the TCA cycle	Prince (1998); see also Parales and Resnick (Chap. 8)
	Polycyclic aromatic hydrocarbons (PAHs) (aerobic) – Bacteria: complex initial PAH ring cleavage reactions involving monooxygenases or combinations of mono- and dioxygenases – Fungi: ligninolytic and non-ligninolytic PAH co-oxidation	Cerniglia (1992, 1997); Paszczynski and Crawford (1995); see also Aitken and Long (Chap. 5)
	Anaerobic Degradation – Conversions of alkanes, monoaromatics, some PAHs and hydrocarbon mixtures under Fe(III), Mn(IV) or sulfate-reducing or -denitrifying conditions or using other terminal electron acceptors including CO_2 – Initial activation mechanisms involving carboxylation, methylation, hydroxylation, dehydrogenation and addition reactions (especially fumarate addition)	Lovely (2000); Widdel and Rabus, (2001); see also Coates (Chap. 4)
Chlorinated hydrocarbons	*Chlorinated aliphatics* – Aerobic transformations by oxidations, typically by hydroxylation or substitution of a hydroxyl group on the molecule or, in the case of unsaturated chlorinated aliphatics, by epoxidation – Anaerobic transformations are typically reductions such as hydrogenolysis, substitution of a chlorine atom by a hydrogen atom or elimination of chlorine atoms from adjacent carbons with formation of a C=C bond	Janssen et al. (1994); Peyton et al. (1995); see also Habash et al. (Chap. 9) and Mohn (Chap. 6)

	Mono- and poly-chlorinated phenols (chlorinated aromatics) Pentachlorophenol (PCP) – Aerobic production of tetrachlorohydroquinone mediated by NADPH-dependent oxygenolytic dechlorination or hydrolytic mechanisms, followed by its further catabolism and mineralization – Non-specific degradation by fungal ligninases – Anaerobic conversion of chlorinated phenols via one or more reductive dehalogenation reactions, leading to phenol which may be mineralized to methane and CO_2	Kennes et al. (1996); McAllister et al. (1996); see also Habash et al. (Chap. 9) and Mohn (Chap. 6)
	Polychlorinated biphenyls (chlorinated aromatics) – Aerobic bacterial dioxygenase-mediated attacks, most commonly by 2,3 dioxygenase attack at an *ortho* chlorinated carbon with the resulting dihydroxy metabolites transformed via *meta*-cleavage products to chlorobenzoate. Attacks at other unsubstituted positions, by different dioxygenases, for example 3, 4, have also been characterized – Non-specific degradation by fungal ligninases – Extensive stepwise removal of *meta* and *para* chlorines by reductive dechlorination under anaerobic conditions typically produces mono- to trichlorinated biphenyls that are amenable to aerobic bacterial degradation	Abramowicz (1990); Paszczynski and Crawford (1995); see also Mohn (Chap. 6)
	2,4-D (2,4-dichlorophenoxy)acetate and related compounds – Aerobic degradation involves removal of acetate producing 2,4-dichlorophenol, which is converted to a dichlorocatechol followed by ring cleavage and mineralization – Fungal ligninolytic mineralization – Anaerobic degradation is initiated by reductive dechlorination	Singh et al. (1999); see also Mohn (Chap. 6)
	DDT [1,1,1-trichloro-2,2-bis(p-chlorophenyl)ethane] and related compounds – Aerobic degradation with initial ring hydroxylation – Anaerobic dechlorination	Foght et al. (2001); see also Mohn (Chap. 6)
Nitroaromatics	*Aerobic degradation* – Aromatic ring dioxygenation with release of nitrite and production of dihydroxy intermediates – Monooxygenation to epoxides – Formation of hydride-Meisenheimer complexes – Partial reduction of nitro group forms NH_3 and hydroxylaminobenzene, which rearranges to a catechol with release of another NH_3 molecule	Nishino et al (2000); Ye et al. (2003)

Table 1. *Continued*

Contaminant	Catabolism	References
	Nitrobenzene – Aerobically metabolized by a partial reduction route, being converted to hydroxylaminobenzene, 2-aminophenol with dioxygenase-mediated ring cleavage or by an initial dioxygenation, producing catechol – An anaerobic reduction route produces aniline	Hawari et al. (2000); Zhao et al. (2001)
	Trinitrotoluene (TNT) Initial reductive attack mechanisms appear to predominate (the nitro group makes TNT vulnerable to reductive attack and resistant to oxygen attack from aerobes), facilitating subsequent mono- and dioxygenase attack leading to ring fission	Esteve-Nunez et al. (2001); Popesku et al. (2003)
Organophosphate derivatives	Hydrolytic conversion of parathion and methyl parathion to dialkyl phosphates and 4-nitrophenol	Ye et al. (2003)
Nitrate esters	– Sequential esterase-initiated hydrolytic denitration conversions with nitrate release – Nitroreductase transformations with production of nitrite or – A glutathione transferase-like conversion with production of nitric oxide and nitrite	Christodoulatos et al. (1997)
Nitrogen heterocycles	*s-Triazine herbicides* Biodegradation by oxidative N-dealkylation of the side chain, dechlorination and ring cleavage	Ralebitso (2002); see also Ye et al. (Chap. 7)
	Hexahydro-1,3,5-trinitro-triazine – Aerobic or anaerobic denitration followed by hydrolytic ring cleavage and further metabolism or autodecomposition – Anaerobic reductive nitroso route followed by hydrolytic ring cleavage – Direct anaerobic hydrolytic ring cleavage	Halasz et al. (2002)
Plastics	– *Polyester:* many bacterial esterases degrade these polymers in nature – *Polyhydroxyalkanoates (PHA)*: degraded by bacterial PHA depolymerases – *Polylactic acid(PLA)*: biodegradation with protease/esterase activity – *Polyurethane(PUR)*: incomplete biodegradation widely reported, PUR does not support growth – *Polyvinyl alcohol(PVA)*: completely degraded by some *Pseudomonas* strains mediated by either dehydrogenase or oxidase-catalyzed reactions attacking the main C–C chain links – *High molecular weight Nylon 66*: degraded by white-rot fungi using Mn-peroxidase	Prijamabada et al. (1995); Deguchi et al. (1997); Li (1999); Shimao et al. (2000); Shimao (2001)

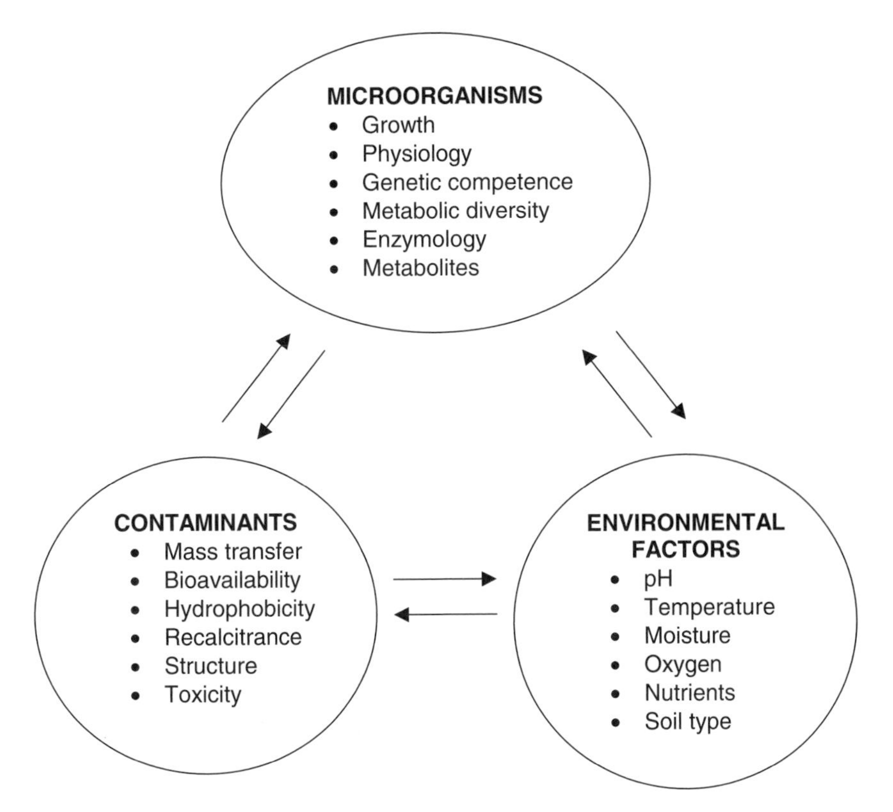

Fig. 1. The biodegradation system in bioremediation

very effective in removing chlorine atoms from PCBs and, indeed, in degrading other halogenated contaminants. These metabolic processes are described in Chapter 6 and one of the enzyme groups involved (reductive dehalogenase) is discussed in Chapter 9. Some anaerobic processes for degradation of nitrogen-containing contaminants are discussed in Chapter 7.

The inability of microbes to mineralize particular contaminants, but to partially transform them, typically means that these organisms require other substrates to support their growth. In such situations, the contaminants are transformed, i.e. 'cometabolism'. The diverse range of structures of PAH molecules requires the PAH-degrading microbes to have a range of enzymes capable of accepting the different PAHs as substrates or, alternatively, having a lesser number of enzymes having broad substrate specificity to PAHs. Consequently, in a particular bioremediation system, some of the more complex PAHs may not get transformed at all or may be partially catabolized into dead-end products. This topic is discussed with reference to PAHs in Chapter 5. Cometabolic aerobic and anaerobic processes also participate in the degradation of halogenated organic compounds. The early oxygenase enzymes for the nonhalogenated analogue appear often to accommodate the halogenated contaminants because of their broad substrate specificity. The capacities of biphenyl dioxygenases to co-oxidize certain PCB congeners and the roles of mono- and di-oxygenases in TCE and PCP degradation are described in Chapter 6.

While mechanisms for aerobic degradation of nitrogen-containing synthetic chemical contaminants have not been completely elucidated, much research is ongoing to determine the degradative mechanisms involved. Participating catalytic reaction types include deaminations, nitroreductions, N-dealkylations, deesterifications, dechlorinations and hydrolysis. There is considerable involvement of monooxygenases and dioxygenases, nitroreductases and esterases in the early stages of catabolism. Some of the main pathways are discussed in Chapter 7.

5 Properties of Some Important Catabolic Enzymes

Mono- and/or dioxygenases participate in many of the early reactions in aerobic bacterial transformations of alkanes, monoaromatics, PAHs, chlorinated hydrocarbons and the nitroaromatics (Wackett and Hershberger 2001; Parales et al. 2002). Many of the monocyclic and polycyclic aromatic compounds, including ring substitute derivatives, are also subject to attack by fungal ligninolytic or non-ligninolytic mechanisms. Many of these compounds may also be degraded under reducing or denitrifying conditions. Anaerobic transformations of chlorinated aliphatic and aromatic compounds, including highly chlorinated PCBs, are usually mediated by reductive dechlorination reactions. Nitroaromatic compounds and nitrogen heterocycles are also amenable to early stage reductive denitration attacks. For heavily substituted aromatics and heterocycles, a sequential anaerobic/aerobic treatment may yield maximum degradation. For example, highly chlorinated PCBs are more effectively remediated by preferential dechlorination under anaerobic conditions followed by oxygenase ring cleavage of the resulting lower chlorinated species under aerobic conditions.

As indicated in Table 1, dioxygenases catalyze the initial steps of biodegradation of an array of mono- and polycyclic aromatic hydrocarbons and related compounds. These multicomponent cofactor-requiring biocatalysts have been classified into different groups based on sequence alignments with further classifications based on phylogenetic analysis. Detailed studies on the biocatalytic mechanisms, structure–activity relationships, substrate specificities and approaches to engineering of aromatic hydrocarbon dioxygenases are described in Chapter 8.

Many toxic industrial halogenated chemicals, including polychlorinated biphenyls (PCBs), pentachlorophenol (PCP), chloroethenes and dichloromethane, threaten the environment and human health. Bacterial reductive, hydrolytic, dehydro- and thiolytic dehalogenases, together with some mono- and dioxygenases, methyl transferases constitute a family of enzymes which participate in the degradation of these halo-organic compounds. One of these groups, the reductive dehalogenases, catalyses the removal of either one halogen atom from the substrate replacing it with a hydrogen or removes two adjacent halogen atoms with double-bond formation. The molecular and functional aspects of these enzymes are discussed in Chapter 9.

6 Designing Microorganisms for Bioremediation

Combined physiological and genetic approaches provide us with a deeper understanding of so many aspects of the biodegradation of xenobiotic compounds. We are clearly in the midst of a biological revolution in this area of environmental microbiology (Head and Bailey 2003). Such research yields new information on the metabolic diversity of the biodegradative organisms, leading to the discovery of new metabolic routes and approaches to optimizing metabolic activities through manipulation of rate-limiting steps.

In order to design improved contaminant-degrading microbes, we need to understand current metabolic processes and perhaps create new metabolic routes which are characterized by improving substrate flux through the pathways so as to avoid accumulation of inhibitory intermediates. In addition, substrate specificities of the catabolic enzymes need to be broadened; cells need to have enhanced mechanisms for accessing some of the more hydrophobic contaminants. Also, newly created strains need to be genetically stable with respect to their biodegradative abilities. Such a rational optimization of known catabolic pathways and the creation of new pathways will rely on the rapidly growing knowledge based on biochemical/pathway genetics. This includes gaining more in-depth understanding of the factors regulating transcription and translation for biosynthesis of the key degradative enzymes and other participating proteins, as well as further characterization of enzyme-catalysed reaction rates and substrate specificities. The increasing information on the structure–function relationship of important catabolic enzymes, using protein engineering techniques, offers further possibilities to improve their activities. For xenobiotics, one strategy for designing superior biocatalysts is the rational combination of catabolic segments from different organisms within one recipient to create complete catabolic routes (Pieper and Reineke 2000).

Novel strategies are being investigated with a view to designing new biocatalysts for bioremediation. Microorganisms may be engineered for heavy metal bioremediation through expression of metal-binding proteins or enzymes that transform these metals into less soluble or less toxic forms. Examples of biocatalytic strategies being explored to enhance biodegradation of contaminants include enzymatic approaches to accelerating the degradation of organophosphates, engineering of recombinant strains for desulfurization of fossil fuels and for improved degradation of specific hydrocarbons. Approaches may be directed to increasing expression or modifying the specificity of a target enzyme or could involve optimization of an entire pathway through directed evolution. These topics are discussed in more detail in Chapter 10.

The introduction to or disposal of recombinant organisms in soil is a related concern, given the potential for recombinant strain proliferation, modification or recombinant gene transfer processes in the soil. One small outcome arising from this concern is that the potential to use recombinant organisms in bioremediation processes has been constrained. There is surely a need to understand better the nature of soils as host media for retention, survival and propagation of these biological organisms/vectors.

7 Combined Chemical/Physical and Biological Treatments

The inability of the degradative enzymes to catabolize specific target xenobiotic compounds is an important metabolic limitation of certain microbes and represents a perceived limitation of bioremedial technologies. However, wood-rotting fungi have evolved biocatalytic systems for transformation of the complex lignin structures in wood where enzymes may not participate directly in the initial transformation of these molecules. Rather, the fungi produce hydroxyl radicals or other reactive chemical species which attack the lignin by mechanisms which are poorly understood. In addition to degrading lignin, these fungal systems are able to transform a range of biologically recalcitrant synthetic chemicals, producing reaction products which are more amenable to further direct transformation by microorganisms. Some of the biocatalytic systems involved in these indirect transformations are described in Chapters 6, 7 and 11. Non-ligninolytic enzymes also participate in these types of reactions.

A variety of agents, promoting initial redox reactions with contaminants rendering them more amenable to further bio-oxidation or bioreduction, can also be used in combined chemical (or physical) and biological remediation strategies. These agents include ozone, Fenton's reagent, potassium permanganate or ferrate, zero valent iron and electrons generated, for example, by electron beam methods. The underlying principles of these methods and their applications are discussed in detail in Chapter 11.

8 Measuring Biodegradation Potential

The monitoring and assessment of a biological process are essential parts of bioremediation. Traditional methods are generally based on differential morphological, physiological and metabolic characteristics of microorganisms. These methods include isolation and cultivation of microbes on solid media, most probably number (MPN) assays, and more recently, BIOLOG substrate utilization patterns. The rates of contaminant degradation and formation of products are monitored. Respirometric tests and pollutant mineralization studies involving measurements of oxygen consumption or CO_2 production can provide useful information on the biodegradability potential of microorganisms. Bioassays can be used to measure the toxicity of a contaminant or inhibition of soil enzymes such as dehydrogenases, lipases, ureases, acid and alkaline phosphatases, arylsulfatases and catalases, which can be considered indicators for monitoring impacts of pollutants soils.

Pure culture experiments in the laboratory are essential for detailed analysis of the physiology, metabolism and genetics of microorganisms. However, physiological conditions in the laboratory culture differ generally from those in the natural environment. Molecular phylogenetic approaches such as denaturing gradient gel electrophoresis (DGGE), terminal restriction fragment length polymorphism (T-RFLP), fluorescence in situ hybridization (FISH) and polymerase chain reaction (PCR)-amplified 16sRNA have been used for the screening and isolation of dominant phylogenetic signatures from the contaminated sites (Watanabe and Hamamura 2003). Fragments of genes coding catabolic enzymes in the degradation pathways can be amplified from environmental DNA samples for analysis of pollutant-degrading microbial populations. Use of catabolic genes is considered an attractive method for analyzing the diversity of potential catabolic microbes and monitoring in situ pollutant degradation. Reverse sample genome probing (RSGP) has been used to track specific microorganisms in a quantitative manner (Hubert et al. 1999).

Microorganisms utilize various cooperative interactions with other species for their growth in the natural ecosystem. One population or a few different populations that best adapt to the environmental conditions become predominant, while others either exist as minorities or eventually die off. In situ biodegradation of contaminants can also occur through the detection of contaminant-specific transformation products, metabolic biomarkers (Smets and Pritchard 2003). Metabolic biomarkers have been detected during transformation of alkanes, aromatic hydrocarbons and explosives such as RDX (Gieg and Suflita 2002; Reusser et al. 2002). The detection of specific catabolic genes using real-time PCR also provides a possible biomarker, which can be correlated with pollutant transformation rates (Johri et al. 1999).

Microbiological and molecular methods used in the assessment of bioremediation systems are discussed in Chapter 12.

9 Conclusions

The significant role of microbial methods for soil and hazardous waste remediation is evident with the changing environmental regulations and social acceptance of bio-based technologies (Zehnder and Wulff 1999; Harayama 2001; Grommen and Verstraete 2002). Limiting factors in biodegradation can be overcome by improving activities of the indigenous organisms, isolating or engineering better degrading strains, and constructing microbial consortia of suitable organisms in addition to overcoming contaminant bioavailability. Advances in molecular techniques and the availability of sophisticated methods for physiological manipulation of microorganisms will provide opportunities for further strain and process improvement. However, the implementation of bioremediation technologies will require a combination of expertise from microbiologists, biochemists, geneticists and chemical and environmental engineers.

We are confident that current research on biochemical aspects of bioremediation will provide the foundation for the development and implementation of new, improved and more dependable processes. This, in turn, will lead to a much greater level of acceptance of this technology as a soil remediation option.

References

Abramowicz DA (1990) Aerobic and anaerobic biodegradation of PCBs: a review. Crit Rev Biotechnol 10:241–251

Atlas RM, Cerniglia CE (1995) Bioremediation of petroleum pollutants: diversity and environmental aspects of hydrocarbon biodegradation. BioScience 45:332–338

Cerniglia CE (1992) Biodegradation of polycyclic aromatic hydrocarbons. Biodegradation 3:351–368

Cerniglia CE (1997) Fungal metabolism of polycyclic aromatic hydrocarbons: past, present and future applications in bioremediation. J Ind Microbiol Biotechnol 19: 324–333

Christodoulatos C, Bhaumik S, Brodman BW (1997) Anaerobic biodegradation of nitroglycerine. Water Res 31:1462–1470

Deguchi T, Kakezawa M, Nishida T (1997) Nylon biodegradation by lignin-degrading fungi. Appl Environ Microbiol 63:329–331

Dua M, Singh A, Sethunathan N, Johri AK (2002) Biotechnology and bioremediation: successes and limitations. Appl Microbiol Biotechnol 59:143–152

Esteve-Nunez A, Caballero A, Ramos JL (2001) Biological degradation of 2,4,6-trinitrotoluene. Microbiol Mol Biol Res 65:335–352

Foght J, April T, Bigger K, Aislebie J (2001) Bioremediation of DDT-contaminated soil – a review. Biorem J 5:225–246

Gieg LM, Suflita JM (2002) Detection of anaerobic metabolites of saturated and aromatic hydrocarbons in petroleum-contaminated aquifers. Environ Sci Technol 36: 3755–3762

Grommen R, Verstraete W (2002) Environmental biotechnology: the ongoing quest. J Biotechnol 98:113–123

Halasz A, Spain J, Paquet L, Beaulieu C, Hawari J (2002) Insights into the formation and degradation mechanisms of methylenedinitramine during the incubation of RDX with anaerobic sludge. Environ Sci Technol 36:633–638

Harayama S (2001) Environmental biotechnology. Curr Opin Biotechnol 12:229–230

Hawari J, Beaudet S, Halasz A, Ampleman G, Thiboutot S (2000) Microbial degradation of explosives: biotransformation versus mineralization. Appl Microbiol Biotechnol 54:605–618

Head IM, Bailey MJ (2003) Environmental biotechnology. Methodological advances spawn new concepts in environmental biotechnology 14:247–247

Hubert C, Shen Y, Voordouw G (1999) Composition of toluene-degrading microbial communities from soil at different concentrations of toluene. Appl Environ Microbiol 65:3064–3070

Iwamoto T, Nasu M (2001) Current bioremediation practice and perspective. J Biosci Bioeng 92:1–8

Janssen DB, Pries F, van der Ploeg JR (1994) Genetics and biochemistry of dehalogenating enzymes. Annu Rev Microbiol 48:161–191

Johri AK, Dua M, Singh A, Sethunathan N, Legge RL (1999) Characterization and regulation of catabolic genes. CRC Crit Rev Microbiol 25:245–273

Kennes C, Wu WM, Bhatnagar L, Zeikus JG (1996) Anaerobic dechlorination and mineralization of pentachlorophenol and 2,4,6-trichlorophenol by methanogenic pentachlorophenol-degrading granules. Appl Microbiol Biotechnol 44:801–806

Li S (1999) Hydrolytic degradation characteristics of aliphatic polyesters derived from lactic acid and glycolic acids. J Biomed Mater Res 48:342–353

Lovely DR (2000) Anaerobic benzene degradation. Biodegradation 11:107–116

McAllister KA, Lee H, Trevors JT (1996) Microbial degradation of pentachlorophenol. Biodegradation 7:1–40
Nishino SF, Spain JC, He Z (2000) Strategies for aerobic degradation of nitroaromatic compounds by bacteria: process discovery to field application. In: Spain JC, Highes JB, Knackmuss H-J (eds) Biodegradation of nitroaromatic compound and explosives, Lewis, Boca Raton, pp 9–61
Parales RE, Bruce NC, Schmid A, Wackett LP (2002) Biodegradation, biotransformation and biocatalysis. Appl Environ Microbiol 68:4699–4709
Paszczynski A, Crawford RL (1995) Potential for bioremediation of xenobiotic compounds by the white-rot fungus *Phanerochaete chrysosporium*. Biotechnol Prog 11: 368–379
Peyton BM, Truex MJ, Skeen RS, Hooker BS (1995) In: Hinchee RE, Leeson A, Semprini L (eds) Bioremediation of chlorinated solvents, Battelle Press, Columbus, Ohio, pp 111–116
Pieper DH, Reineke W (2000) Engineering bacteria for bioremediation. Curr Opin Biotechnol 11:262–270
Popesku JT, Singh A, Zhao J-S, Hawari J, Ward OP (2003) High TNT-transforming activity by a mixed culture acclimated and maintained on crude oil-containing media. Can J Microbiol 49:362–366
Prijambada ID, Negoro S, Yomo T, Urabe I (1995) Emergence of nylon oligomer degradation enzymes in Pseudomonas aeruginosa PAO through experimental evolution. Appl Environ Microbiol 61:2020–2022
Prince RC (1998) Bioremediation. In: Kirk-Othmer Encyclopedia of chemical technology. Supplement to the 4th edn, Wiley, New York, pp 48–89
Providenti MA, Lee H, Trevors JT (1993) Selected factors limiting the microbial degradation of recalcitrant compounds. J Ind Microbiol 12:379–395
Ralebitso TK, Senior E, van Verseveld HW (2002) Microbial aspects of atrazine degradation in natural environments. Biodegradation 13:11–19
Reusser DE, Istok JD, Beller HR, Field JA (2002) In situ transformation of deutered toluene and xylene to benzylsuccinic acid analogues in BTEX-contaminated aquifers. Environ Sci Technol 36:4127–4134
Rogers JE (1998) Soil bioremediation: review of land treatment and composting options. Land Contam Reclam 6:215–222
Shimao M (2001) Biodegradation of plastics. Curr Opin Biotechnol 12:242–247
Shimao M, Tamogami T, Kishida S, Harayama S (2000) The gene pvaB encodes oxidized polyvinyl alcohol hydrolase of *Pseudomonas* sp. Strain M15C and forms an operon with the polyvinyl alcohol dehydrogenase gene *pvaA*. Microbiology 146: 649–657
Singh BK, Kuhad RC, Singh A, Lal R, Tripathi KK (1999) Biochemical and molecular basis of pesticide degradation. Crit Rev Biotechnol 19:197–225
Smets BF, Pritchard PH (2003) Elucidating the microbial component of natural attenuation. Curr Opin Biotechnol 14:283–288
Stapleton RD, Ripp S, Jimenez L, Cheol-Koh S, Fleming JT, Gregory IR, Sayler GS (1998) Nucleic acid analytical approaches in bioremediation: site assessment and characterization. J Microbiol Meth 32:165–178
Top EM, Springael D (2003) The role of mobile genetic elements in bacterial adaptation to xenobiotic organic compounds. Curr Opin Biotechnol 14:262–269
van Beilen JB, Wubbolts MG, Witholt B (1994) Genetics of alkane oxidation by *Pseudomonas oleovorans*. Biodegradation 5:161–174
Van der Meer JR, Senchilo V (2003) Genomic islands and the evolution of catabolic pathways in bacteria. Curr Opin Biotechnol 14:248–254

van Elsas JD, Duarte GF, Rosado AS, Smalla K (1998) Microbiological and molecular biological methods for monitoring microbial inoculants and their effects in the soil environment. J Microbiol Meth 32:133–154
Van Hamme JD, Singh A, Ward OP (2003) Recent advances in petroleum microbiology. Microbiol Mol Biol Rev 7:503–549
Verstraete W (2002) Environmental biotechnology for sustainability. J Biotechnol 94: 93–100
Wackett LP, Hershberger CD (2001) Biocatalysis and biodegradation: microbial transformation of organic compounds. ASM Press, Washington, DC, pp 39–69
Ward OP, Singh A, Van Hamme J (2003) Accelerated biodegradation of petroleum hydrocarbon waste. J Ind Microbiol Biotechnol 30:260–270
Watanabe K (2001) Microorganisms relevant to bioremediation. Curr Opin Biotechnol 12:237–241
Watanabe K, Hamamura N (2003) Molecular and physiological approaches to understanding the ecology of pollutant degradation. Curr Opin Biotechnol 14:289–295
Widada J, Nojiri H, Omori T (2002) Recent developments in molecular techniques for identification and monitoring of xenobiotic-degrading bacteria and their catabolic genes in bioremediation. Appl Microbiol Biotechnol 60:45–59
Widdel F, Rabus R (2001) Anaerobic biodegradation of saturated and aromatic hydrocarbons. Curr Opin Biotechnol 12:259–276
Ye J, Singh A, Ward OP (2003) Biodegradation of nitroaromatics and other nitrogen-containing xenobiotics. World J Microbiol Biotechnol (in press)
Zehnder AJB, Wulff H (1999) Business and the environment. Nature Biotechnol 17:25
Zhao J-S, Ward OP, Lubicki P, Cross JD, Huck P (2001) Process for degradation of nitrobenzene: combining electron beam irradiation with biotransformation. Biotechnol Bioeng 73:306–312

2 Microbial Community Dynamics During Bioremediation of Hydrocarbons

E. Anne Greene[1] and Gerrit Voordouw[2]

1 Introduction

Hydrocarbons are introduced into environments through constant, low-level releases and large-scale accidental spills. Widespread use of hydrocarbons has resulted in a variety of environments being exposed. Shifts in microbial populations occur when hydrocarbons enter an environment, and during their subsequent degradation (MacNaughton et al. 1999). Biodegradation requires appropriate environmental conditions and an active microbial population. Physical parameters that affect microbial growth can be measured. There is currently no single method to easily assess microbial community dynamics during hydrocarbon degradation. A broad overview of the various methods used for studying community dynamics will be presented here, followed by a discussion of some specific research results.

2 Methods for Characterization of Microbial Communities

Microbial community dynamics encompass changes in the structure of a community, as defined by the activity and phylogeny of the species in an environment. The complex nature of environmental microbial communities makes their complete assessment difficult. Molecular and nonmolecular techniques can be used to track community dynamics throughout hydrocarbon exposure and degradation (Table 1). Each method has certain advantages and limitations. The approach to monitoring communities will vary depending on the information,

[1]Department of Biological Sciences, University of Calgary, 2500 University Dr. NW, Calgary, Alberta, T2N 1N4, Canada
[2]Department of Biological Sciences, University of Calgary, 2500 University Dr. NW, Calgary, Alberta, T2N 1N4, Canada, email: voordouw@ucalgary.ca, Tel: +1-403-2206388, Fax: +1-403-2899311

Soil Biology, Volume 2
Biodegradation and Bioremediation
(ed. by. A. Singh and O. P. Ward)

Table 1. Summary of techniques used to study microbial community dynamics

Technique	Assay type	Sample components required for assays	Culturing		PCR		Detects		Quantification of community		References
			Required	Improves detection	Required	Improves detection	Activity	Potential	Specific	Total	
Non-molecular techniques											
Culturing	Growth[a]	Inoculum	Yes	Yes	No	No	In situ	Ex situ	Yes[b]	No	Atlas (1984), Jones (1977), Tchelet et al. (1999)
C source profile	Growth[a]	Inoculum	Yes	Yes	No	No	No	Yes[a]	No	No	Carpenter-Boggs et al. (1998), Garland (1997)
Hydrocarbon use	Chemical	Hydrocarbons (and inoculum)[c]	No	Yes	No	No	In situ	Ex situ	Yes	No	Becker (1999), Gieg et al. (1999)
Terminal electron acceptor	Chemical	Electron acceptors or reduced products[d]	No	Yes	No	No	In situ	Ex situ	Yes[e]	No	Christensen et al. (1994), Gieg et al. (1999)
Stable isotopes	Isotope	Hydrocarbons or metabolic products	No	Yes	No	No	In situ	Ex situ	Yes	No	Grossman (1997)
Molecular techniques											
Low molecular weight RNA pattern	Phylogenetic	RNA	No	Yes	No	No	No	No	Yes	No	Stahl (1997)
Probes											
Targeted oligonucleotide probes	Physiologic, phylogenetic	DNA or RNA	No	Yes	No	Yes	RNA	DNA	Yes	Yes	Brockman (1995), Stahl (1997), Koizumi et al. (2002)
Fluorescence in situ hybridiztion	Physiologic, phylogenetic	DNA or RNA	No	Yes	No	Yes	RNA	DNA	No	No	Amann et al. (1990), Tresse et al. (2002)
Nucleic acid amplification techniques											
Polymerase chain reaction (PCR)	Physiologic, phylogenetic	DNA	No	Yes	Yes	Yes	No	Yes	No	No	Stahl (1997)
Reverse-transcription PCR (RT-PCR)	Physiologic, phylogenetic	RNA	No	Yes	Yes	Yes	Yes	Yes	Yes	No	Stahl (1997)
Repetitive extragenic palindromic PCR (REP-PCR)	Physiologic, phylogenetic	DNA	No	Yes	Yes	Yes	No	No	No	No	Matheson et al. (1997)
Competitive-quantitative PCR	Physiologic, phylogenetic	DNA	No	Yes	Yes	Yes	No	Yes	Yes	No	Mesarch et al. (2000)

Arbitrarily primed AP-PCR (RAPD)	Phylogenetic	DNA	No	Yes	Yes	Yes	No	No	Yes	No	Pan et al. (2001)
Denaturing gradient gel electrophoresis (DGGE)	Phylogenetic	DNA	No	Yes	Yes	Yes	No	No	No	No	Kurisu et al. (2002), Tresse et al. (2002)
Terminal restriction fragment length polymorphism (T-RFLP)	Phylogenetic	DNA	No	Yes	Yes	Yes	No	No	No	No	Stahl (1997), Marsh et al. (2000)
Amplified ribosomal DNA restriction analysis (ARDRA)	Phylogenetic	DNA	No	Yes	Yes	Yes	No	No	No	No	Fernandez et al. (1999), Fries et al. (1997)
Single strand conformation polymorphism (SCCP)	Phylogenetic	DNA	No	Yes	Yes	Yes	No	No	No	No	Schwieger and Tebbe (1998)
Microarrays and macroarrays											
DNA microarrays and macroarrays	Physiologic, phylogenetic	DNA or RNA	No	Yes	No	Yes	RNA	DNA	Yes[f]	No	Cho and Tiedje (2001), Greer et al. (2001), Greene and Voordouw (2003)
Reverse sample genome probing (RSGP)	Physiologic, phylogenetic	DNA or RNA	Yes[g]	Yes	No	No	RNA	DNA	Yes	No	Greene and Voordouw (2003), Voordouw\ (1998)
Non-nucleic acid techniques											
Phospholipid fatty acid (PLFA) analysis	Physiologic, phylogenetic	PLFA	No	Yes	No	No	No	No	Yes	Yes	Hellman et al. (1997), Tresse et al. (2002), White et al. (1997), Doumenq et al. (1999)
Quinone profile	Physiologic, phylogenetic	Quinones	No	Yes	No	No	No	No	Yes	No	Kurisu et al. (2002), Hu et al. (2001), White et al. (1997)
Ether lipids	Archaeal phylogeny	Ether lipids	No	Yes	No	No	No	No	Yes	No	White et al. (1997)

[a] May not assess the hydrocarbon-degrading potential, depending on the carbon sources used.
[b] Depending on culturing conditions, a specific culturable community may be enumerated.
[c] Culturing, particularly when appropriate nutrients are provided, may improve detection of hydrocarbon-degrading activity. Requirements for culturing shown in brackets.
[d] Detection of O_2, NO_3^-, Mn^{4+}, Fe^{3+}, SO_4^{2-}, CO_2 and reduced products; steady-state H_2 measurements can be used to confirm these results in situ.
[e] Measurement of terminal electron acceptors will allow quantification, if the most probable number or plate count assays are done under conditions selecting for particular hydrocarbon-degrading activities. This requires culturing.
[f] Quantification is possible if an internal standard is used.
[g] Culturing is typically used to generate the macroarray, however, this could be done using PCR and cloning techniques.

samples and time available. Several methods may be used to define the same community, in order to overcome the limitations of a single method.

In cases where changes from the time of exposure to hydrocarbons cannot be assessed, a nearby uncontaminated location with a similar hydrogeological profile can be used as a reference to estimate the microbial community present at the contaminated site before hydrocarbon exposure (Gieg et al. 1999). Sampling at various locations within and downstream from a contaminant plume may also allow reconstruction of what has occurred at the plume source over time.

2.1 Non-molecular Methods

Culture-based ex situ techniques, e.g., plate counts and most probable number (MPN) assays, are useful for enumerating specific culturable populations and introduced strains (Tchelet et al. 1999). However, 12% or less of the total microbial population are as yet culturable (Jones 1977; Atlas 1984). Carbon source utilization profiles (e.g., as determined by Biolog plates or the use of specific compounds) can distinguish between microbial communities or changes within a single community, and evaluate metabolic potential (Garland 1997; Carpenter-Boggs et al. 1998). These tests detect changes in community profile based on biochemical activity, but provide no information on phylogeny.

Steady-state hydrogen measurements can be used to monitor in situ microbial physiology. This information is confirmed by the presence of terminal electron acceptors or their reduced products (Gieg et al. 1999). Hydrocarbon use and laboratory assays demonstrating metabolic potential for hydrocarbon degradation can be correlated with these measurements. Stable isotope ratios can also serve as an indicator of microbial activity (Grossman 1997). The ratio of $^{12}C:^{13}C$ in the environment is constant and unaffected by abiotic processes. Microbial activity can be isotopically selective, resulting in shifts in the $^{12}C:^{13}C$ ratio in substrates and products, depending on the type of molecule and microbial activity (e.g., anabolism vs. catabolism; Grossman 1997).

2.2 Molecular Methods

Culture-independent nucleic acid samples are more representative of whole communities, improving phylogenetic evaluation considerably (Stahl 1997). The advantages are that samples can be frozen without being disturbed and nucleic acids can be extracted from most of the microbial population (Brockman 1995). However, as RNA is relatively short-lived, it belongs to the active component of the microbial community and can be used to determine activity (Brockman 1995). Active cells with more RNA are distinguishable from inactive ones (Roane and Pepper 2000). Semiquantitative assessment of microbial activity is possible by comparing the ratio of rRNA:rDNA; rRNA will increase with increasing activity (Stahl 1997).

A phylogenetic perspective of a community can be gained using 16S rRNA-targeted methods. The 16S rRNA gene has both highly conserved and species-dependent variable regions, therefore the relative abundance of specific populations can be estimated by comparing the amounts of total community and specific 16S rRNA (Koizumi et al. 2002). Different regions of 16S rDNA will give different results; some variable regions are less diverse than others (Schmalenberger et al. 2001). Complicating factors are that some organisms have multiple 16S rDNA gene sequences (Stahl 1997), and phylogenetically similar microbial strains may be physiologically different. Genes that determine physiology can also be detected. However, important degradative pathways including genes significantly divergent from the DNA sequences being used for detection may be missed.

Nucleic acid hybridization is one molecular approach for defining communities. The lower detection limits are typically 10^4 to 10^6 DNA or RNA copies g^{-1} of soil. Whole cells or extracted nucleic acids can be detected (Brockman 1995). Quantification is possible with visualization of cells (Wilson et al. 1999); e.g., by fluorescence in situ hybridization (FISH) using 16S rRNA-targeting probes linked to a fluorescent dye. Organisms can be enumerated using microscopy or flow cytometry (Amann et al. 1990), but relatively inactive cells (e.g., in biofilms) may not be detected by FISH (Tresse et al. 2002).

Many community analysis techniques use the polymerase chain reaction (PCR) to amplify specific DNA sequences. Reverse transcription (RT-PCR) results in amplification of specific RNA sequences. Several methods of PCR amplification are described in Table 1. Amplification of repetitive extragenic palindromic elements (REP-PCR) results in strain-specific fragments that hybridize with DNA from the strain of interest but not from other samples (Matheson et al. 1997). In a test of this method, the detection limit was 10^4 *Burkholderia cepacia* cells in a background of 10^5 nontarget cells. The technique worked in groundwater and soil backgrounds, and with other environmental isolates (Matheson et al. 1997). In competitive-quantitative PCR, competitor DNA is amplified from the same primers as the gene of interest, allowing quantification of the latter. Mesarch et al. (2000) developed primers targeting the gene for catechol 2,3-dioxygenase, which is involved in hydrocarbon degradation pathways in environments such as petroleum-contaminated soil. The detection limit was 10^2 to 10^3 gene copies g^{-1} soil; this was decreased to 10^0 to 10^1 gene copies g^{-1} soil by hybridizing the resulting PCR products with a labeled probe. Arbitrarily primed PCR (AP-PCR, also known as randomly amplified polymorphic DNA or RAPD) uses a single primer at low stringency to generate multiple amplicons for strain fingerprinting and is particularly effective for distinguishing closely related strains (Pan et al. 2001).

Mixtures of rDNA fragments are generated by PCR-amplification of partial 16S rDNA sequences in extracted total community DNA using conserved primers. Denaturing gradient gel electrophoresis (DGGE) allows separation of amplified 16S rDNA sequences at the species level (Kurisu et al. 2002) and has proven a useful tool for measuring bacterial diversity (Tresse et al. 2002). Because this method avoids lengthier cloning procedures, it has been widely employed (e.g., Röling et al. 2002). Each separated band can be extracted and sequenced to identify specific community members (Greer et al. 2001). The entire, mixed 16S rDNA amplicon can also be analyzed by terminal restriction fragment length polymorphism (T-RFLP), which separates these using terminal fragments of 16S rDNA after digestion with restriction endonucleases (Stahl 1997). All known sequences in the Ribosomal Database Project web site (http://www.cme.msu.edu/DRP/html/analyses.html) can be used for T-RFLP analysis (Marsh et al. 2000). Other techniques for separating 16S rDNA amplicons include amplified ribosomal DNA restriction analysis (ARDRA), which is based on gel separation of restriction endonuclease-digested PCR products (Fries et al. 1997; Fernandez et al. 1999), and single-strand conformation polymorphism (SSCP). A method for analyzing environmental communities by SSCP is reported by Schwieger and Tebbe (1998).

Micro- and macroarrays are powerful tools for phylogenetic or physiologic genomics; with microarrays samples can be screened with up to 10^5 probes (Cho and Tiedje 2001). Amplification of sample genes is not necessary (Small et al. 2001). Specificity varies; probes based on PCR amplicons can detect sequences with 80% identity or less, however, 25 to 30 base pair oligonucleotide probes may not hybridize to sequences with a singe base-pair mismatch (Greer et al. 2001). Detection limits for microarrays are not as low as with some PCR methods. For instance, an oligonucleotide microarray consisting of species-specific 16S rRNA probes was used to detect *Geobacter* and *Desulfovibrio* spp. in soil. The sensitivity limit was 7.5×10^6 *Geobacter* cell equivalents (Small et al. 2001), which is similar to the detection limit for nucleic acid hybridization (Brockman, 1995). Detection is more sensitive with macroarrays, as a larger amount of each probe is present (Greene and Voordouw

2003). A combination of PCR and microarray technologies can improve sensitivity (Greer et al. 2001). Randomly generated genome fragments can be used as probes to identify microorganisms; individual species can be detected with approximately 100 randomly generated fragments, thus a typical microarray could accommodate 1,000 species (Cho and Tiedje 2001). Lambda DNA can be used as a reference to normalize variations in hybridization and spot size within DNA arrays, allowing the amount of each target gene present to be estimated (Cho and Tiedje 2002). Analysis of environmental samples using microarrays has been shown to give similar results to DGGE and membrane hybridization using various probes (Koizumi et al. 2002).

Reverse sample genome probing (RSGP) has also been used for monitoring microbial diversity (Voordouw 1998). Community DNA is labeled and hybridized to macroarrays consisting of genomic DNA spots representing bacterial strains that have limited cross-hybridization (Voordouw et al. 1991). A large amount of microbial diversity in a sample (i.e., as represented in total soil DNA without prior enrichment) can result in overly complex analyses (Chao et al. 1997; Bagwell and Lovell 2000; Greene and Voordouw 2003). In environments with more limited diversity the method has proved useful. Oil field populations have been successfully monitored by RSGP (e.g., Voordouw et al. 1991; Telang et al. 1997). The technique has also been used to monitor the effect of various hydrocarbon compounds on soil microbial communities (Shen et al. 1998; Hubert et al. 1999; Greene et al. 2000, 2002).

Phospholipid fatty acids (PLFA) extracted from samples can be used to quantitate and characterize viable biomass and community components (Hellman et al. 1997; White et al. 1997; Tresse et al. 2002). Specific PLFAs can indicate the phylogeny and physiology of microbial community members (Hellman et al. 1997), however, individual strains may not be distinguishable (White et al. 1997). PLFA turn over quickly; their analyses therefore indicate active microbial populations (Carpenter-Boggs et al. 1998). Hydrocarbon exposure causes changes in PLFA (Doumenq et al. 1999), which may suggest community shifts that have not actually occurred. Similar analyses can be done on archaea using ether lipids. Other cellular lipids also act as signature biomarkers (White et al. 1997). For example, respiratory quinones can be used to quantify biomass and identify specific phylogenetic groups (White et al. 1997; Hu et al. 2001; Kurisu et al. 2002). Quinone profiling has been used successfully in activated sludge and soil communities (Hu et al. 2001).

As is clear from the above compilation, a wide variety of techniques are used to study microbial communities. Even in cases where similar samples, experiments and methods for observing and defining microbial communities were used, the results obtained are often divergent, limiting meaningful discussion of findings on community dynamics. Cases in which results are somewhat comparable will be presented. These represent specific examples, rather than the entire body of work on microbial community dynamics during hydrocarbon degradation.

3 Microbial Community Dynamics Following Hydrocarbon Exposure

3.1 Monitoring Microbial Strains Added to Environments

The fate of seeded microbial strains can be monitored using unique characteristics to distinguish the introduced strain from background populations. Although introduced strains can survive and may be active, amendment with microorganisms does not usually improve bioremediation in the field (Atlas 1995). However, in laboratory communities, sufficient numbers for useful activity may remain. Watanabe et al. (1998) examined the population dynamics of two phenol-degrading strains, a *Pseudomonas* and a *Comomonas* spp., in an activated sludge community and demonstrated that the growth of both strains was able to compensate for predation in the system. An introduced 1,2,4-trichlorobenzene (TCB)-degrading *Pseudomonas* strain washed out of sludge bioreactors in both the presence and absence of TCB, but colonized a soil column and remained active there (Tchelet et al. 1999). Up to 10^3 to 10^4 cells ml^{-1} in a background of 10^9 cells ml^{-1} were detected. In contrast, MacNaughton et al. (1999) found that an oil-degrading mixed consortium enriched from a beach and added to a nearby oiled test plot was soon lost from the oiled plot, suggesting an inability to compete with the indigenous population. Bachoon et al. (2001a) added an *Acinetobacter* strain to oil-amended salt marsh sediment microcosms. The strain remained detectable, but oil degradation was not improved.

3.2 Correlation of Hydrocarbon Degradation and Gene Expression

Nucleic acid techniques have been used to correlate the degradation of specific hydrocarbon compounds with the activity or presence of genes encoding known hydrocarbon-degrading enzymes. Fleming et al. (1993) demonstrated that mRNA from the *nahA* gene encoding a naphthalene-degrading enzyme was found concomitantly with decreases in naphthalene concentration and ^{14}C-naphthalene mineralization in hydrocarbon-contaminated microcosms. In another study, enumeration of 2,4-dichlorophenoxyacetic acid (2,4-D)-degrading organisms in soil microcosms was positively correlated with biodegradation of 2,4-D and the presence of *tfd* genes, which encode enzymes involved in 2,4-D degradation, (Holben et al. 1992). However, active microbial populations can be overlooked if the gene products being assayed are not employed in degradation pathways. Ka et al. (1994) used gene probes for two different 2,4-D degradation genes (*tfdA* and *spa*) to assess 2,4-D-degrading populations in different soil microcosms. The *tfdA* probe is typically used to detect 2,4-D degraders (Ka et al. 1994). However, depending on the soil sample, either *tfdA* or *spa* genes were dominant, thus the commonly used probe alone would not have detected active microbial populations in some soil samples. Although gene probe techniques are useful in determining microbial potential in a sample, it is important that the limitations of the method be realized.

3.3 Shifts in Microbial Community Composition Due to Hydrocarbon Exposure

Most researchers observe shifts in microbial community structure after exposure to hydrocarbons. Due to the diversity of methods used, environments examined and microbial communities observed, a meaningful comparison of these results is often difficult. Molecular biology techniques allow communities to be defined by phylogenetic parameters, facilitating more detailed discussion. Nonmolecular analyses provide no phylogenetic information, however, the physiologically relevant level of detail can offer more process-relevant data.

Overall, the number of culturable heterotrophs in environmental samples does not change significantly following oil amendment, although specific physiological subgroups such as hydrocarbon degraders may be directly affected. A typical microbial community consists of less than 1% hydrocarbon degraders, but this can increase to 10% upon exposure to oil pollutants (Atlas 1995). Many examples of this exist in the literature. Lindstrom et al. (1999) found similar numbers of culturable heterotrophs in pristine and oil-contaminated soils. Mesarch and Nies (1997) demonstrated that enumeration of organisms that use benzoate reflected hydrocarbon-degrading activity in samples, while the total number of heterotrophs in contaminated and uncontaminated soils was similar. Duncan et al. (1997) demonstrated similar numbers of heterotrophs in oil-exposed soils with or without nutrient treatment or in unexposed soils after 3 years, although the numbers of certain groups (hydrocarbon degraders and sulfate-reducing bacteria) were higher in the oil-contaminated samples. During bioremediation treatments, heterotrophic plate counts of salt marsh sediment microcosms were the same for oiled and unoiled sediments (Bachoon et al. 2001b). Treatment of the same microcosms with nutrients or an oil-degrading bacterial inoculum resulted in increased numbers of oil-degrading bacteria but did not necessarily correspond with increased oil degradation rates (Bachoon et al. 2001a). The number of alkane and polyaromatic hydrocarbon degraders at an experimental oil spill site initially increased in oil-amended plots, but slowly declined over 14 weeks (MacNaughton et al. 1999).

Community-level physiology profiling is a useful tool for following community dynamics (Garland 1997). The addition of diesel to each of the three soil communities enriched a population capable of using a wide variety of carbon sources (Bundy et al. 2002). Biolog profiles were strongly affected by diesel contamination, however, this did not correspond to changes in the community as determined by PLFA analysis. Convergent community profiles did not develop as a result of diesel exposure in different soils. Becker (1999) also found that for communities with similar pollutant-degrading capabilities, the molecular characterization and substrate utilization (Biolog and specific hydrocarbon utilization) patterns were quite different. In contrast, Wünsche et al. (1995) found that the substrate utilization profiles of hydrocarbon-exposed soils tended to converge and proposed community convergence on a metabolic level. The number of compounds utilized decreased in uncontaminated soil, possibly due to attrition of the population in the absence of nutrients (Wünsche et al. 1995).

Microbial community structure in relation to hydrogeochemistry has also been evaluated. A landfill leachate-polluted aquifer was examined to determine the relationship between hydrochemistry and community structures measured by DGGE (Röling et al. 2001). Groundwater samples showed differences in community structure based on the processes occurring in a given location, while sediment community structures did not. Less than 1% of the dissolved organic carbon consisted of xenobiotic compounds, thus the presence of other heterotrophic organisms may have masked community changes due to the presence of these compounds. Gieg et al. (1999) compared communities in a petroleum condensate-contaminated aquifer and uncontaminated reference area over a 4-year period. The contaminated region was depleted in oxygen and sulfate, and had a higher population of sulfate-reducing bacteria. Hydrogen concentrations and the presence of metabolites suggested sulfate-reducing and methanogenic populations. Laboratory-based assays demonstrated microbial hydrocarbon utilization under these conditions. A hydrocarbon-degrading sulfate-reducing and methanogenic microbial community likely developed as higher energy terminal electron acceptors became depleted. Other studies have also demonstrated that as terminal electron acceptors are depleted, the active population shifts accordingly (e.g., Christensen et al. 1994). Hess et al. (1997) used FISH to demonstrate that the structure of a microbial community varies depending on the electron acceptors available.

A common finding of many studies is a decrease in microbial diversity after hydrocarbon exposure (Atlas et al. 1991). The use of hydrocarbons as a carbon source will enrich only certain community members. For instance, Hanson et al. (1999) demonstrated that 96% of a microbial community incorporated some ^{13}C isotope from glucose, while only 27% incorporated ^{13}C-labeled toluene. Lindstrom et al. (1999) demonstrated a decrease in diversity in oiled soils. The remaining microbial population were metabolic generalists, and more hydrocarbon-degrading organisms were present. However, even in closed systems such as bioreactors, enrichment of some community members due to the presence of hydrocarbons may result in an apparent rather than an actual decrease in diversity (Stoffels et al. 1998; Greene et al. 2002). An exception to the more common observation of decreased diversity upon hydrocarbon exposure is a study by Juck et al. (2000), where the same or increasing diversity was found at a hydrocarbon-contaminated site.

Enrichment of specific community members does not necessarily depend on the carbon source provided. Stoffels et al. (1998) used a car-painting facility wastewater inoculum enriched in α- and β-proteobacteria to inoculate a fermentor and monitored community dynamics using FISH. In this study, incubation conditions were the major force effecting change in the microbial community. Dominant bacterial species in the fermentor were closely related to *Pseudomonas putida* or *P. mendocina* (γ-proteobacteria), and community diversity appeared to decrease. Transfer to a trickle-bed reactor resulted in increased diversity, with α- and β-proteobacteria being dominant, thus diversity was not lost in the fermentor. Greene et al. (2000) also observed that while dominant microbial strains may vary during enrichment, phylogenetic diversity is not lost. *Pseudomonas* spp. (γ-proteobacteria) were dominant over the first 13 weeks of incubation in the presence of benzene, toluene, *m*-xylene, styrene, DCPD, naphthalene or all six compounds

(Greene et al. 2000, 2002). Longer-term incubations (45 to 60 weeks) resulted in a population shift to *Alcaligenes* spp. (β-proteobacteria). The level of soil contamination and type of compound present had little effect on the microbial population; incubation time appeared to be the most important factor. Two-week incubations of hydrocarbon-contaminated soil with toluene or with DCPD also resulted in enrichment of *Pseudomonas* spp. (Shen et al. 1998; Hubert et al. 1999). These results were interpreted in terms of a food web in which an interdependent community develops irrespective of the nature of the hydrocarbon input (Greene et al. 2002). In contrast, Colores et al. (2000) reported that *Pseudomonas* and *Alcaligenes* spp. increase after the amendment of soils with hexadecane and a surfactant (both compounds served as carbon and energy sources); hexadecane alone resulted in the enrichment of *Rhodococcus* and *Nocardia* spp. (both gram positive). Thus the nature of the carbon and energy source may affect community dynamics in some but not all cases.

In uncontaminated shallow sand aquifer material, α-, β- and γ-proteobacteria were equally represented (Shi et al. 1999). Fuel-contaminated aquifer material had β- and γ-proteobacteria as dominant strains. The addition of toluene to microcosms resulted in an increase in α-proteobacteria at the expense of β and γ-proteobacterial phyla. However, more than 60% of the microorganisms present were unaccounted for by phylogenetic probes. The authors postulated that these were mainly δ-proteobacteria, as anaerobic activity was detected (Shi et al. 1999).

MacNaughton et al. (1999) monitored the bacterial community at a site that underwent an experimental oil spill and several types of bioremediation treatment. Oil exposure resulted in a substantial change in the dominant components of the microbial community from eukaryotes to gram-negative bacteria. Plots amended with nutrients or inoculated with an oil-degrading mixed consortium had similar microbial populations, but were different from oil-only or control plots. Oil treatment resulted in the enrichment of α-proteobacteria and the *Flexibacter-Cytophaga–Bacteroides* phylum. While the PLFA profile of oiled plots regained similarity to the unoiled control plots after 14 weeks, DGGE analysis suggested that significant differences between the communities remained. It is thus important to use multiple methods to document community shifts.

Replicate samples do not necessarily have identical microbial communities. Community dynamics in microcosms emulating an oil-contaminated intertidal zone were monitored by DGGE (Röling et al. 2002). Replicate samples taken from the same microcosm at the same time were 95% similar, however, samples from a replicate microcosm were as similar (68%) as samples that had undergone different bioremediation treatments (63%). Communities with different microbial populations had similar abilities to remove hydrocarbons (Röling et al. 2002). Bundy et al. (2002) demonstrated that the community profiles of three pristine soils diverged after diesel contamination. Oil amendment resulted in a large decrease in community diversity after 6 days. By the 26th day, initial diversity levels had returned, but the community was then comprised of a different population. While representing only 8% of the community before oil contamination, α-proteobacteria became dominant afterwards. Some γ-proteobacteria, present in unamended oiled communities, were supplanted by γ-proteobacteria in nutrient-amended microcosms. Grossman et al. (2000) and Kasai et al. (2001) found γ-proteobacteria after oil spills in the Norwegian Arctic and in Japan, respectively. Increases in eubacterial DNA corresponding to increased degradation have often been observed (Atlas and Bartha 1972; Atlas 1992; MacNaughton et al. 1999; Bachoon et al. 2001a). Overall, γ-proteobacteria are frequently enriched after relatively short-term exposure to hydrocarbons in the absence of additional nutrients. However, different results have also been observed. Bachoon et al. (2001a), using domain- and group-specific oligonucleotide probes to analyze microbial community dynamics in salt marsh sediment microcosms during oil exposure and bioremediation treatments, found degradative activity associated with *Streptomyces* spp. (gram positive) and uncharacterized bacteria but not with *Pseudomonas* spp. (γ-proteobacteria).

Microbial communities with different or constantly shifting compositions can perform the same activities. Fernández et al. (1999) found that phylogenetic changes occurred constantly in a physiologically stable, glucose-fed, methanogenic bioreactor using ARDRA. Changes in carbon and electron flows occurred during the 605-day incubation, which may have affected the community composition without affecting the bioreactor performance. ARDRA, Biolog and specific hydrocarbon utilization were used by Becker (1999) to monitor aerobic heterotrophic communities from hydrocarbon-contaminated sites. Communities with similar pollutant-degrading capabilities had different species profiles and substrate utilization patterns. The DGGE analysis of a styrene-degrading biotrickling filter showed that the biofilm community was always more diverse than the planktonic community derived from the biofilm (Tresse et al. 2002). Of the bands present in the styrene-enriched inoculum, only 50% was detected in the biofilm. Acclimation took 35 days, after which variation in the DGGE patterns decreased; none was detected between days 82 and 184, unlike the continual changes described by Fernández et al. (1999).

Some community structures are extremely sensitive to hydrocarbon inputs. The response of marine microbial mats to hydrocarbons was monitored by DGGE (Abed et al. 2002). Most DGGE bands were replaced during the treatment, and diversity decreased, thus the community structure shifted considerably. The cyanobacterial population was strongly affected. Cyanobacteria are sensitive to pollution, thus may be good indicators of contamination.

4 Conclusions

In conclusion, it appears that although some comparisons can be made between studies on community dynamics, on the whole, the results from this type of research are wide-ranging, even when similar techniques for determining community structure are used. This emerging field will benefit greatly from the development of standardized tests and means by which microbial communities and their activities are defined. Standard incubation practices will also allow a better comparison between different results. However, given the variety of hydrocarbon-contaminated environments, and the ability of microbial species to rapidly adapt to changes, it is likely that microbial community dynamics during hydrocarbon degradation will always prove to be as diverse as the microbial communities themselves.

Acknowledgements. We would like to thank V. Brunelle for assistance in manuscript preparation, and C.R. Hubert, S.A. Haveman, M.S. Hicks and P.H. Beatty for helpful comments. E.A. Greene was supported by a Natural Sciences and Engineering Research Council of Canada Postdoctoral Fellowship.

References

Abed RMM, Safi NMD, Köster J, de Beer D, El-Nahhal Y, Rullkötter J, Garcia-Pichel F (2002) Microbial diversity of a heavily polluted microbial mat and its community changes following degradation of petroleum compounds. Appl Environ Microbiol 68:1674–1683

Amann RI, Binder BJ, Olson RJ, Chisholm SW, Devereux R, Stahl DA (1990) Combination of 16S rRNA-targeted oligonucleotide probes with flow cytometry for analyzing mixed microbial populations. Appl Environ Microbiol 56:1919–1925

Atlas, RM (1984) Diversity of microbial communities. Adv Microb Ecol 7:1–48

Atlas RM (1992) Petroleum microbiology. Encycl Microbiol 3:363–369

Atlas RM (1995) Bioremediation of petroleum pollutants. Int Biodeterior Biodegrad: 317–327

Atlas RM, Bartha R (1972) Degradation and mineralization of petroleum in seawater: limitation by nitrogen and phosphorus. Biotech Bioeng 14:309–317

Atlas RM, Horowitz A, Krichevsky M (1991) Response of microbial populations to environmental disturbance. Microbial Ecol 22:287–338

Bachoon DS, Araujo R, Molina M, Hodson RE (2001a) Microbial community dynamics and evaluation of bioremediation strategies in oil-impacted salt marsh sediment microcosms. J Ind Microbiol Biotech 27:72–79

Bachoon DS, Hodson RE, Araujo R (2001b) Microbial community assessment in oil-impacted salt marsh sediment microcosms by traditional and nucleic acid-based indices. J Microbiol Meth 46(11):35–47

Bagwell CE and Lovell CR (2000) Persistence of selected *Spartina alterniflora* rhizoplane diazotrophs exposed to natural and manipulated environmental variability. Appl Environ Microbiol 66:4625–4633

Becker PM (1999) About the order in aerobic heterotrophic microbial communities from hydrocarbon-contaminated sites. Int Biodeterior Biodegrad 43:135–146

Brockman FJ (1995) Nucleic-acid-based methods for monitoring the performance of in situ bioremediation. Mol Ecol 4:567–578

Bundy JG, Paton GI, Campbell CD (2002) Microbial communities in different soil types do not converge after diesel contamination. J Appl Microbiol 92:276–288

Carpenter-Boggs L, Kennedy AC, Reganold JP (1998) Use of phospholipid fatty acids and carbon source utilization patterns to track microbial community succession in developing compost. Appl Environ Microbiol 64:4062–4064

Chao W-L, Tien C-C, Chao C-C (1997) Investigation of the effect of different kinds of fertilizers on the compositions of a soil microbial community using the molecular biology technique. J Chin Agric Chem Soc 35:252–262

Cho J-C, Tiedje JM (2001) Bacterial species determination from DNA-DNA hybridization by using genome fragments and DNA microarrays. Appl Environ Microbiol 67: 3677–3682

Cho J-C, Tiedje JM (2002) Quantitative detection of microbial genes by using DNA microarrays. Appl Environ Microbiol 68:1425–1430

Christensen TH, Kjeldsen P, Albrechtsen H-J, Heron G, Nielsen PH, Bjerg PL, Holm PE (1994) Attenuation of landfill leachate pollutants in aquifers. Crit Rev Environ Sci Technol 24:119–202

Colores GM, Macur RE, Ward DM, Inskeep WP (2000) Molecular analysis of surfactant-driven microbial population shifts in hydrocarbon-contaminated soil. Appl Environ Microbiol 66(7):2959–2964

Doumenq P, Acquaviva M, Asia L, Durbec JP, Dréau YL, Mille G, Bertrand JC (1999) Changes in fatty acids of *Pseudomonas nautica*, a marine denitrifying bacterium, in response to *n*-eicosane as a carbon source and various culture conditions. FEMS Microbiol Ecol 28:151–161

Duncan K, Levetin E, Wells H, Jennings E, Hettenbach S, Bailey S, Lawlor K, Sublette K, Fisher JB (1997) Managed bioremediation of soil contaminated with crude oil: soil chemistry and microbial ecology three years later. Appl Biochem Biotech 63–65: 879–889

Fernández A, Huang S, Seston S, Xing J, Hickey R, Criddle C, Tiedje J (1999) How stable is stable? Function versus community composition. Appl Environ Microbiol 65: 3697–3704

Fleming JT, Sanseverino J, Sayler GS (1993) Quantitative relationship between naphthalene catabolic gene frequency and expression in predicting PAH degradation in soils at town gas manufacturing plants. Environ Sci Technol 27:1068–1074

Fries MR, Hopkins GD, McCarty PL, Forney LJ, Tiedje JM (1997) Microbial succession during a field evaluation of phenol and toluene as the primary substrates for trichloroethene cometabolism. Appl Environ Microbiol 63(4):1515–1522

Garland JL (1997) Analysis and interpretation of community-level physiological profiles in microbial ecology. FEMS Microbiol Ecol 24:289–300

Gieg LM, Kolhatkar RV, McInerney MJ, Tanner RS, Harris SH Jr, Sublette KL, Suflita JM (1999) Intrinsic bioremediation of petroleum hydrocarbons in a gas condensate-contaminated aquifer. Environ Sci Technol 33:2550–2560

Greene EA, Voordouw G (2003) Analysis of environmental microbial communities by reverse sample genome probing. J Microbiol Meth 53:211–219

Greene EA, Kay JG, Jaber K, Stehmeier LG, Voordouw G (2000) Composition of soil microbial communities enriched on a mixture of aromatic hydrocarbons. Appl Environ Microbiol 66:5282–5289

Greene EA, Kay JG, Stehmeier LG, Voordouw G (2002) Microbial community composition at an ethane pyrolysis plant site at different hydrocarbon inputs. FEMS Microbiol Ecol 40:233–241

Greer CW, Whyte LG, Lawrence JR, Masson L, Brousseau R (2001) Genomics technologies for environmental science. Environ Sci Technol 35:360A–366A

Grossman EL (1997) Stable carbon isotopes as indicators of microbial activity in aquifers. In: Hurst LJ, Knudsen GR, McInerney MJ, Stetzenbach LD, Walter MV (eds) Manual of Environmental Microbiology. ASM Press, Washington, pp 565–576

Grossman MJ, Prince RC, Garrett RM, Garrett KK, Bare RE, O'Neil KR, Sowlay SM, Hinton KL, Sergy GA, Owens EH, Guenette CC (2000) Microbial diversity in oiled and un-oiled shoreline sediments in the Norwegian Arctic. In: Bell CR, Brylinsky M, Johnson-Green (eds) Microbial biosystems: new frontiers. Proceedings of the 8th International Symposium on Microbial Ecology. Atlantic Canada Society for Microbiology, Kentville, NS, Canada, pp 775–787

Hanson JR, MacAlady JL, Harris D, Scow KM (1999) Linking toluene degradation with specific microbial populations in soil. Appl Environ Microbiol 65:5403–5408

Hellman B, Zelles L, Palojärvi A, Bai Q (1997) Emission of climate-relevant trace gases and succession of microbial communities during open-windrow composting. Appl Environ Microbiol 63:1011–1018

Hess A, Zarda B, Hahn D, Häner A, Stax D, Höhener P, Zeyer J (1997) In situ analysis of denitrifying toluene- and *m*-xylene-degrading bacteria in a diesel fuel-contaminated laboratory aquifer column. Appl Environ Microbiol 63:2136–2141

Holben WE, Schroeter BM, Calabrese VGM, Olsen RH, Kukor JK, Biederbeck VO, Smith AE, Tiedje JM (1992) Gene probe analysis of soil microbial populations selected by amendment with 2,4-dichlorophenoxyacetic acid. Appl Environ Microbiol 58:3941–3948

Hu H-Y, Lim B-R, Goto N, Fujie K (2001) Analytical precision and repeatability of respiratory quinones for quantitative study of microbial community structure in environmental samples. J Microbiol Meth 47:17–24

Hubert C, Shen Y, Voordouw G (1999) Composition of toluene-degrading microbial communities from soil at different concentrations of toluene. Appl Environ Microbiol 65:3064–3070

Jones GJ (1977) The effect of environmental factors on estimated viable and total populations of planktonic bacteria in lakes and experimental enclosures. Freshwater Biol 7:61–97

Juck D, Charles T, Whyte LG, Greer CW (2000) Polyphasic microbial community analysis of petroleum hydrocarbon-contaminated soils from two northern Canadian communities. FEMS Microbiol Ecol 33:241–249

Ka JO, Holben WE, Tiedje JM (1994) Use of gene probes to aid in recovery and identification of functionally dominant 2,4-dichlorophenoxyacetic acid-degrading populations in soil. Appl Environ Microbiol 60:1116–1120

Kasai Y, Kishira H, Syutsubo K, Harayama S (2001) Molecular detection of marine bacterial populations on beaches contaminated by the *Nakhodka* tanker oil-spill accident. Environ Microbiol 3(4):246–255

Koizumi Y, Kelly JJ, Nakagawa T, Urakawa H, El-Fantroussi S, Al-Muzaini S, Fukiu M, Urushigawa Y, Stahl DA (2002) Parallel characterization of anaerobic toluene- and ethylbenzene-degrading microbial consortia by PCR-denaturing gradient gel electrophoresis, RNA-DNA membrane hybridization and DNA microarray technology. Appl Environ Microbiol 68:3215–2115

Kurisu F, Satoh H, Mino T, Matsuo T (2002) Microbial community analysis of thermophilic contact oxidation process by using ribosomal RNA approaches and the quinone profile method. Wat Res 36:429–438

Lindstrom JE, Barry RP, Braddock JF (1999) Long-term effects on microbial communities after a subarctic oil spill. Soil Biol Biochem 31:1677–1689

MacNaughton SJ, Stephen JR, Venosa AD, Davis GA, Chang Y-J, White DC (1999) Microbial population changes during bioremediation of an experimental oil spill. Appl Environ Microbiol 65:3566–3574

Marsh TL, Saxman P, Cole J, Tiedje J (2000) Terminal restriction fragment length polymorphism analysis program, a web-based research tool for microbial community analysis. Appl Environ Microbiol 66(8):3616–3620

Matheson VG, Munakata-Marr J, Hopkins GD, McCarty PL, Tiedje JM, Forney LJ (1997) A novel means to develop strain-specific DNA probes for detecting bacteria in the environment. Appl Environ Microbiol 63:2863–2869

Mesarch MB, Nies L (1997) Modification of heterotrophic plate counts for assessing the bioremediation potential of petroleum-contaminated soils. Environ Technol 18: 639–646

Mesarch MB, Nakatsu CH, Nies L (2000) Development of catachol 2,3-dioxygenase-specific primers for monitoring bioremediation by competitive quantitative PCR. Appl Environ Microbiol 66:678–683

Pan YP, Li Y, Caufield PW (2001) Phenotypic and genotypic diversity of *Streptococcus sanguis* in infants. Oral Microbiol Immunol 16:235–242

Roane TM, Pepper IL (2000) Microscopic techniques. In: Maier RM, Pepper IL, Gerba CP (eds) Environmental microbiology. Academic Press, New York, pp 195–211

Röling WFM, van Breukelen BM, Braster M, Lin B, van Verseveld HW (2001) Relationships between microbial community structure and hydrochemistry in a landfill leachate-polluted aquifer. Appl Environ Microbiol 67:4619–4629

Röling WFM, Milner MG, Jones DM, Lee K, Daniel F, Swannell RJP, Head IM (2002) Robust hydrocarbon degradation and dynamics of bacterial communities during nutrient-enhanced oil spill bioremediation. Appl Environ Microbiol 68: 5537–5548

Schmalenberger A, Schwieger F, Tebbe CC (2001) Effect of primers hybridizing to different evolutionarily conserved regions of the small-subunit rRNA gene in PCR-based microbial community analyses and genetic profiling. Appl Environ Microbiol 67:3557–3563

Schwieger F, Tebbe CC (1998) A new approach to utilize PCR-single-strand-conformation polymorphism for 16S rRNA gene-based microbial community analysis. Appl Environ Microbiol 64:4870–4876

Shen Y, Stehmeier LG, Voordouw G (1998) Identification of hydrocarbon-degrading bacteria in soil by reverse sample genome probing. Appl Environ Microbiol 64: 637–645

Shi Y, Zwolinski MD, Schreiber ME, Bahr JM, Sewell GW, Hickey WJ (1999) Molecular analysis of microbial community structures in pristine and contaminated aquifers: field and laboratory microcosm experiments. Appl Environ Microbiol 65:2143–2150

Small J, Call DR, Brockman FJ, Straub TM, Chandler DP (2001) Direct detection of 16S rRNA in soil extracts by using oligonucleotide microarrays. Appl Environ Microbiol 67:4708–4716

Stahl DA (1997) Molecular approaches for the measurement of density, diversity and phylogeny. In: Hurst LJ, Knudsen GR, McInerney MJ, Stetzenbach LD, Walter MV (eds) Manual of environmental microbiology. ASM Press, Washington, pp 102–114

Stoffels M, Amann R, Ludwig W, Hekmat D, Schleifer K-H (1998) Bacterial community dynamics during start-up of a trickle-bed bioreactor degrading aromatic compounds. Appl Environ Microbiol 64:930–939

Tchelet R, Meckenstock R, Steinle P, Van der Meer JR (1999) Population dynamics of an introduced bacterium degrading chlorinated benzenes in a soil column and in sewage sludge. Biodegradation 10:113–125

Telang AJ, Ebert S, Foght JM, Westlake DWS, Jenneman GE, Gevertz D, Voordouw G (1997) The effect of nitrate injection on the microbial community in an oil field as monitored by reverse sample genome probing. Appl Environ Microbiol 63:1785–1793

Tresse O, Lorrain M-J, Rho D (2002) Population dynamics of free-floating and attached bacteria in a styrene-degrading biotrickling filter analyzed by denaturing gradient gel electrophoresis. Appl Microbiol Biotechnol 59:585–590

Voordouw G (1998) Reverse sample genome probing of microbial community dynamics. ASM News 64:627–633

Voordouw G, Voordouw JK, Karkhoff-Schweizer RR, Fedorak PM, Westlake DWS (1991) Reverse sample genome probing, a new technique for identification of bacteria in environmental samples by DNA hybridization, and its application to the identification of sulfate-reducing bacteria in oil field samples. Appl Environ Microbiol 57:3070–3078

Watanabe K, Yamamoto S, Hono S, Harayama S (1998) Population dynamics of phenol-degrading bacteria in activated sludge determined by *gyrB*-targeted quantitative PCR. Appl Environ Microbiol 64:1203–1209

White DC, Pinkart HC, Ringelberg DC (1997) Biomass measurements: biochemical approaches. In: Hurst LJ, Knudsen GR, McInerney MJ, Stetzenbach LD, Walter MV (eds) Manual of environmental microbiology. ASM Press, Washington, pp 91–101

Wilson VL, Tatford BC, Yu X, Rajki SC, Walsh MM, LaRock P (1999) Species-specific detection of hydrocarbon-utilizing bacteria. J Microbiol Meth 39:59–78)

Wünsche L, Brüggemann L, Babel W (1995) Determination of substrate utilization patterns of soil microbial communities: an approach to assess population changes after hydrocarbon pollution. FEMS Microbiol Ecol 17:295–306

3 Bioavailability and Biodegradation of Organic Pollutants – A Microbial Perspective

Jonathan D. Van Hamme[1]

Biodegradation – "... the biologically catalyzed reduction in complexity of chemicals ..."
Mineralization – "... conversion of an organic substrate to inorganic products ..."

Alexander (1999)

Bioavailability – for the purpose of this chapter, bioavailability refers to the capacity of an ion, element or compound to interact with and induce a response (e.g. metabolism or intoxication) in a living organism.

1 Why Do Microorganisms Biodegrade Organic Pollutants?

Environmental pollutants are a global issue due to direct contamination from growing industrialized centers, application of pesticides, herbicides and insecticides, and indirect contamination resulting from long-range atmospheric transport that distributes persistent pollutants such as polychlorinated biphenyls (PCBs) around the world (Meijer et al. 2003). Pollutant characteristics, environmental conditions, soil and vegetation type, and proximity to source create a complex set of conditions influencing pollutant lifecycles. Soils are key reservoirs for environmental pollutants with deposition and persistence being dependent on factors such as atmospheric exchange, formation of bound residues, burial, and biodegradation (Mackay 2001).

It is well known that microorganisms play the central role in nutrient cycling on this planet (Ehrlich 1995) and, at one time, their vast enzymatic capacity was considered to be infallible (Alexander 1999). However, fallibilities are evident as anthropogenic wastes accumulate in natural environments. So, in order to understand and enhance biodegradation, the question is often asked: why does a particular microorganism expend valuable energy resources to transform or metabolize an organic pollutant? Philosophical arguments aside, observations of microorganisms in natural and laboratory-based environments have revealed a number of reasons.

[1]Department of Biology, The University College of the Cariboo, 900 McGill Road, Kamloops B.C., V2C 5N3, Canada, e-mail: jvanhamme@cariboo.bc.ca, Tel: +1-250-3776064, Fax: +1-250-8285450

Soil Biology, Volume 2
Biodegradation and Bioremediation
(ed. by. A. Singh and O. P. Ward)

First, in any growth or survival situation, there will most likely be a limiting nutrient. For example, in soil environments, an organism may be limited in the amount of utilizable carbon present with respect to other key nutrients such as nitrogen, phosphorus, sulfur and trace metals important for the production of cellular structures and energy generation. Thus, if a new carbon source is introduced in the form of an organic pollutant, cells capable of extracting the carbon from such molecules will thrive (Annweiler et al. 2000; Prenafeta-Boldu et al. 2001). Indeed, it has been shown that, after such an influx, the proportion of organisms capable of metabolizing such compounds can increase from near non-detectable levels to sheer dominance (Leahy and Colwell 1990). Dominance may persist for long periods after the initial influx, as was found for hydrocarbon degrading microorganisms almost 20 years after a subarctic oil spill (Lindstrom et al. 1999). It is important to keep in mind that carbon is not the only nutrient available in organic pollutants. Nitrogen (Kilbane et al. 2000; Stamper et al. 2002), sulfur (Oldfield et al. 1998; Greene et al. 2000; Schleheck et al. 2003), phosphorus (Ramanathan and Lalithakumari 1999) and metals are other examples where pollutants can be the source of biogenic materials. In some cases, limited biotransformations take place before an organic pollutant becomes part of important cellular structures (Rodgers et al. 2000) or storage bodies (Ishige et al. 2002). For example, nocardioform bacteria such as the rhodococci possess complex cell envelopes containing mycolic acids, which are large 2-alkyl 3-hydroxy branched fatty acids (Sutcliffe 1998). These mycolic acids play an important role in protection from extracellular insults and dictate how these organisms partition with respect to hydrophobic-hydrophilic interfaces, such as those found at the junction of oil and water (Van Hamme and Ward 2001). Recent studies have shown that alkanes may be completely incorporated into membrane phospholipids in rhodococci (Rodgers et al. 2000). The route by which these compounds are added to the cell membrane is not well understood. That is, are these hydrocarbons transported first into the cell for processing followed by subsequent export to the cell surface, or do the enzymatic transformations and addition occur at the surface of the cell?

A second reason for microbial degradation of pollutants is that an organic pollutant may exert toxic pressure on a cell (Sikkema et al. 1995) and specific enzymatic modifications may be used to alter pollutant properties to allow for sequestration within, or removal from, the cell.

Third, a biodegradation process may be completely fortuitous from an anthropomorphic point of view. That is, either we have no understanding of the process, or the process is indeed without a true purpose. This may occur when an organic pollutant, apparently by chance, resembles a natural substrate of the enzyme system, when an enzyme system has a broad or relaxed substrate specificity to allow an organism to degrade a wide variety of substrates, or when extracellular enzyme systems are involved that generate long-armed free radicals to achieve reaction (Bogan and Lamar 1996; McKinzi and Dichristina 1999; Rodríguez et al. 1999; Inoue et al. 2003).

2 Microbial Metabolic Capabilities

Microorganisms are able to transform pollutants via the production of enzymes, chelating agents, vesicles and storage bodies, and cell surface agents. As described throughout this volume, organic species such as hydrocarbons, pesticides and explosives, or inorganic species such as heavy metals may be targeted substrates.

Reactions catalyzed by microorganisms may produce biomass, transient or stable metabolites, or, in the case of mineralization, inorganic products. The implications of microbially driven reactions are usually evaluated from a toxicity perspective, with mineralization being the ultimate form of detoxication. When metabolism does not lead to mineralization, newly formed compounds will have different properties with respect to bioavailability (biodegradability and toxicity) and mobility. Detoxication may occur due to decreased bioavailability following reactions that form soil-bound complexes (Kästner et al. 1999) or polymerizations that form high molecular weight compounds with unknown fates (Bressler and Fedorak 2001). Alternatively, metabolites may have greater toxicity (Maymo-Gatell et al. 1997) or inhibit biodegradation of other pollutants in a mixture (Kazunga and Aitken 2000).

The enzymes involved in these processes may be intracellular, membrane bound or extracellular, which presents unique challenges in terms of bioavailaility. For example, for intracellular and membrane-bound enzymes (e.g. the *alk* genes for alkane metabolism; van Beilen et al. 2001), there must necessarily be a mechanism for bringing the substrate into the cell to contact the enzyme. Extracellular enzymes work outside of the cell so uptake of the original pollutant is not necessarily required. For example, fungi produce a range of extracellular enzymes such as peroxidases (Pickard et al. 1991; Bogan et al. 1996; Novotný et al. 1999) and laccases (Thurston 1994) that have activity against environmental pollutants. Typically, extracellular enzymes are in the fungal realm and, while there are bacterial examples (Liao and McCallus 1998;

Ruijssenaars et al. 1999; Nankai et al. 1999), they are not often related to pollutants. Recently, however, Inoue et al. (2003) isolated three extracellular pyoverdins, normally produced in response to iron starvation, from *Pseudomonas chlororaphis* capable of cleaving tin-carbon bonds in a variety of organotin compounds.

Overall, the diversity of reactions catalyzed by microbial enzymes can be appreciated by examining single, broad-substrate enzymes. Naphthalene dioxygenase, for example, can catalyze over 76 types of reactions including mono- and di-hydroxylations, desaturations, *O*- and *N*-dealkylations and sulfoxidations against mono- and heterocyclic compounds (Ellis et al. 2000; Resnick et al. 1996). Also, individual microorganisms are known to possess more than one set of genes for the metabolism of a given pollutant, such as alkanes (Whyte et al. 2002) and aromatic hydrocarbons (Ferrero et al. 2002), have metabolic machinery to degrade more than one pollutant (Kahng et al. 2001; Story et al. 2000, 2001), or use both extracellular and intracellular enzymes for metabolism of a single substrate (Van Hamme et al. 2003). Finally, in any given ecosystem, a multitude of microorganisms with a diversity of metabolic capabilities will interact to carry out pollutant biotransformations (Trzesicka-Mlynarz and Ward 1995; Boonchon et al. 2000; Kanaly et al. 2000; Van Hamme and Ward 2001; Marchesi and Weightman 2003). Even with this metabolic arsenal, there are still times when pollutant accumulation becomes problematic.

3 Impediments to Microbial Biodegradation

In the environment, pollutants may be either resistant to biodegradation, biodegraded at very low rates, or degraded to an extent where a low residual concentration remains (Nocentini et al. 2000). Impediments to biodegradation are varied, intimately connected, and generally result from physical phenomena limiting substrate and cofactor bioavailability or the lack of appropriate biochemical machinery.

To begin, appropriate biochemical machinery may include not only catabolic enzymes that catalyze direct and indirect reactions against pollutants, but also include machinery for accessory functions for metabolism and protection from toxic effects. These include sensing and chemotaxis, biosurfactant production, changes to cell surface properties, and energy-dependent uptake and efflux. Generally, if a pollutant is a naturally occurring compound (as in the case of petroleum hydrocarbons), or closely resembles such a compound, existing biochemical machinery is able to carry out biodegradation. On the other hand, in cases where xenobiotic compounds have novel structures, existing machineries may not be sufficiently relaxed to accommodate.

When biochemical machinery is known to exist and biodegradation still does not occur, it may be a simple case of appropriate microorganisms or groups of microorganisms not being present in the contaminated environment. Microbes able to degrade pollutants are often widespread in soil and water environments under a range of redox conditions, and, if present, existing microbial populations and communities will be enriched upon the introduction of a biodegradable pollutant. Indeed, there are examples of biodegradative potential being present even in previously uncontaminated environments (Kent and Triplett 2002; Marchesi and Weightman 2003). However, there are situations when microorganisms with the appropriate characteristics are simply not available for enrichment. A striking example is when a site is contaminated with radioactive materials releasing constant lethal doses of ionizing radiation. *Deinococcus radiodurans*, a radiation tolerant bacterium, is being evaluated (Fredrickson et al. 2000) and engineered to remediate metals (Brim et al. 2000; Lloyd and Lovely 2001) and organic pollutants (Daly et al. 2000) in such environments. However, other environmental factors such as pH and temperature variabilities within a field site (Bending et al. 2003; Braddock and McCarthy 1996), or the absence of an appropriate co-substrate (Billingsley et al. 1999a; Smith et al. 2003) may limit biodegradation. While there are exceptions (Mishra et al. 2001), efforts to inoculate allochthonous microorganisms into contaminated environments often fail due to competition with adapted native organisms and, in cases where biodegradation is enhanced, improvements can be attributed to the introduction of a limiting nutrient rather than the inoculum itself (Venosa et al. 1992; Margesin and Schinner 1997).

Pollutants, depending on their properties, may enter the environment as solids, liquids or gases and be transported into different compartments depending on environmental conditions. Pollutants may remain as bulk plumes (NAPL or DNAPL) in groundwater, as undissolved solids, or be transported through dissolution processes. Flux and partitioning between air, water, soil and biological materials (plants, animals, microorganisms) will determine in part the bioavailability, or effective concentration, of any given pollutant (Harms and Bosma 1997).

Pollutant concentration is an important factor for two key reasons: toxicity and induction. Many pollutants are lipophilic and tend to partition naturally into cell wall structures. Since these pollutants possess solvent properties, toxic effects may include disruption of membrane structure resulting in leakage of important cellular constituents, destruction of membrane-based energy generation processes through dissipation of electrical potential and pH gradients, or inhibition of membrane proteins (Sikkema et al. 1995). The higher the concentration of pollutant, the more likely an organism will be unable to defend itself from toxic effects. On the other end of the spectrum, many enzymes involved in biodegradation are only expressed when contaminant, and in some cases cofactor, levels are above certain concentrations. Of course, enzyme activity depends on enzyme-substrate affinity constants and substrate concentration. For example, trichloroethylene (TCE) is degraded by methanotrophs expressing either a soluble, cytoplasmic methane monoxygenase (sMMO), or a membrane-associated particulate MMO (pMMO). Cells expressing pMMO typically degrade TCE 10 to 100 times slower than sMMO-expressing cells but are able to achieve greater extents of biodegradation due to a higher enzyme-substrate affinity. Interestingly, the bioavailability of copper in soil, dictated by soil binding sites, determines the rate of TCE biodegradation by *Methylosinus trichosporium* OB3b, as sMMO is expressed at low copper-to-biomass ratios, while pMMO is expressed at high ratios (Morton et al. 2000).

In soil systems, it is generally believed that pollutants first partition onto soil surfaces through reversible adsorption and, over time, bioavailability decreases as sorption to soil matter increases or as pollutants become trapped in micropores that are too small for microbial penetration. Here, hydrophobic contaminants may partition into soil organic matter or onto air-water interfaces and hydrophilic pollutants may adsorb to soil minerals (Haderlein and Schwarzenbach 1993). Generally, it is believed that desorption must necessarily precede biodegradation, but there are examples where sorbed substrates are degraded directly by attached cells (Park et al. 2001). Alternatively, a pollutant may become unavailable following a chemical transformation to a new form. For example, covalent linkage of chlorinated phenols and anilines to humic substances or soil phenols has been observed following reactions catalyzed by enzymes such as peroxidases, or by inorganic substances such as iron, silica and clay oxides and oxyhydroxides (Harms and Bosma 1997). In any case, whenever a biological or chemical reaction alters a pollutant, a new compound with different properties is produced.

Microorganisms have developed mechanisms to overcome both negative effects of toxicants and mass transfer limitations for metabolizable substrates and are often observed, in natural and laboratory environments, to have substantial effects on the state of environmental pollutants. Chemotaxis, cell surface properties and biosurfactants, and uptake and efflux mechanisms are discussed here.

4
Chemotaxis

In situations where biodegradation occurs, it is necessary for the capable microorganism and the pollutant to be in close proximity before other challenges such as solubility, concentration, partitioning, and transport can be overcome. In reactor-based systems, a primary consideration is proper mixing to enhance contact between microbial catalysts and their substrate. In natural environments, a pollutant may be spilled directly into an ecosystem home to capable microorganisms, or it may travel through as a plume. Bacterial motility driven by tactic responses may play an important role in these systems.

As a behavior, taxis may be metabolism-dependent energy taxis, metabolically independent, or driven by transport of compounds across membranes. Energy taxis dictated by fluctuating energy levels in the cell may be in response to oxygen (aerotaxis) and other electron acceptors, light (phototaxis), or a carbon source (chemotaxis) (Taylor et al. 1999).

Taxis can be beneficial to microorganisms capable of biodegrading a pollutant as it allows movement towards higher concentrations of substrate that may be stationary, adsorbed to a solid surface such as soil particles, or traveling within a plume. Alternatively, a microorganism may move away from toxic levels of a pollutant or its metabolites. Taxis also allows for microorganisms to associate with one another, often as a biofilm, or with other organisms such as plants which may increase biodegradation through exchange of catabolic plasmids, growth factors and metabolites. Many of these possible benefits have not been well established experimentally as pollutants and chemotaxis has not been well studied due to technical difficulties associated with performing chemotaxis assays with sparingly soluble and volatile compounds (Parales and Harwood 2002).

Recently, methods have been developed and chemotaxis towards aromatic hydrocarbons such as naphthalene and toluene, chloroaromatics such as 2,4-dichlorophenoxyacetate, chlorinated aliphatics such as DCE, TCE and PCE, and nitroaromatics such as *p*-nitrophenol have been observed in the laboratory (Pandey and Jain 2002; Parales and Harwood 2002). Chemoreceptors for salicylate and naphthalene (*nahY* in *Pseudomonas putida* G7) (Grimm and Harwood 1999), 4-hydroxybenzoate (*PcaK* in *P. putida* PRS2000) (Harwood et al. 1994), and 2,4-D [*tfdK* in *Ralstontia eutropha* JMP134(pJP4)] (Hawkins and Harwood 2002) have been discovered. In the latter two cases, the chemoreceptors are also required for active uptake (Leveau et al. 1998; Ditty and Harwood 1999). These few examples have shown that chemotaxis, uptake and pollutant biodegradation are linked as, in these cases, chemoreceptor genes are located within co-regulated biodegradative gene clusters (Parales and Harwood 2002). Gene-sequencing projects and more detailed analyses of known biodegradative gene clusters will yield new, and undoubtedly diverse, pollutant chemoreceptors. For example, van Beilen et al. (2001) have postulated that the *alkN* gene, found between the alkane-degrading *alkBFGHJKL* and *alkST* operons in *P. putida* GPo1, may be involved in chemotaxis. Sequence analysis has shown that *alkN* encodes a protein with 30% sequence similarity to methyl-accepting transducer proteins such as the one found in strain G7 for naphthalene (Grimm and Harwood 1999).

Evaluating the impact of chemotaxis on bioavailability may be difficult in the field, but excellent indirect evidence for enhanced biodegradation due to bacterial taxis has been provided in one laboratory study (Marx and Aitken 2000). Here, it was shown that, in unmixed aqueous systems, strain G7 was able to degrade naphthalene faster than both a chemotactic mutant and a non-motile mutant of the same strain. A new technique using magnetic resonance imaging and immunomagnetic labeling for observing the transport of live bacteria through porous media (Sherwood et al. 2003) will be useful for evaluating the impact of motility on biodegradation in the environment.

A better understanding of the genetic regulation of chemotaxis to environmental pollutants is required if a dynamic picture of how motile bacteria influence, and are influenced by, contaminant partitioning, biosurfactant production, soil properties, microbe–microbe, and microbe–plant interactions is to be drawn.

5
Cell Surface Properties and Biosurfactants

Microorganisms interact with their environments via their cell surfaces, and how a microorganism interacts with a pollutant will be determined by physical properties that are a function of environmental conditions, pollutant properties, and cell surface chemical composition. When examining bioavailability in natural environments, interfaces rule (van Loosdrecht et al. 1990). Air, water, soil, plants, macro- and microorganisms all create interfaces where transport and exchange take place to fulfill the requirements for biodegradation or intoxication.

Environmental pollutants are often hydrophobic in nature and tend to partition into soil organic matter or accumulate at hydrophobic interfaces such as air-water interfaces. To give hydrophobic, sparingly soluble compounds as an example, three microbial methods for accessing hydrophobic compounds are generally discussed. These are the uptake of sparingly soluble compounds from water, direct adhesion to bulk liquids or solids, and the production of biosurfactants to pseudosolubilize hydrophobic compounds in the aqueous phase. Note that these mechanisms are usually discussed under headings such as "hydrocarbon uptake mechanisms" (Volkering 1998). However, for terminology, "accession" is preferred as the steps to decrease the distance between microbe and pollutant precede internalization or uptake (Van Hamme and Ward 2001).

Focusing first on adherence to hydrophobic phases, it is apparent that if a microbe can control its cell surface properties then it will be better able to cope with accessing or avoiding pollutants that are either potential substrates or toxicants. There are many examples of microorganisms possessing highly hydrophobic cell surfaces (Hanson et al. 1994; Smits et al. 2003; Stelmack et al. 1999; Whyte et al. 1999), especially when grown in the presence of bulk hydrophobic phases. An example of interaction of hydrophobic and hydrophilic cells with crude oil (*Rhodococcus* sp. strain F9-D79 and *Pseudomonas* sp. strain JA5-B45) is shown in Fig. 1 (Van Hamme and Ward 2001). Increased hydrophobicity can be attributed to cell bound surface-active compounds (e.g. lipopolysaccharide, lipoteichoic acids, lipoglycans, glycolipids, fatty acids, neutral lipids, phospholipids), polymers (e.g. emulsan, liposan)

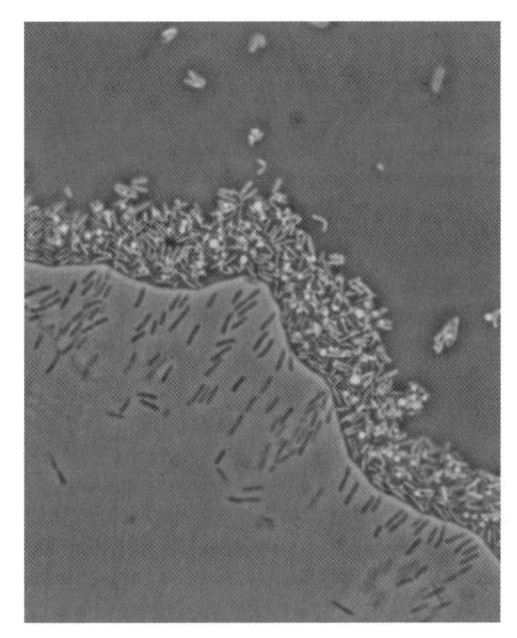

Fig. 1. Phase contrast micrograph showing the hydrophobic *Rhodococcus* sp. strain F9-D79 adhered to Bow River crude oil (*right*) with the hydrophilic *Pseudomonas* sp. strain JA5-B45 in the aqueous phase adjacent to the oil–water interface (*left*)

or proteins (Neu 1996). Adhesion may be required for pollutant metabolism as is observed when exogenous (bio)surfactants disrupt biodegradation by hydrophobic bacteria (Foght et al. 1989; Stelmack et al. 1999; Van Hamme and Ward 2001). Furthermore, the release of hydrophobic polymers from the cell surface has been observed to release bacteria from hydrocarbon droplets exhausted of utilizable carbon (Rosenberg 1993). Alterations in cell surface properties are also key determinants in natural biofilm colonization and spread. Recent studies have shown the importance of biofilms in a variety of processes and, under environmental conditions, participation in differentiated, functional biofilms is how microorganisms spend the majority of their time between transient planktonic adventures (Stoodley et al. 2002). As a physiological form, biofilms have been shown to be advantageous to microbial survival for many reasons including protection from toxic effects, metabolite exchange, nutrient accession and lateral gene exchange. Thus, microbial alterations of cell surface properties will impact pollutant bioavailability directly through interfacial phenomena and indirectly through biofilm colonization.

Biosurfactants released into the extracellular milieu serve various physiological functions and greatly impact pollutant bioavailability (Neu 1996; Desai and Banat 1997). Cell surface biosurfactants may be shed during bacterial de-adhesion from a hydrophobic surface as discussed above, or may be released to serve a direct role in pollutant accession. That is, biosurfactants may mediate pollutant micellar solubilization in order to enhance cell-substrate contact and subsequent uptake. Miller and Bartha (1989) performed some elegant experiments showing that encapsulation of a solid alkane (n-C_{36}) in liposomes greatly enhanced mass transport and allowed for bacterial metabolism to occur. Solubilization or desorption by surfactants has been observed for metals as well as organic pollutants (Christofi and Ivshina 2002). Alternatively, an excreted biosurfactant may serve as a conditioning film in order to alter a surface from hydrophobic to hydrophilic which allows for the adherence of a hydrophilic organism. Finally, biosurfactants play a role in gliding motility, which is a continuous bacterial desorption on a two-dimensional interface driven by surface–tension gradients (Neu 1996).

Historically, the literature has offered conflicting reports on the effect of (bio)surfactants on pollutant biodegradation. In fact, different effects are expected depending on toxic effects, diauxie, and the mechanism by which a microorganism accesses a pollutant. That is, a (bio)surfactant may be toxic, serve as a preferred carbon source, sequester substrate too strongly (Van Hamme and Ward 1999), improve mass transport due to desorption from surfaces and pseudosolubilization (Billingsley et al.

1999b), or disrupt cell-substrate hydrophobic interactions (Van Hamme and Ward 2001). Of course, these effects depend on environmental conditions, soil type, contaminant properties, and the presence of other microorganisms. For example, in one study, rhamnolipids and Triton X-100 were found to increase pesticide desorption from soils above the critical micellization concentration (cmc). Below the cmc, surfactant sorbed to soil resulting in a doubling of pesticide sorption due to the increased surface hydrophobicity of soil particles (Mata-Sandoval et al. 2002).

At this time, much of the evidence for the various roles of biosurfactants vis-à-vis pollutant bioavailability and biodegradation has been obtained in liquid culture. However, as the genetic systems regulating biosurfactant production are unraveled (Sullivan 1998), more direct studies are being undertaken. Holden et al. (2002) used green fluorescent protein (GFP) expression to show that, unlike in liquid culture, expression of genes for rhamnolipid and PA bioemulsifying protein did not improve biodegradation rates or end points for *n*-hexadecane in an unmixed sand culture. In sand cultures, it appeared that adherence to the hydrocarbon-water interface was more important for biodegradation than biosurfactant production. More studies of this type will probably not change the fundamentals of how biosurfactants impact bioavailability and bidegradation, but will allow for a better understanding of their impact in specific field situations.

6 Active Uptake and Efflux

Imagine a situation where a bacterial cell is in a position to internalize a pollutant for a given purpose. That is, all requirements to bring the cell and the potential substrate into contact have been met – be that through alterations in cell surface hydrophobicity, pollutant alteration by extracellular enzymes or other microbes in a consortium, biosurfactant production or chemotaxis. In addition, assuming the substrate is not passively internalized via equilibrium diffusion or partitioning, the question remains: are there specific mechanisms by which the cell can actively and specifically select and transport the pollutant into the intracellular milieu?

Specific, carrier-mediated transport systems (active transport and group translocation) driven by ATP or proton motive force are well known for polar molecules such as amino acids, sugars and ions. These systems are essential for cells to accumulate the necessary materials for cells to maintain homeostasis. For environmental pollutants, the story is beginning to emerge as microbial internalization of compounds such as 4-toluene sulfonate (Locher et al. 1993) and 4-chlorobenzoate (Groenewegen et al. 1990) into microbial cells are being linked to active transport.

To give petroleum hydrocarbons as an example, it is known that *n*-alkane degradation rates can exceed the rates of diffusive flux into microbial cells (Leahy and Colwell 1990) and that microbes can accumulate internal, unit membrane-bound vesicles of pure and partially oxidized alkanes (Kennedy et al. 1975; Scott and Finnerty 1976; Kim et al. 2002). In addition, it is known that *n*-alkane degradation often precedes the degradation of aromatic hydrocarbons when present as a complex mixture. Taken together, these observations suggest energy-driven, selective transport mechanisms. Indeed, recent evidence has shown that *Rhodococcus erythropolis* S+14He accumulates unaltered alkanes as inclusions via an unknown energy-dependent mechanism (Kim et al. 2002). Strikingly, S+14He is able to preferentially accumulate *n*-hexadecane from mixtures of structurally similar alkanes that would be expected to partition similarly, presenting an intriguing scenario whereby preferential alkane degradation may arise not only from enzyme-substrate affinities, but also from the presence of specific active uptake systems (Fig. 2).

A similar scenario has been observed in *Sphingomonas herbicidovorans* MH, a strain that degrades 2,4-D, and the enantiomeric herbicides (*RS*)-2-(2,4-dichlorophenoxy)propanoic acid (dichlorprop) and (*RS*)-2-(4-chloro-2-methylphenoxy)acetic acid (mecoprop) (Nickel et al. 1997; Zipper et al. 1998). Preferential degradation of (*S*)-mecoprop can be partly explained by the constitutive expression of a (*S*) enantiomer-specific α-ketoglutarate-dependent dioxygenase and the induction requirements for an (*R*)-specific dioxygenase (Nickel et al. 1997). Furthermore, MH has been found to posses separate uptake systems for 2,4-D, (*S*)-dichlorprop and Mecoprop, and (*R*)-dichlorprop and mecoprop. These uptake systems, probably constructed from enantiomeric-specific binding proteins, are driven by proton motive force (Zipper et al. 1998).

In the cases described above, specific pumps have not yet been identified by genetic methods and this is an area that deserves more attention. As mentioned earlier, the chemotactic bacterium *Ralstonia eutropha* JMP134 has a gene (*tfdK*), which is important for chemotaxis to 2,4-D. This gene, found on a large plasmid with other genes for 2,4-D utilization, is also involved in nonessential 2,4-D active transport. The predicted hydrophobic protein product (Leveau et al. 1998) is similar to transmembrane proteins of the major facilitator superfamily responsible for catalyzing unimport, symport and antiport functions in

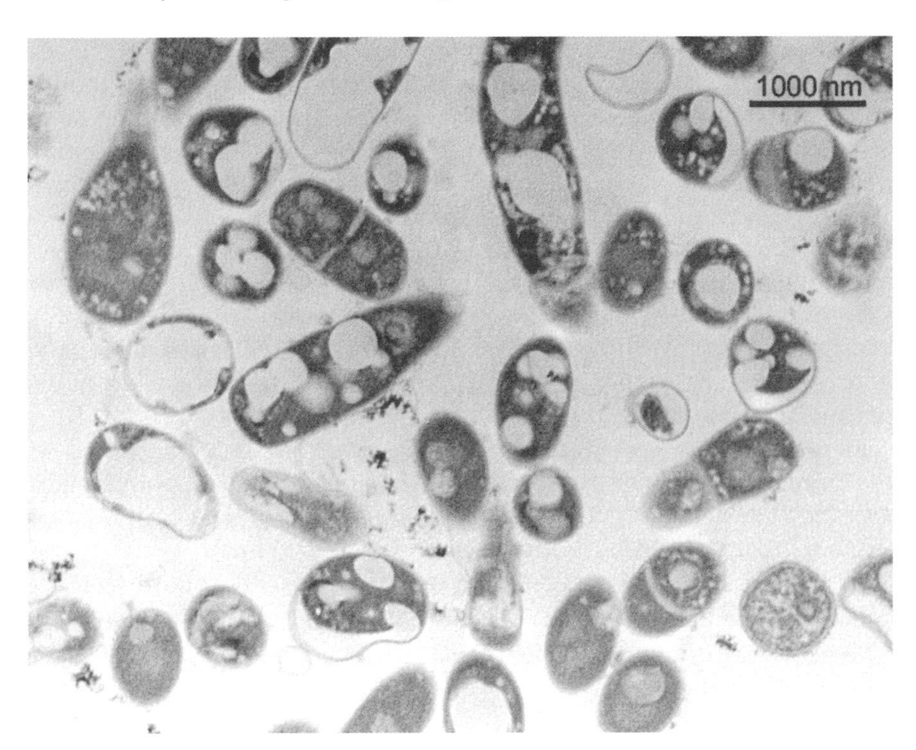

Fig. 2. Round hydrocarbon inclusions in *R. erythropolis* S+14He cells following growth on *n*-hexadecane

Archaea, *Bacteria* and *Eukarya* (Marger and Saier 1993; Pao et al. 1998).

Now consider a situation where a pollutant has been internalized, either within the leaflets of the cell membrane or into the cytoplasm, and is exerting a toxic effect or is simply no longer desirable. At this point, a microorganism may want to expel the pollutant in an active and specific manner. Since some microorganisms can survive in high concentrations of organic solvents that can have disastrous effects on biological membranes (Sikkema et al. 1995), and metabolite exchange is known to be important in the biodegradation of many pollutants, specific efflux mechanisms for lipophilic compounds were thought to exist. Indeed, exclusion of lipophilic compounds such as toluene, styrene, propylbenzene, ethylbenzene, anthracene and fluoranthene has been linked to active efflux mechanisms (Bugg et al. 2000; Ramos et al. 2002).

Inroads into the molecular basis for pollutant efflux have been recently made with aromatic hydrocarbons. Closely related to antibiotic efflux pumps, specific hydrocarbon efflux systems for toluene have been characterized in Pseudmonads (Kieboom et al. 1998; Ramos et al. 1998). In general, these pumps have been divided into three classes: those that only pump antibiotics, those that only pump hydrocarbons, and those that pump both antibiotics and hydrocarbons. These efflux pumps belong to the resistance-nodulation-division (RND) family (Ramos et al. 2002) which are constructed from a cytoplasmic-membrane pump protein, a peripheral cytoplasmic membrane linker protein, and an outer membrane periplasmic channel protein (Hancock and Brinkmann 2002). Their physiological role is to extrude potentially toxic compounds from the cytoplasm or, perhaps even directly from the periplasmic space, to the extracellular milieu.

In one of the few strains that have been well characterized at the molecular level, *P. putida* DOT-T1E, multiple efflux pumps exist and, in one case, the genes for toluene efflux are linked to the genes for toluene metabolism (Ramos et al. 1998, 2002). Thus, it is also possible that these pumps serve a homeostatic role and counter-balance specific uptake mechanisms to allow for metabolism while avoiding lethal effects at dangerous substrate concentrations. This is the case for a variety of transition metals that are essential for cell function at low levels and toxic at higher concentrations (Rosen 2002).

At this time, it is important to identify similar pumps in other organisms and for other environmental pollutants in order to start building an understanding of how pump specificity is determined. It is intriguing to consider if and how these pumps may be involved in metabolite efflux and how these phenomena impact pollutant biodegradation in microbial consortia.

7 Conclusions: Microbial Communities and Their Neighborhoods

This chapter has served as a brief introduction to some of the fundamentals of bioavailability and biodegradation of environmental pollutants. For a more detailed treatment of the subject, the reader is directed to Alexander (1999). As a final thought, it is useful to recall the words of W.T. Sedgwick from the inaugural issue of the Journal of Bacteriology, "It was an acute observation of the late C.S. Pierce that some of the most fruitful of modern sciences have been bred by the crossing of older ones" (Sedgwick 1916). For the future, careful consideration of advances in the understanding of microbial communities, biofilm structure and function, microbial mobility and metabolism, and pollutant mass transfer will allow for a deeper understanding of pollutant bioavailability and biodegradation in the environment.

Acknowledgements. Special thanks to Melissa Elsa Haveroen for critical reading of the manuscript.

References

Alexander M (1999) Biodegradation and bioremediation, 2nd edn. Academic Press, San Diego, pp 1–7

Annweiler E, Richnow HH, Antranikian G, Hebenbrock S, Garms C, Franke S, Francke W, Michaelis W (2000) Naphthalene degradation and incorporation of naphthalene-derived carbon into biomass in the thermophile *Bacillus thermoleovorans*. Appl Environ Microbiol 66:518–523

Bending GD, Lincoln SD, Sorensen SR, Morgan JAW, Aamand J, Walker A (2003) In-field spatial variability in the degradation of the phenyl-urea herbicide isoproturon is the result of interactions between degradative *Sphingomonas* spp. and soil pH. Appl Environ Microbiol 69:827–834

Billingsley KA, Backus SM, Ward OP (1999a) Production of metabolites from chlorobiphenyls by resting cells of *Pseudomonas* strain LB400 after growth on different carbon sources. Can J Microbiol 45:178–184

Billingsley KA, Backus SM, Ward OP (1999b) Effect of surfactant solubilization on biodegradation of polychlorinated biphenyl congeners by *Pseudomonas* LB400. Appl Microbiol Biotechnol 52:255–260

Bogan BW, Lamar RT (1996) Polycyclic aromatic hydrocarbon-degrading capabilities of *Phanerochaete laevis* HHB-1625 and its extracellular ligninolytic enzymes. Appl Environ Microbiol 62:1597–1603

Bogan BW, Lamar RT, Hammel KE (1996) Fluorene oxidation in vivo by *Phanerochaete chrysosporium* and in vitro during manganese peroxidase-dependent lipid peroxidation. Appl Environ Microbiol 62:1788–1792

Boonchan S, Britz ML, Stanley GA (2000) Degradation and mineralization of high-molecular-weight polycyclic aromatic hydrocarbons by defined fungal-bacterial cocultures. Appl Environ Microbiol 66:1007–1019

Braddock JF, McCarthy KA (1996) Hydrologic and microbiological factors affecting persistence and migration of petroleum hydrocarbons spilled in a continuous-permafrost region. Environ Sci Technol 30:2626–2633

Bressler DC, Fedorak PM (2001) Purification, stability, and mineralization of 3-hydroxy-2-formylbenzothiophene, a metabolite of dibenzothiophene. Appl Environ Microbiol 67:821–826

Brim H, McFarlan SC, Fredrickson JK, Minton KW, Zhai M, Wackett LP, Daly NJ (2000) Engineering *Deinococcus radiodurans* for metal remediation in radioactive mixed waste environments. Nat Biotechnol 18:85–90

Bugg T, Foght JM, Pickard MA, Gray MR (2000) Uptake and active efflux of polycyclic aromatic hydrocarbons by *Pseudomonas fluorescens* LP6a. Appl Environ Microbiol 66:5387–5392

Christofi N, Ivshina IB (2002) Microbial surfactants and their use in field studies of soil remediation. J Appl Microbiol 93:915–929

Daly MJ (2000) Engineering radiation-resistant bacteria for environmental biotechnology. Curr Opin Biotechnol 11:280–285

Desai JD, Banat IM (1997) Microbial production of surfactants and their commercial potential. Microbiol Mol Biol Rev 61:47–64

Ditty JL, Harwood CS (1999) Conserved cytoplasmic loops are important for both the transport and chemotaxis functions of PcaK, a protein from *Pseudomonas putida* with 12 membrane-spanning regions. J Bacteriol 181:5068–5074

Ehrlich HL (1995) Geomicrobiology. Marcel Dekker, New York.

Ellis LBM, Hershberger D, Wackett LP (2000) The Univeristy of Minnesota Biocatalysis/Biodegradation Database: microorganisms, genomics and prediction. Nucleic Acids Res 28:377–379

Ferrero M, Llobet-Brossa E, Lalucat J, Garciá-Valdés E, Rosselló-Mora R, Bosch R (2002) Coexistence of two distinct copies of naphthalene degradation genes in *Pseudomonas* strains isolated from the western Mediterranean region. Appl Environ Microbiol 68:957–962

Foght JM, Gutnick DL, Westlake DWS (1989) Effect of emulsan on biodegradation of crude oil by pure and mixed bacterial cultures. Appl Environ Microbiol 55:36–42

Fredrickson JK, Kostandarithes HM, Li SW, Plymale AE, Daly MJ (2000) Reduction of Fe(III), Cr(VI), U(VI), and Tc(VIII) by *Deinococcus radiodurans* R1. Appl Environ Microbiol 66:2006–2011

Greene EA, Beatty PH, Fedorak PM (2000) Sulfonlane degradation by mixed cultures and a bacterial isolate identified as a *Variovorax* sp. Arch Microbiol 174:111–119

Grimm AC, Harwood CS (1999) NahY, a catabolic plasmid-encoded receptor required for chemotaxis of *Pseudomonas putida* to the aromatic hydrocarbon naphthalene. J Bacteriol 181:3310–3316

Groenewegen PEJ, Driessen AJM, Konings WN, de Bont JAM (1990) Energy-dependent uptake of 4-chlorobenzoate in the coryneform bacterium NTB-1. J Bacteriol 172: 419–423

Haderlein SB, Schwarzenbach RP (1993) Adsorption of substituted nitrobenzenes and nitrophenols to mineral surfaces. Environ Sci Technol 37:316–326

Hancock REW, Brinkman FSL (2002) Function of *Pseudomonas* porins in uptake and efflux. Annu Rev Microbiol 56:17–38

Hanson KG, Kale VC, Desai AJ (1994) The possible involvement of cell surface and outer membrane proteins of *Acinetobacter* sp. A3 in crude oil degradation. FEMS Microbiol Lett 122:275–280

Harms H, Bosma TNP (1997) Mass transfer limitation of microbial growth and pollutant degradation. J Ind Microbiol Biotechnol 18:97–105

Harwood CS, Nichols NN, Kim M-K, Ditty JL, Parales RE (1994) Identification of the *pcaRKF* gene cluster from *Pseudomonas putida*: involvement in chemotaxis, biodegradation, and transport of 4-hydroxybenzoate. J Bacteriol 176: 6479–6488

Hawkins AC, Harwood CS (2002) Chemotaxis of *Ralstonia eutropha* JMP134 (pJP4) to the herbicide 2,4-dichlorophenoxyacetate. Appl Environ Microbiol 68:968–972

Holden PA, LaMontagne MG, Bruce AK, Miller WG, Lindow SE (2002) Assessing the role of *Pseudomonas aeruginosa* surface-active gene expression in hexadecane biodegradation in sand. Appl Environ Microbiol 68:2509–2518

Inoue H, Takimura O, Kawaguchi K, Nitoda T, Fuse H, Murakami K, Yamaoka Y (2003) Tin-carbon cleavage of organotin compounds by pyoverdine from *Pseudomonas chlororaphis*. Appl Environ Microbiol 69:878–883

Ishige T, Tani A, Takabe K, Kawasaki K, Sakai Y, Kato N (2002) Wax ester production from *n*-alkanes by *Acinetobacter* sp. Strain M-1: ultrastructure of cellular inclusions and role of acyl coenzyme A reductase. Appl Environ Microbiol 68:1192–1195

Kahng H-Y, Malinverni JC, Majko MM, Kukor JJ (2001) Genetic and functional analysis of the *tbc* operons for catabolism of alkyl- and chloroaromatic compounds in *Burkholderia* sp. Strain JS150. Appl Environ Microbiol 67:4805–4816

Kanaly RA, Bartha R, Watanabe K, Harayama S (2000) Rapid mineralization of benzo[*a*]pyrene by a microbial consortium growing on diesel fuel. Appl Environ Microbiol 66:4205–4211

Kästner M, Streibich S, Beyrer M, Richnow HH, Fritsche W (1999) Formation of bound residues during microbial degradation of [^{14}C]anthracene in soil. Appl Environ Microbiol 65:1834–1842

Kazunga C, Aitken MD (2000) Products from the incomplete metabolism of pyrene by polycyclic aromatic hydrocarbon-degrading bacteria. Appl Environ Microbiol 66:1917–1922

Kennedy RS, Finnerty WR, Sudarsanan K, Young RA (1975) Microbial assimilation of hydrocarbons. I. The fine structure of a hydrocarbon oxidizing *Acinetobacter* sp. Arch Microbiol 102:75–83

Kent AD, Triplett EW (2002) Microbial communities and their interactions in soil and rhizosphere ecosystems. Annu Rev Microbiol 56:211–236

Kieboom J, Dennis JJ, de Bont JAM, Zylstra GJ (1998) Identification and molecular characterization of an efflux pump involved in *Pseudomonas putida* S12 solvent tolerance. J Biol Chem 273:85–91

Kilbane II JJ, Ranganathan R, Cleveland L, Kayser KJ, Ribiero C, Linhares MM (2000) Selective removal of nitrogen from quinoline and petroleum by *Pseudomonas ayucida* IGTN9m. Appl Environ Microbiol 66:688–693

Kim IS, Foght JM, Gray MR (2002) Selective transport and accumulation of alkanes by *Rhodococcus erythropolis* S+14He. Biotechnol Bioeng 80:650–659

Leahy JG, Colwell RR (1990) Microbial degradation of hydrocarbons in the environment. Microbiol Rev 54:305–315

Leveau JH, Zehnder AJB, van der Meer JR (1998) The *tfdK* gene product facilitates the uptake of 2,4-dichlorphenoxyacetate by *Ralstonia eutropha* JMP134(pJP4). J Bacteriol 180:2237–2243

Liao C-H, McCallus DE (1998) Biochemical and genetic characterization of an extracellular protease from *Pseudomonas fluorescens* CY091. Appl Environ Microbiol 64: 914–921

Lindstrom JE, Barry RP, Braddock JF (1999) Long-term effects on microbial communities after a subarctic oil spill. Soil Biol Biochem 31:1677–1689

Lloyd JR, Lovely DR (2001) Microbial detoxification of metals and radionuclides. Curr Opin Biotechnol 12:248–253

Locher HH, Poolman B, Cook AM, Konings WN (1993) Uptake of 4-toluene sulfonate by *Comamonas testosteroni* T-2. J Bacteriol 175:1075–1080

Mackay D (2001) Multimedia environmental models: the fugacity approach. Lewis/ CRC, Boca Raton

Marchesi JR, Weightman AJ (2003) Diversity of α-halocarboxylic acid dehalogenases in bacteria isolated from a pristine soil after enrichment and selection on the herbicide 2,2-dichloropropionic acid (Dalapon). Environ Microbiol 5:48–54

Marger MD, Saier MJ (1993) A major superfamily of transmembrane facilitators that catalyse uniport, symport and antiport. Trends Biochem Sci 18:13–20

Margesin R, Schinner F (1997) Efficiency of indigenous and inoculated cold-adapted soil microorganisms for biodegradation of diesel oil in alpine soils. Appl Environ Microbiol 63:2660–2664

Marx RB, Aitken MD (2000) Bacterial chemotaxis enhances naphthalene degradation in a heterogenous aqueous system. Environ Sci Technol 34:3379–3383

Mata-Sandoval JC, Karns J, Torrents A (2002) Influence of rhamnolipids and Triton X-100 on the desorption of pesticides from soils. Environ Sci Technol 36:4669–4675

Maymo-Gatell X, Chien Y, Gossett JM, Zinder SH (1997) Isolation of a bacterium that reductively dechlorinates tetracloroethene to ethene. Science 276:1568–1571

McKinzi AM, Dichristina TJ (1999) Microbially driven Fenton reaction for transformation of pentachlorophenol. Environ Sci Technol 33:1886–1891

Meijer SN, Ockenden WA, Sweetman A, Breivik K, Grimalt JO, Jones KC (2003) Global distribution and budget of PCBs and HCB in background surface soils: implications for sources and environmental processes. Environ Sci Technol 37:667–672

Miller RM, Bartha R (1989) Evidence from liposome encapsulation for transport-limited microbial metabolism of solid alkanes. Appl Environ Microbiol 55:269–274

Mishra S, Jyot J, Kuhad RC, Lal B (2001) Evaluation of inoculum addition to stimulate *in situ* bioremediation of oily-sludge-contaminated soil. Appl Environ Microbiol 67: 1675–1681

Morton JD, Hayes KF, Semrau JD (2000) Bioavailability of chelated and soil-adsorbed copper to *Methylosinus trichosporium* OB3b. Environ Sci Technol 34:4917–4922

Nankai H, Hashimoto W, Miki H, Kawai S, Murata K (1999) Microbial system for polysaccharide depolymerization: enzymatic route for xanthan depolymerization by *Bacillus* sp. Strain GL1. Appl Environ Microbiol 65:2520–2526

Neu TR (1996) Significance of bacterial surface-active compounds in interaction of bacteria with interfaces. Microbiol Rev 60:151–166

Nickel K, Suter MJ-F, Kohler H-PE (1997) Involvement of two α-ketoglutarate-dependent dioxygenases in enantioselective degradation of (*R*)- and (*S*)-mecoprop by *Sphingomonas herbicidovorans* MH. J Bacteriol 179:6674–6679

Nocentini M, Pinelli D, Fava F (2000) Bioremediation of a soil contaminated by hydrocarbon mixtures: the residual concentration problem. Chemosphere 41:1115– 1123

Novontný A, Erbanová P, Šašek V, Kubátová A, Cajthaml T, Lang E, Krahl J, Zadrazil F (1999) Extracellular oxidative enzyme production and PAH removal in soil by exploratory mycelium of white rot fungi. Biodegradation 10:159–168

Oldfield C, Wood NT, Gilbert SC, Murray FD, Faure FR (1998) Desulphurization of benzothiophene and dibenzothiophene by actinomycete organisms belonging to the genus *Rhodococcus*, and related taxa. Antonie van Leeuwenhoek 74:119–132

Pandey G, Jain RK (2002) Bacterial chemotaxis towards environmental pollutants: role in bioremediation. Appl Environ Microbiol 68:5789–5795

Pao SS, Paulsen IT, Saier Jr. MH (1998) Major facilitator superfamily. Microbiol Mol Biol Rev 62:1–34

Parales RE, Harwood CS (2002) Bacterial chemotaxis to pollutants and plant-derived aromatic molecules. Curr Opin Microbiol 5:266–273

Park J-H, Zhao X, Voice TC (2001) Biodegradation of non-desorbable naphthalene in soils. Environ Sci Technol 35:2734–2740

Pickard MA, Kadima TA, Carmichael RD (1991) Chloroperoxidase, a peroxidase with potential. J Ind Microbiol 7:235–242

Prenafeta-Boldú F, Luykx DMAM, Vervoort J, de Bont JAM (2001) Fungal metabolism of toluene: monitoring of fluorinated analogs by ^{19}F nuclear magnetic resonance spectroscopy. Appl Environ Microbiol 67:1030–1034

Ramanathan MP, Lalithakumari D (1999) Complete mineralization of methylparathion by *Pseudomonas* sp. A3. Appl Biochem Biotechnol 80:1–12

Ramos JL, Duque E, Gallegos M-T, Godoy P, Ramos-González MI, Rojas A, Terán W, Segura A (2002) Mechanisms of solvent tolerance in Gram-negative bacteria. Annu Rev Microbiol 56:743–768

Ramos JL, Duque E, Godoy P, Segura A (1998) Efflux pumps involved in toluene tolerance in *Pseudomonas putida* DOT-T1E. J Bacteriol 180:3323–3329
Raushel FM (2002) Bacterial detoxification of organophosphate nerve agents. Curr Opin Microbiol 5:288–295
Resnick SM, Lee K, Gibson DT (1996) Diverse reactions catalyzed by naphthalene dioxygenase from *Pseudomonas* ap. Strain NCIB 9816. J Ind Microbiol Biotechnol 7:438–456
Rodríguez E, Pickard MA, Vazquez-Duhalt R (1999) Industrial dye decolorization by laccases from ligninolytic fungi. Curr Microbiol 38:27–32
Rodgers RP, Blumer EN, Emmett MR, Marshall AG (2000) Efficacy of bacterial bioremediation: demonstration of complete incorporation of hydrocarbons into membrane phospholipids from *Rhodococcus* hydrocarbon degrading bacteria by electrospray ionization Fourier transform ion cyclotron resonance mass spectrometry. Environ Sci Technol 34:535–540
Rosen BP (2002) Transport and detoxification systems for transition metals, heavy metals and metalloids in eukaryotic and prokaryotic microbes. Compar Biochem Physiol Part A 133:689–693
Rosenberg E (1993) Exploiting microbial growth on hydrocarbons – new markets. Trends Biotechnol 11:419–424
Ruijssenaars HJ, de Bont JAM, Hartman S (1999) A pyruvated mannose-specific xanthan lyase involved in xanthan degradation by *Paenibacillus alginolyticus* XL-1. Appl Environ Microbiol 65:2446–2452
Schleheck D, Lechner M, Schönenberger R, Suter MJ-F, Cook AM (2003) Desulfonation and degradation of the disulfodiphenylethercarboxylates from linear alkyldiphenyletherdisulfonate surfactants. Appl Environ Microbiol 69:938–944
Scott CLL, Finnerty WR (1976) Characterization of intracytoplasmic hydrocarbon inclusions from hydrocarbon-oxidizing *Acinetobacter* species HO1-N. J Bacteriol 127:481–489
Sedgwick WT (1916) The genesis of a new science – bacteriology. J Bacteriol 1:1–4
Sherwood JL, Sung JC, Ford RM, Fernandez EJ, Maneval JE, Smith JA (2003) Analysis of bacterial random motility in a porous medium using magnetic resonance imaging and immunomagnetic labeling. Environ Sci Technol 37: 781–785
Sikkema J, de Bont JAM, Poolman B (1995) Mechanisms of membrane toxicity of hydrocarbons. Microbiol Rev 59:201–222
Smith CA, O'Reilly KT, Hyman MR (2003) Characterization of the initial reactions during the cometabolic oxidation of methyl *tert*-butyl ether by propane-grown *Mycobacterium vaccae* JOB5. Appl Environ Microbiol 69:796–804
Smits THM, Wick LY, Harms H, Keel C (2003) Characterization of the surface hydrophobicity of filamentous fungi. Environ Microbiol 5:85–91
Stamper DM, Radosevich M, Hallberg KB, Traina SJ, Tuovinen OH (2002) *Ralstonia basilensis* M91-3, a denitrifying soil bacterium capable of using *s*-triazines as nitrogen sources. Can J Microbiol 48:1089–1098
Stelmack PL, Gray MR, Pickard MA (1999) Bacterial adhesion of soil contaminants in the presence of surfactants. Appl Environ Microbiol 65:163–168
Stoodley P, Sauer K, Davies DG, Costerton JW (2002) Biofilms as integrated differentiated communities. Annu Rev Microbiol 56:187–209
Story SP, Parker SH, Hayasaka SS, Riley MB, Kline EL (2001) Convergent and divergent points in catabolic pathways involved in utilization of fluoranthene, naphthalene, anthracene, and phenanthrene by *Sphingomonas paucimobilis* var. EPA505. J Ind Microbiol Biotechnol 26:369–382

Story SP, Parker SH, Kline JD, Tzeng T-RJ, Mueller JG, Kline EL (2000) Identification of four structural genes and two putative promoters necessary for utilization of naphthalene, phenanthrene, and fluoranthene by *Sphingomonas paucimobilis* var. EPA505. Gene 260:155–169

Sullivan ER (1998) Molecular genetics of biosurfactant production. Curr Opin Biotechnol 9:263–269

Sutcliffe IC (1998) Cell envelope composition and organisation in the genus *Rhodococcus*. Antonie van Leeuwenhoek 74:49–58

Taylor BL, Zhulin IB, Johnson MS (1999) Aerotaxis and other energy-sensing behaviour in bacteria. Annu Rev Microbiol 53:103–128

Thurston CF (1994) The structure and function of fungal laccases. Microbiology 140: 19–26

Trzesicka-Mlynarz D, Ward OP (1995) Degradation of polycyclic aromactic hydrocarbons (PAHs) by a mixed culture and its component pure cultures, obtained from PAH-contaminated soil. Can J Microbiol 41:470–476

van Beilen JB, Panke S, Lucchini S, Franchini AG, Röthlisberger M, Witholt B (2001) Analysis of *Pseudomonas putida* alkane-degradation gene clusters and flanking insertion sequences: evolution and regulation of the *alk* genes. Microbiology 147: 1621–1630

Van Hamme JD, Ward OP (1999) Influence of chemical surfactants on the biodegradation of crude oil by a mixed bacterial culture. Can J Microbiol 45:130–137

Van Hamme JD, Ward OP (2001) Physical and metabolic interactions of *Pseudomonas* sp. Strain JA5-B45 and *Rhodococcus* sp. Strain F9-D79 during growth on crude oil and effect of a chemical surfactant on them. Appl Environ Microbiol 67:4874–4879

Van Hamme JD, Wong ET, Dettman H, Gray MR, Pickard MA (2003) Dibenzyl sulfide metabolism by white rot fungi. Appl Environ Microbiol 69:1320–1324

van Loosdrecht MCM, Lyklema J, Norde W, Zehnder AJB (1990) Influence of interfaces on microbial activity. Microbiol Rev 54:75–87

Venosa AD, Haines JR, Allen DM (1992) Efficacy of commercial inocula in enhancing biodegradation of weathered crude oil contaminating a Prince William Sound beach. J Ind Microbiol Biotechnol 10:1–11

Volkering F, Breure AM, Rulkens WH (1998) Microbiological aspects of surfactant use for biological soil remediation. Biodegradation 8:401–417

Whyte LG, Smits THM, Labbé D, Witholt B, Greer CW, van Beilen JB (2002) Gene cloning and characterization of multiple alkane hydroxylase systems in *Rhodococcus* strains Q15 and NRRL B-16531. Appl Environ Microbiol 68:5933–5942

Whyte LG, Slagman SJ, Pietrantonio F, Bourbonnière L, Koval SF, Lawrence JR, Inniss WE, Greer CW (1999) Physiological adaptations involved in alkane assimilation at a low temperature in *Rhodococcus* sp. Strain Q15. Appl Environ Microbiol 65: 2961–2968

Zipper C, Bunk M, Zehnder AJB, Kohler H-PE (1998) Enantioselective uptake and degradation of the chiral herbicide dichlorprop [(*RS*)-2,4-dichlorophenoxy)proanoic acid] by *Sphinogomonas herbicidovorans* MH. J Bacteriol 180:3368–3374

4 Anaerobic Biodegradation of Hydrocarbons

John D. Coates[1]

1 Introduction

Hydrocarbons are a generic group of compounds composed exclusively of hydrogen and carbon. They represent one of the most important groups of chemicals to mankind because of their natural abundance, their industrial importance, their extensive use as a primary energy source throughout the world, and their toxicity. Benzene, for example, has a broad range of industrial uses and represents one of the top 20 production volume chemicals produced in the United States, which represents 35% of the worldwide production. In addition to use in petroleum-based fuels, benzene is used for the manufacture of a diversity of other chemicals, rubbers, lubricants, dyes, detergents, drugs, and pesticides. Alternative sources, including volcanoes, forest fires, and cigarette smoke, also contribute significantly to benzene in the environment. Benzene is considered one of the most prevalent organic contaminants in groundwater (Anderson and Lovley 1997) and poses a significant health risk due to its toxicity and relatively high solubility. It is ranked fifth on the US National Priorities List (NPL), and has been found in more than 50% of the 1428 current or former NPL sites (URL: http://www.atsdr.cdc.gov/cxcx3.html). Benzene is highly toxic and is a known human carcinogen and the United States Environmental Protection Agency (EPA) has set the maximum permissible level of $5\,\mu g\,l^{-1}$ of benzene in drinking water with an ultimate goal of zero tolerance.

Microbial interaction with hydrocarbons such as benzene has been a major focus of study over the last 50 years. Although initial research efforts focused on the involvement of microorganisms in the formation of petroleum deposits (Zobell 1945, 1946, 1949, 1950; Stone and Zobell 1952), there has been a recent redirection towards the microbial catab-

[1]Department of Plant and Microbial Biology, University of California Berkeley, Berkeley, California 94720, USA, e-mail: jcoates@nature.berkeley.edu, Tel: +1-510-6438455, Fax: +1-510-6424995

Soil Biology, Volume 2
Biodegradation and Bioremediation
(ed. by. A. Singh and O. P. Ward)

olism of hydrocarbons. This was primarily motivated by the growing need for remediation of environments contaminated as a result of crude and fuel oil spills (Atlas 1981, 1995).

Prior to the 1980s, microbial hydrocarbon catabolism studies were all performed under aerobic conditions and anaerobic hydrocarbon degradation by microorganisms was not considered significant (Atlas 1981). It was not until the late 1980s that it was conclusively demonstrated that hydrocarbons could be biodegraded in the absence of oxygen (Vogel and Grbic-Galic 1986; Grbic-Galic and Vogel 1987). These studies focused on the degradation of toluene and benzene using strict anaerobic techniques under methanogenic conditions (Vogel and Grbic-Galic 1986). Since then, many studies have demonstrated the anaerobic biodegradability of a diverse range of hydrocarbons including both aliphatic and aromatic structures (Lovley et al. 1994, 1995; Coates et al. 1996a–c, 1997, 2001a, b; Burland and Edwards 1999).

Hydrocarbon degradation has now been demonstrated under the dominant terminal electron accepting processes known to occur in anaerobic environments including nitrate (Burland and Edwards 1999; Coates et al. 2001b), Fe(III) (Lovley et al. 1994, 1996; Anderson et al. 1998; Anderson and Lovley 1999), sulfate (Edwards et al. 1992; Lovley et al. 1995; Coates et al. 1996a, b; Phelps et al. 1996; Kazumi et al. 1997; Weiner and Lovley 1998a; Kropp et al. 2000), and CO_2 (Vogel and Grbic-Galic 1986; Grbic-Galic and Vogel 1987; Kazumi et al. 1997; Weiner and Lovley 1998b). Many pure cultures of anaerobic organisms that can degrade hydrocarbons such as benzene, toluene, ethylbenzene, xylene, hexadecane, and naphthalene have been described (Lovley and Lonergan 1990; Rabus et al. 1993; Beller et al. 1996; Coates et al. 1996b, 2001a, b; Galushko et al. 1999; Harms et al. 1999a, b) and many new and novel catabolic pathways and genetic systems have been elucidated.

2 Monoaromatic Compounds

Monoaromatic hydrocarbons such as benzene, toluene, ethylbenzene and xylene (Fig. 1), collectively know as BTEX, are commonly found in gasoline and are highly volatile substances (Coates et al. 2002). Due to their relatively high solubility and toxicity, they represent a significant health risk and primarily enter the environment through processes associated with manufacture, storage, and handling of gasoline and petroleum fuels. Considerable emissions also result from aqueous discharge of a broad range of industrial effluents including metal, paint, and textile manufacture, wood processing, chemical production, and tobacco products.

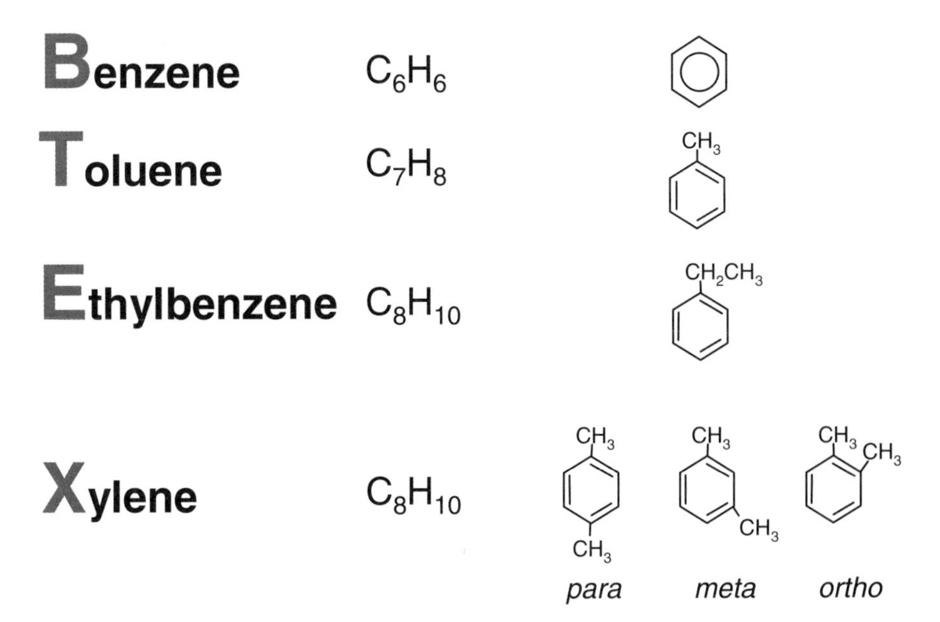

Fig. 1. Molecular formula and structure of the respective BTEX components

2.1 Toluene Biodegradation

Of the BTEX components, the anaerobic biodegradation of toluene has been most intensely studied and is probably the most comprehensively understood. Toluene is biodegradable with nitrate, Mn(IV), Fe(III), sulfate, or CO_2 (methanogenesis) as terminal electron acceptors (Vogel and Grbic-Galic 1986; Grbic-Galic and Vogel 1987; Lovley et al. 1989; Dolfing et al. 1990; Evans et al. 1991a, b; Edwards et al. 1992; Rabus et al. 1993; Fries et al. 1994; Coates et al. 1996b; Langenhoff et al. 1997; Meckenstock 1999). More recently, it has been demonstrated that anaerobic toluene degradation can also be coupled to the reduction of humic substances (Cervantes et al. 2001) or chlorine oxyanions such as chlorate or perchlorate (Coates et al. 2001b), or it can be assimilated as a carbon source by anoxygenic phototrophs (Zengler et al. 1999a).

Geobacter metallireducens strain GS-15 was the first organism in pure culture demonstrated to be capable of the anaerobic oxidation of toluene which it completely oxidized to CO_2 with the reduction of Fe(III) (Lovley et al. 1989). *G. metallireducens* was incapable of oxidizing other BTEX components (Lovley et al. 1989; Lovley and Lonergan 1990;

Lonergan and Lovley 1991). Other *Geobacter* species such as *G. grbiciae* have since been isolated and described which can also oxidize toluene with Fe(III) (Coates et al. 1996c, 2001a; Coates and Lovley 2003).

Several organisms are now known to couple anaerobic toluene degradation to various other forms of anaerobic respiration including nitrate reduction by *Thauera aromatica* K172 (Anders et al. 1995), *Thauera aromatica* T1 (Evans et al. 1991a), *Azoarcus* sp. strain T (Dolfing et al. 1990), *Azoarcus tolulyticus* Tol4 (Zhou et al. 1995), *Azoarcus tolulyticus* Td15 (Fries et al. 1994), Strain ToN1 (Rabus and Widdel 1995), *Dechloromonas* strain RCB and *Dechloromonas* strain JJ (Coates et al. 2001b), perchlorate reduction by *Dechloromonas* strain RCB (Coates et al. 2001b), and sulfate reduction by *Desulfobacula toluolica* (Rabus et al. 1993) and *Desulfobacterium cetonicum* (Harms et al. 1999b).

All of the toluene-oxidizing nitrate reducers are facultative anaerobes and are members of the beta subclass of the Proteobacteria. Several of the *Azoarcus* and *Thauera* species were originally described as *Pseudomonas* species (Dolfing et al. 1990; Schocher et al. 1991), but have subsequently been reclassified into their current taxonomic positions (Anders et al. 1995; Zhou et al. 1995). Toluene-degrading *Azoarcus* and *Thauera* species are readily isolated from anaerobic sludge or creek sediments with nitrate or nitrous oxide as the electron acceptor and various electron donors (Schocher et al. 1991; Anders et al. 1995).

The biochemistry and genetics of toluene degradation by the *Azoarcus* and *Thauera* species have been intensely investigated over the last decade (Biegert et al. 1996; Beller and Spormann 1997a, 1999; Coschigano and Young 1997; Heider et al. 1998; Coschigano 1999; Krieger 1999) and comprehensively reviewed elsewhere (Spormann and Widdel 2000; Widdel and Rabus 2001). Briefly, these studies revealed that the first step in the catabolism of toluene is the addition of fumarate onto the toluene methyl group to form benzylsuccinate (Fig. 2). This reaction is mediated by a novel glycyl radical enzyme benzylsuccinate synthase (BSS) (Leuthner et al. 1998). Free benzylsuccinate is often found in culture broths of *T. aromatica* as a transient intermediate when grown on toluene (Biegert et al. 1996), and is now believed to be an inducer of the genetic pathway for toluene catabolism (P.W. Coschigano, pers. comm.).

Although the BSS-based toluene pathway was first identified in *Thauera* and *Azoarcus* species growing under nitrate-reducing conditions, this is now recognized as the common mechanism for the activation of toluene by phylogenetically diverse organisms growing under a range of alternative anaerobic conditions. These include dissimilatory Fe(III)-reduction by *Geobacter metallireducens* (Kane et al. 2002) and

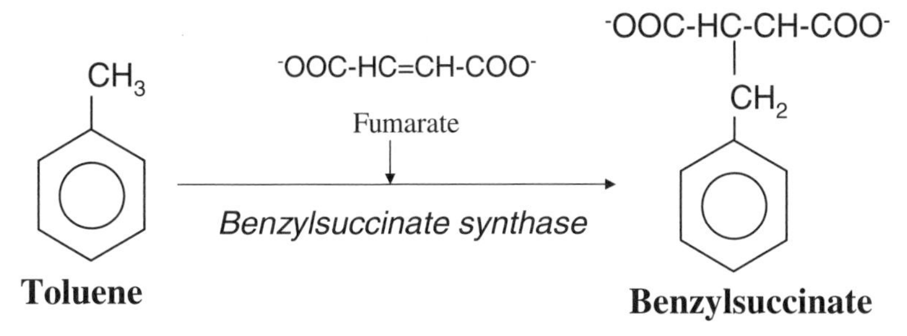

Fig. 2. Activation of toluene through addition of fumarate to the methyl group to form benzylsuccinate. This activation step appears to be highly conserved amongst all known anaerobic toluene-degrading organisms

dissimilatory sulfate reduction by *Desulfobacula toluolica* (Beller and Spormann 1997b) both of which are members of the delta subclass of the Proteobacteria. Similarly, this mechanism is also utilized during the anoxic phototrophic assimilation of toluene by *Blastochloris sulfoviridis*, of the alpha subclass of the Proteobacteria (Zengler et al. 1999a).

2.2 Benzene Biodegradation

Although isolates capable of the anaerobic oxidation of toluene and the metabolic pathways involved have been extensively documented in recent years, the anaerobic oxidation of benzene has been comparatively more difficult. Organisms capable of anaerobic benzene degradation have, until recently, been elusive and this metabolism has only been observed in sediment studies (Lovley et al. 1994, 1995, 1996; Coates et al. 1996b, 1997; Anderson et al. 1998; Weiner and Lovley 1998a, b; Anderson and Lovley 1999) or with microbial enrichments (Vogel and Grbic-Galic 1986; Grbic-Galic and Vogel 1987; Weiner and Lovley 1998a; Burland and Edwards 1999). Although anaerobic benzene degradation has been demonstrated under nitrate-reducing (Burland and Edwards 1999), Fe(III)-reducing (Lovley et al. 1994, 1996), sulfate-reducing (Lovley et al. 1995; Coates et al. 1996a, 1997), and methanogenic (Grbic-Galic and Vogel 1987; Weiner and Lovley 1998b) conditions, until recently, no specific organisms or genera have been associated with this ability. The first two organisms, strains RCB and JJ, capable of anaerobic benzene degradation were isolated and described by Coates and coworkers (Coates et al. 2001b). These organisms were closely related to each other and were members of the newly described

Dechloromonas genus in the beta subclass of the Proteobacteria (Achenbach et al. 2001; Coates et al. 2001b). Both strains completely oxidized benzene to CO_2 in the absence of oxygen and coupled this metabolism to the reduction of nitrate (Coates et al. 2001b). Members of the *Dechloromonas* genus are generally recognized for their ability to grow by dissimilatory perchlorate reduction (Bruce et al. 1999; Coates et al. 1999; Achenbach and Coates 2000; Achenbach et al. 2001; Coates 2003), another common groundwater contaminant associated with the activity of the munitions industry (Urbansky 1998) and when tested, one of the isolates, *Dechloromonas* strain RCB, could also couple growth and benzene oxidation to the reduction of perchlorate (Coates et al. 2001b). As such, this organism offers great potential for the bioremediation of environments, co-contaminated with both BTEX and perchlorate in a single treatment strategy. Previous studies have demonstrated that these organisms are, in general, metabolically versatile and can use a broad range of alternative electrons donors (Coates 2003). As such, the selective pressures for *Dechloromonas* species in the environment may be based on the diversity of their metabolic capabilities rather than on any individual metabolism. In support of their potential role in the nitrate-dependent anaerobic biodegradation of benzene a recent molecular analysis of a benzene-degrading nitrate-reducing enrichment culture indicated that the microbial population was dominated (70% of the cloned 16S rRNA gene sequences) by an organism 93% identical to *Dechloromonas* strain JJ (Ulrich and Edwards 2003). Interestingly, this organism was equally related (93% identical based on 16S rDNA sequence homology) to *Azoarcus* species (Ulrich and Edwards 2003), which are well known for their ability to anaerobically degrade other BTEX components.

The biochemical pathway of anaerobic benzene degradation is currently unknown but several possibilities exist for the initial reaction (for a review see Coates et al. 2002) including an initial alkylation to form toluene, a hydroxylation to form phenol, or a carboxylation to form benzoate (Fig. 3). Previous independent studies, performed with sediments or enrichment cultures, have demonstrated the formation of phenol and benzoate as extracellular intermediates, during the anaerobic degradation of benzene (Coates et al. 2002 and references therein).

2.3 Ethylbenzene Biodegradation

In contrast to toluene and benzene, there is relatively little known about the anaerobic biodegradation of ethylbenzene and, until recently, only

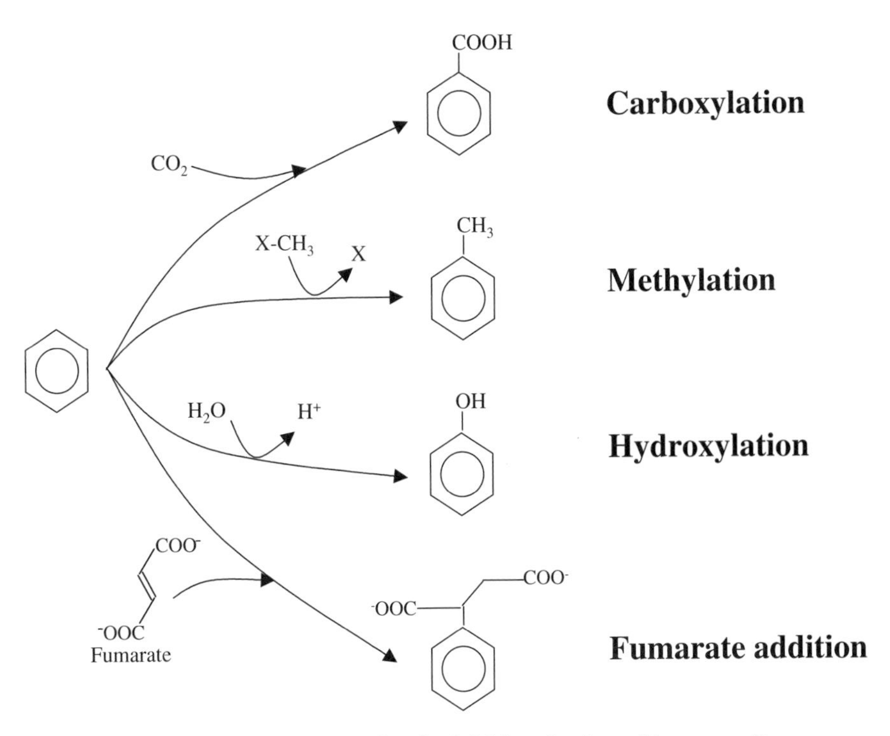

Fig. 3. Four putative mechanisms for the initial activation of benzene. *X* may represent tetrahydrofolate, S-adenosyl-methionine, or cobalamin protein

three organisms had been described that were capable of this metabolism (Rabus and Widdel 1995; Ball et al. 1996). These organisms, strains EbN1, PbN1 (Rabus and Widdel 1995) and EB1 (Ball et al. 1996), were facultative anaerobes and coupled ethylbenzene oxidation to the reduction of nitrate. The three isolates were closely related to each other and to the previously described *Thauera* species in the beta subclass of the Proteobacteria. Strains EbN1 and PbN1 were isolated, with ethylbenzene and propylbenzene, respectively, from enrichments prepared with homogenized mixtures of river and ditch mud samples (Rabus and Widdel 1995), while strain EB1 was isolated from ethylbenzene-degrading enrichments, prepared with sediment from an oil refinery treatment pond (Ball et al. 1996). In general, ethylbenzene was completely mineralized to CO_2 by these isolates with the reduction of nitrate, the transitory production of nitrite, and ultimate formation of N_2 (Rabus and Widdel 1995; Ball et al. 1996). Ethylbenzene is initially activated for degradation by these organisms through a dehydrogenation reaction of the methylene group of the ethyl side chain to form 1-phenylethanol, which is further oxidized to form the aromatic ketone acetophenone

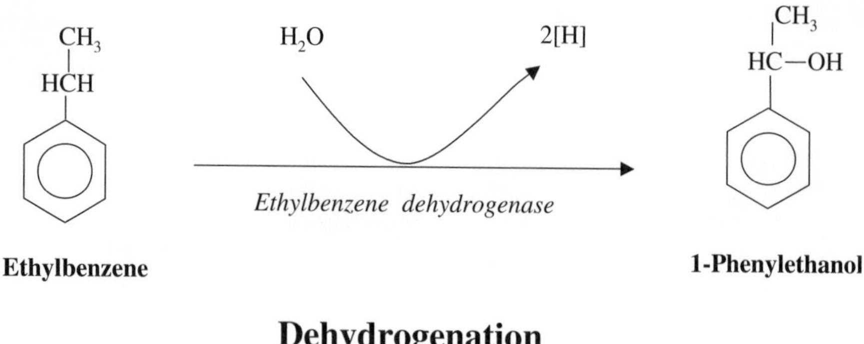

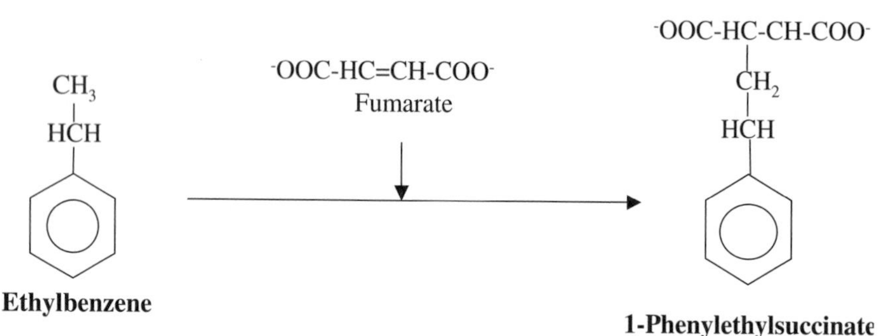

Fig. 4. Activation of ethylbenzene with nitrate as the electron acceptor is effected through a dehydrogenation reaction at the methylene group of the ethyl side chain to form 1-phenylethanol. With sulfate as the electron acceptor ethylbenzene is activated by a fumarate addition reaction at the secondary carbon atom of the ethyl group to form 1-phenylethylsuccinate

(Ball et al. 1996; Rabus and Heider 1998; Fig. 4). Stable isotope labeling studies indicated that the hydroxyl group of the 1-phenylethanol formed in the initial dehydrogenation reaction is derived from water (Ball et al. 1996) and the reaction is mediated by ethylbenzene dehydrogenase, a novel member of the dimethyl sulfoxide reductase family of molybdoterin-containing enzymes (Johnson et al. 2001). In strain EbN1, complete ethylbenzene mineralization involves a complex genetic system with a two component regulatory mechanism (Rabus et al. 2002).

More recent studies on anaerobic ethylbenzene degradation under sulfate-reducing conditions have resulted in the enrichment and isolation of a novel organism, strain EbS7 from enrichments prepared with marine sediments (Kniemeyer et al. 2003). Strain EbS7 completely oxidizes ethylbenzene to CO_2 with the dissimilatory reduction of sulfate to sulfide (Kniemeyer et al. 2003). Strain EbS7 is a member of the delta subclass of the Proteobacteria and is most closely related to strains NaphS2 and mXyS1, which can anaerobically oxidize naphthalene and *m*-xylene, respectively (Galushko et al. 1999; Harms et al. 1999b; Kniemeyer et al. 2003). In contrast to the initial dehydrogenation reaction, used by denitrifying ethylbenzene degraders, activation of ethylbenzene for catabolism by strain EbS7 is achieved by a fumarate addition reaction at the secondary carbon atom of the ethyl group to form 1-phenylethyl-succinate (Kniemeyer et al. 2003; Fig. 4). This alternative catabolic pathway for aromatic ethylbenzene degradation, which is similar to the activation step for toluene outlined above, is explained by the difference in the redox potential of the respective electron acceptors, nitrate and sulfate (Kniemeyer et al. 2003). A similar type of activation reaction at the secondary carbon group has previously been observed in the anaerobic catabolism of *n*-alkanes under nitrate-reducing conditions (Rabus et al. 2001).

2.4 Xylene Biodegradation

Anaerobic biodegradation of the three structural isomers of dimethylbenzene (*meta*-, *ortho*-, and *para*-xylene) has been predominantly studied under nitrate- and sulfate-reducing conditions. Although, studies based on sediments or enrichment cultures were demonstrated to biodegrade *para*-xylene in the absence of oxygen (Kuhn et al. 1988; Haner et al. 1995), to date, no pure culture that can mineralize *p*-xylene completely to CO_2 exists. In contrast, several organisms have now been isolated which can completely mineralize *meta*- and *ortho*-xylene coupled to the reduction of nitrate (Rabus and Widdel 1995; Hess et al. 1997). Many of these organisms are closely related to each other and to the previously identified toluene-degrading denitrifiers belonging to the beta-subclass of the Proteobacteria. In addition, several of the known toluene-degrading *Azoarcus* or *Thauera* species are also capable of anaerobic xylene degradation while several of the xylene-degrading isolates could also utilize toluene (Dolfing et al. 1990; Fries et al. 1994). Xylene-degrading denitrifiers have now been isolated from a broad range of environments including aquifer material contaminated with diesel fuel (Hess et al. 1997), freshwater mud samples (Rabus and Widdel 1995), laboratory columns treating toluene (Dolfing et al. 1990), and compost (Fries et al. 1994).

The initial reactions, involved in anaerobic *m*-xylene oxidation, are thought to be similar to those of toluene oxidation under nitrate reducing conditions and involve an initial addition of fumarate onto one of the methyl groups to form 3-methylbenzylsuccinate, which is subsequently oxidized to 3-methylbenzoate (Krieger 1999; Fig. 5). The initial addition reaction is mediated by 3-methylbenzylsuccinate synthase, which similarly to benzylsuccinate synthase retains the abstracted hydrogen atom from the methyl carbon during the reaction (Krieger 1999).

More recently, the marine dimethylbenzene-degrading, sulfate-reducing organisms, strains oXyS1 and mXyS1,have been isolated with *ortho*-xylene and *meta*-xylene as the sole carbon and energy sources, respectively (Harms et al. 1999b). Both strains were members of the Desulfobacteriaceae family in the delta subclass of the Proteobacteria, which is composed of a metabolically versatile group of organisms (Widdel and Bak 1992). While strain oXyS1 was closely related (greater than 98% identity based on 16S rDNA sequence analysis) to the *Desulfobacterium cetonicum* and *Desulfosarcina variabilis* species in this family, strain mXyS1 was only 86.9% similar to its closest relative, *Desulfococcus multivorans*, and represented the first example of a previously unrecognized line of descent within the delta subclass of the Proteobacteria (Harms et al. 1999b). The isolates were obtained from

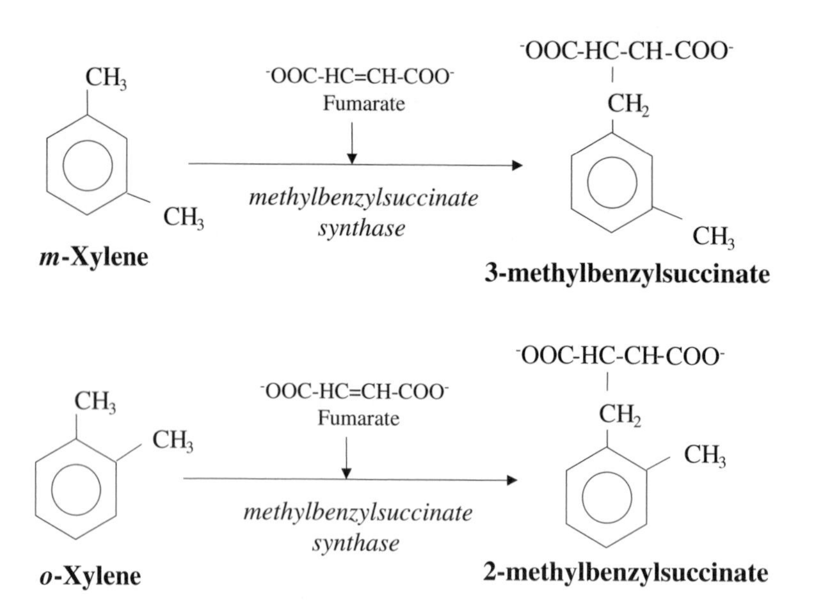

Fig. 5. Anaerobic xylene activation occurs through a fumarate addition reaction at one of the methyl groups to form the respective methylbenzylsuccinate

enrichment cultures growing anaerobically on crude oil and sulfate in a seawater-based medium (Harms et al. 1999b). In addition to degrading dimethylbenzene, these isolates were able to grow on a range of alternative aromatic compounds, including toluene and higher homologues of dimethylbenzene such as ethyltoluene and isopropyltoluene (Harms et al. 1999b). In addition to pure substrates, the isolates were also able to grow with crude oil as the sole carbon and energy source (Harms et al. 1999b).

3 Polycyclic Aromatic Hydrocarbon Compounds

Polycyclic aromatic hydrocarbons (PAH) contain two or more benzene rings in their structure and are components of airborne combustion particulates, petroleum waste, and creosote. They are more hydrophobic and have lower solubility than the monoaromatics, and are particularly persistent in sedimentary environments, due to their low aqueous solubility. PAHs are often introduced into the environment as a result of oil refinery discharges, domestic runoffs, accidental oil spills, or weathering of creosote-impregnated pylons. PAHs are often deposited in the bottom harbor sediments due to their low volatility, and high affinity for sediment particles (McElroy et al. 1989). Once in the sediments, PAHs may persist until they are degraded, resuspended, bioaccumulated, or removed by dredging (Heitkamp and Cerniglia 1987). This is an environmental concern because PAHs may adversely affect bottomfish and sediment infauna (McElroy et al. 1989). In addition, PAHs have been identified by the USEPA as priority pollutants as many are known carcinogens and mutagens (Keith and Telliard 1979).

The ability of aerobic microorganisms to degrade PAHs such as naphthalene, phenanthrene, anthracene, and benzo[*a*]pyrene is well known and has been observed in studies with a variety of soils and sediments (Hambrick et al. 1980; Herbes 1981; Bauer and Capone 1985; Heitkamp and Cerniglia 1987; Heitkamp et al. 1987; Bauer and Capone 1988; Shiaris 1989; Madsen et al. 1992; Dyksterhouse et al. 1995). However, these studies and others (Flyvbjerg et al. 1993; Rueter et al. 1994) have generally indicated that PAHs were not degraded in the absence of oxygen (Cerniglia 1992). A few early studies did suggest that some PAHs were biodegradable in the absence of oxygen if nitrate was available as an electron acceptor, but these same studies also indicated that the PAHs persisted under sulfate-reducing or methanogenic conditions (Mihelcic and Luthy 1988a; Al-Bashir et al. 1990; Leduc et al. 1992). It was only in the last decade that the true extent of the anaerobic catabolism of PAHs, under anything other than nitrate-reducing conditions, was demonstrated (Coates et al. 1996a, b, 1997; Bedessem et al. 1997; Zhang and Young 1997).

The anaerobic oxidation of PAHs under highly reduced anoxic conditions was first demonstrated in studies performed with heavily contaminated sulfate-reducing harbor sediments (Coates et al. 1996a). These and later (Hayes et al. 1999) studies demonstrated that extended prior exposure and adaptation to PAH was imperative to the biodegradation. Sulfate-dependent oxidation of [^{14}C]-naphthalene and phenanthrene to $^{14}CO_2$ was observed without a detectable lag phase in chronically PAH-contaminated sediments, while little or no degradation was observed in relatively pristine sediments from nearby sites (Coates et al. 1996a). PAH biodegradation activity could be conferred onto the pristine sediments by inoculation of these sediments with active PAH-degrading sediments from the more heavily contaminated site.

More recently, PAH degradation has also been observed in sediments and enrichments derived from freshwater aquifers coupled to Fe(III) reduction (Anderson and Lovley 1999) and sulfate reduction (Bedessem et al. 1997), respectively. Of all PAH compounds, the anaerobic biodegradation of naphthalene is probably one of the most often observed and the most comprehensively understood, however, anaerobic degradation of other PAHs including methylnaphthalene, fluorine, fluoranthene, and phenanthrene has also been shown (Coates et al. 1996, 1997; Zhang and Young 1997; Annweiler et al. 2000). In general, the rate and extent of anaerobic PAH degradation is correlated with the number of benzene rings and the presence or absence of side chains (Coates et al. 1997).

Pure cultures of several anaerobic nitrate-reducing organisms (McNally et al. 1998; Rockne et al. 2000) and a sulfate-reducing organism (Galushko et al. 1999), which can degrade PAH compounds, have recently been described. Phylogenetic analyses of the nitrate-reducing PAH-oxidizers indicated that the majority were members of the *Pseudomonas* genus in the gamma subclass of the Proteobacteria (McNally et al. 1998; Rockne et al. 2000). The *Pseudomonas* genus comprises many aerobic naphthalene-degrading bacteria and can account for up to 86.9% of the hydrocarbon-degrading microorganisms found in gasoline-contaminated aquifers (Ridgeway et al. 1990).

The PAH-degrading sulfate-reducer strain NaphS2 was isolated from marine sulfidogenic sediments from the North Sea harbor in Germany. Phylogenetic analysis based on a 16S rRNA gene sequence indicated that this organism belonged to the *Desulfobacterium* genus in the delta subclass of the Proteobacteria and was most closely related (96.9% sequence identity) to *Desulfobacterium* strain mXyS1, an organism isolated for its ability to grow by anaerobic *m*-xylene degradation (Harms et al. 1999b). A recent microbial community sampling analysis of PAH-degrading sediments from a diversity of PAH-contaminated marine environments has indicated that the NaphS2 phylotype was ubiquitous in these environments but was not found in non-PAH-degrading sediments, implying their importance in intrinsic PAH degradation in contaminated marine environments (Hayes and Lovley 2002).

Efforts to elucidate the initial reactions of naphthalene oxidation under anaerobic conditions have yielded contrasting results (Bedessem et al. 1997; Zhang and Young 1997; Meckenstock et al. 2000; Zhang et al. 2000). In one study, performed on sulfidogenic freshwater enrichment cultures, naphthalenol was transiently formed during naphthalene degradation, suggesting hydroxylation as the first activation step of anaerobic naphthalene degradation (Bedessem et al. 1997). In contrast, recent studies on anaerobic naphthalene degradation by sulfidogenic marine enrichments (Zhang and Young 1997; Zhang et al. 2000) and sulfidogenic freshwater enrichments (Meckenstock et al. 2000) have demonstrated the accumulation of 2-naphthoic acid from naphthalene which is indicative of a carboxylation reaction as the activating step in naphthalene catabolism (Fig. 6). A similar carboxylation reaction was

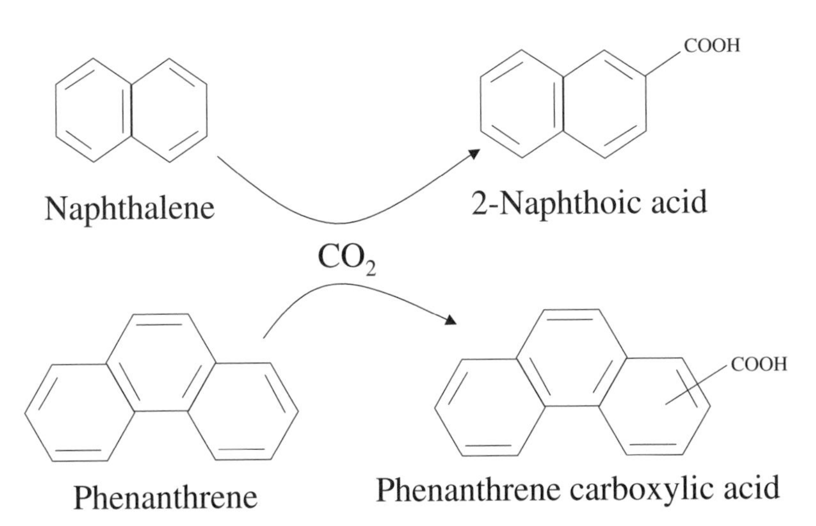

Fig. 6. Initial activation of naphthalene and phenanthrene during anaerobic degradation through a carboxylation reaction to form the respective carboxylic acid

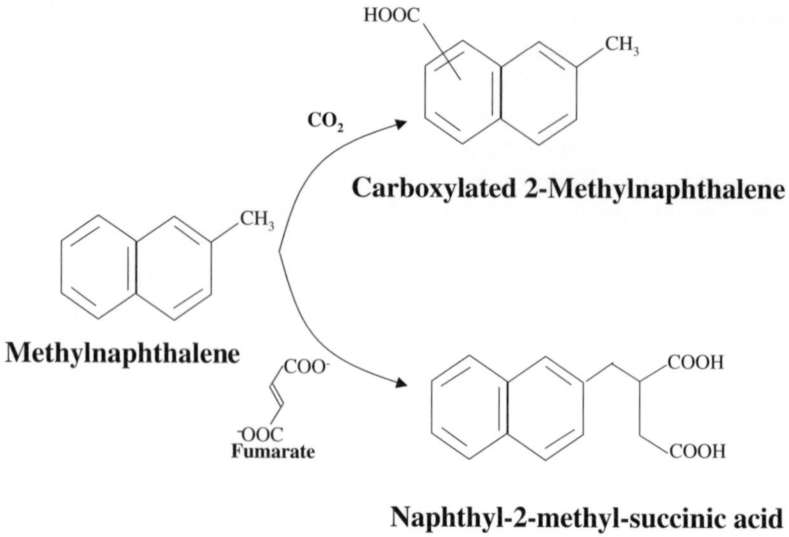

Fig. 7. Two possible mechanisms for the initial activation of methylnaphthalene via a carboxylation reaction with inorganic bicarbonate or through a fumarate addition reaction at the methyl group side chain

also observed during the anaerobic degradation of phenanthrene by sulfidogenic marine enrichments (Zhang and Young 1997).

Similarly to naphthalene, two alternative pathways have also been proposed for the anaerobic degradation of methylnaphthalene (Annweiler et al. 2000, 2002; Sullivan et al. 2001; Fig. 7). Studies of methylnaphthalene degradation by a freshwater sulfidogenic enrichment culture indicated that the first step in the activation of methylnaphthalene is analogous to that for anaerobic toluene degradation and involves a fumarate addition reaction at the methyl group on the benzene ring to form naphthyl-2-methyl-succinic acid which was subsequently oxidized to 2-naphthoic acid (Annweiler et al. 2000, 2002). More recently, in a study which had been performed with a sulfidogenic marine enrichment culture, Sullivan and coworkers (2001) demonstrated the transient formation of carboxylated 2-methylnaphthalene, suggesting that an alternative pathway may also exist for anaerobic sulfate-dependent methylnaphthalene degradation, in which the methylnaphthalene is carboxylated, however, this has yet to be elucidated (Sullivan et al. 2001).

4
Alkanes

Saturated aliphatic hydrocarbons (alkanes) are widespread in both terrestrial and marine environments. They are often produced by living organisms (Birch and Bachofen 1988) or may result from diagenic and catagenic processes (Tissot and Welte 1984). In addition to natural sources, alkanes may enter the environment from anthropogenic activities involving crude and fuel oils. In contrast to many of the aromatic hydrocarbons, alkanes are considered non-toxic although the shorter chain (C_5 to C_{12}) alkanes are known to affect biological membranes (Sikkema et al. 1995). Early studies have suggested the possibility of anaerobic alkane biodegradation under sulfate-reducing (Novelli and Zobell 1944; Davis and Yarbrough 1966) and denitrifying (Traxler and Bernard 1969) conditions, however, the results of these experiments were either irreproducible or the cultures from the studies were not maintained (Swain et al. 1978; Atlas 1981; Griffin and Traxler 1981). More recently, there have been several reports on the anaerobic biodegradability of alkanes either as complex mixtures (Rueter et al. 1994; Coates et al. 1997; Caldwell et al. 1998) or as individual compounds (Aeckersberg et al. 1991, 1998; Rueter et al. 1994; Coates et al. 1997; Caldwell et al. 1998; So and Young 1999b; Zengler et al. 1999b; Anderson and Lovley 2000; Ehrenreich et al. 2000; Kropp et al. 2000) under denitrifying, sulfate-reducing, and methanogenic conditions. In addition, there are now several examples of alkane-degrading sulfate-reducing and denitrifying organisms in pure culture (Aeckersberg et al. 1991; Rueter et al. 1994; So and Young 1999b; Ehrenreich et al. 2000). Aeckersberg and coworkers (1991) provided the first clear evidence of the anaerobic degradation of alkanes by a pure culture with a sulfate-reducing organism they isolated from a saline-water phase of an oil-water separator in an oil production plant (Aeckersberg et al. 1991). This organism, strain Hxd3, was closely related to the *Desulfococcus* genus in the Desulfobacteriaceae family in the delta subclass of the Proteobacteria (Aeckersberg et al. 1998). It grew anaerobically on C_{10} to C_{20} alkanes, which it completely oxidized to CO_2 with sulfate as the sole electron acceptor (Aeckersberg et al. 1991, 1998). Three more alkane-degrading sulfate-reducers have subsequently been isolated (Rueter et al. 1994; Aeckersberg et al. 1998; So and Young 1999b). One of these isolates, strain Pnd3, was phylogenetically similar to strain Hxd3 although it exhibited significant phenotypic differences (Aeckersberg et al. 1998). Strain Pnd3 was isolated from pentadecane-degrading enrichment cultures established with marine mud samples. Caproate was used as a water-soluble electron donor for isolation of this organism in an agar solidified basal medium (Aeckersberg et al. 1998). Strain Pnd3 used a more limited range of *n*-alkanes (C_{14}–C_{17}) than strain Hxd3, although it showed greater versatility in its ability to oxidize fatty acids (Aeckersberg et al. 1998). Both of these isolates grew well on hexadecane (C_{16}) but showed alternative cellular fatty acid composition after growth. Growth on an alkane with an odd number of carbon atoms (C-odd) yielded predominantly cellular fatty acids with an even number of carbon atoms (C-even) and a C-even alkane yielded C-odd fatty acids (Aeckersberg et al. 1998). These results suggested that an alkane was altered by an odd numbered carbon(s) unit during its transformation. In contrast, it was shown for Pnd3 (Aeckersberg et al. 1998), and subsequently for the recently described alkane-degrading sulfate-reducer, strain AK-01 (So and Young 1999a), that C-odd alkanes yielded C-odd fatty acids, while C-even alkanes yielded C-even fatty acids. Isotope labeling studies suggested that the initial attack of the alkanes by Hxd3 included an initial subterminal carboxylation with inorganic bicarbonate at the C-3 position of the alkane followed by the elimination of the two adjacent terminal carbon atoms (So et al. 2003; Fig. 8). In contrast, alkane degradation by strain AK-01 was mediated by an addition of exogenous carbon subterminally at the C-2 position, however, the carbon addition was not direct carboxylation by inorganic bicarbonate (So and Young 1999a), but rather, may be the result of a fumarate addition reaction as has been recently demonstrated for a dodecane-utilizing, sulfate-reducing enrichment culture (Kropp et al. 2000; Fig. 8). A similar type of fumarate addition at the subterminal carbon of the alkane chain has also been demonstrated as the initial reaction for alkane degradation by denitrifying isolates (Rabus et al. 2001).

5
Biomarkers of Anaerobic Hydrocarbon Degradation

The practical application of various metabolic biomarkers to distinguish abiotic losses of hydrocarbons in contaminated sites from biologically mediated degradation processes has recently been proposed (Beller 2000; Elshahed et al. 2001; Pelz et al. 2001; Mancini et al. 2003). Specific intermediates of metabolic pathways can serve as biomarkers in the environment, where their detection is indicative of the presence of an active microbial population utilizing the metabolic substrate concerned. A compound can be used as a biomarker only if certain criteria are met. The compound must be readily detectable in soil, water, and sediment samples; it must be highly specific to a particular pathway (i.e., not a product of commonly utilized substrates by a host of diverse

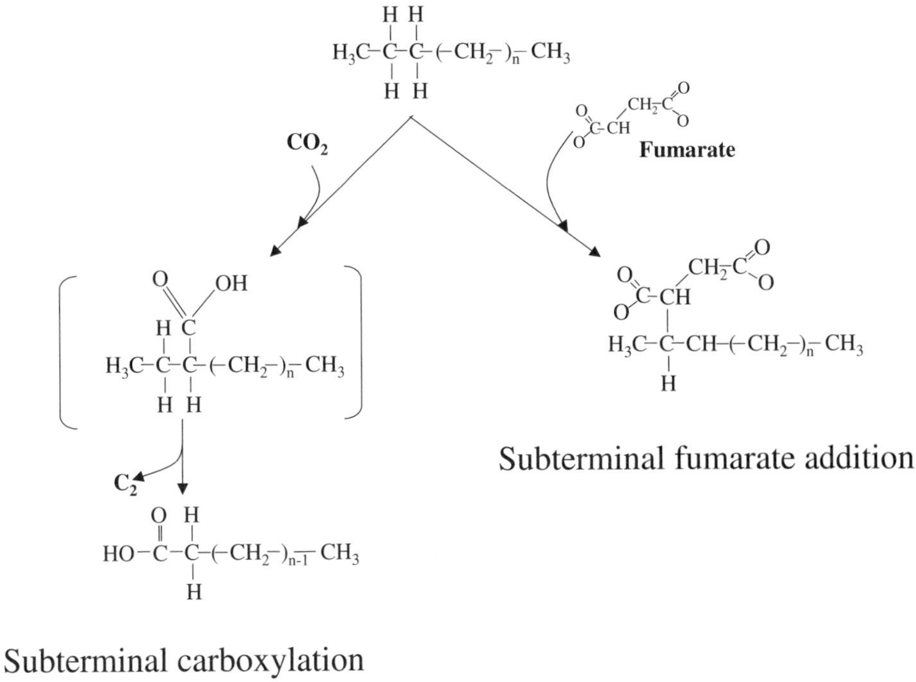

CO_2

Fumarate

Subterminal fumarate addition

Subterminal carboxylation

Fig. 8. Two possible mechanisms for the initial activation of *n*-alkanes during anaerobic biodegradation involving a subterminal fumarate addition reaction at the C-2 carbon atom or a subterminal carboxylation reaction with inorganic bicarbonate at the C-3 carbon atom

organisms), and it must not be found in pristine environments (Beller 2000). Ideally, the presence of the biomarker compound in a particular environment should signify its formation from a known substrate via a specific pathway by a particular group of microorganisms. If these criteria are met, then monitoring the disappearance of the substrate concomitant with the appearance of the biomarker over time is a reliable indication of in situ biodegradation.

For example, benzylsuccinate and *e*-phenylitaconate have been commonly used as biomarkers for anaerobic toluene oxidation (Beller 2000). Similarly, methybenzylsuccinate, the first intermediate of the degradative pathway of dimethylbenzene (Krieger 1999), has also been identified as a suitable biomarker for xylene degradation. Benzylsuccinate, or its methylated analogs, are not anthropogenic compounds, and are highly specific to the anaerobic catabolic pathways of toluene, xylene and ethylbenzene, especially under nitrate-reducing conditions. As such, these compounds represent excellent candidates for use as biomarkers of aromatic hydrocarbon degradation in the environment.

Detailed environmental studies indicated that the concentration of such biomarkers decreased over time concomitantly with a decrease in the concentration of the monoaromatic hydrocarbons (Beller et al. 1995). However, as methylbenzylsuccinate was relatively more recalcitrant in the environment than benzylsuccinate, it was selected as a more effective biomarker for in situ monitoring of intrinsic BTEX biodegradation (Beller et al. 1995). Transiently produced benzylsuccinate and methylbenzylsuccinate were similarly detected in sulfate-reducing and methanogenic sediments which actively degraded toluene and dimethylbenzene (Elshahed et al. 2001). Furthermore, 3-phenyl-1,2-butanedicarboxylic acid was detected and identified as a possible metabolite of ethylbenzene oxidation under sulfate-reducing conditions in the same sediments (Elshahed et al. 2001).

Acknowledgments. The research of JDC is supported by grants from the US Department of Energy NABIR program (DE-FG02–98-ER-62689) and the US Department of Defense SERDP Program (DACA72–00-C-0016).

References

Achenbach LA, Coates JD (2000) Disparity between bacterial phylogeny and physiology. ASM Newsl 66:714–716

Achenbach LA, Bruce RA, Michaelidou U, Coates JD (2001) *Dechloromonas agitata* N.N. gen., sp. nov. and *Dechlorosoma suillum* N.N. gen., sp. nov., two novel environmentally dominant (per)chlorate-reducing bacteria and their phylogenetic position. Int J Syst Evol Microbiol 51:527–533

Aeckersberg F, Bak F, Widdel F (1991) Anaerobic oxidation of saturated hydrocarbons to CO_2 by a new type of sulfate-reducing bacterium. Arch Microbiol 156:5–14

Aeckersberg F, Rainey FA, Widdel F (1998) Growth, natural relationships, cellular fatty acids and metabolic adaptation of sulfate-reducing bacteria that utilize long-chain alkanes under anoxic conditions. Arch Microbiol 170:361–369

Al-Bashir B, Cseh T, Leduc R, Samson R (1990) Effect of soil/contaminant interactions on the biodegradation of naphthalene in flooded soil under denitrifying conditions. Appl Microbiol Biotechnol 34:414–419

Anders H, Kaetzke A, Kaempfer P, Ludwig W, Fuchs G (1995) Taxonomic position of aromatic-degrading denitrifying pseudomonad strains K-172 and KB 740 and their description as new members of the genera *Thauera*, as *Thauera aromatica* sp. nov., and *Azoarcus*, as *Azoarcus evansii* sp. nov., respectively, members of the beta subclass of the Proteobacteria. Int J Syst Bacteriol 45:327–333

Anderson RT, Lovley DR (1997) Ecology and biogeochemistry of in situ groundwater bioremediation. Adv Microbial Ecol 15:289–350

Anderson RT, Lovley DR (1999) Naphthalene and benzene degradation under Fe(III)-reducing conditions in petroleum-contaminated aquifers. Biorem J 3:121–135

Anderson RT, Lovley DR (2000) Biogeochemistry: hexadecane decay by methanogenesis. Nature 404:722–723

Anderson RT, Rooney-Varga J, Gaw CV, Lovley DR (1998) Anaerobic benzene oxidation in the Fe(III) reduction zone of petroleum-contaminated aquifers. Environ Sci Technol 32:1222–1229

Annweiler E, Materna A, Safinowski M, Kappler A, Richnow HH, Michaelis W, Meckenstock RU (2000) Anaerobic degradation of 2-methylnaphthalene by a sulfate-reducing enrichment culture. Appl Environ Microbiol 66: 5329–5333

Annweiler E, Michaelis W, Meckenstock RU (2002) Identical ring cleavage products during anaerobic degradation of naphthalene, 2-methylnaphthalene, and tetralin indicate a new metabolic pathway. Appl Environ Microbiol 68:852–858

Atlas RM (1981) Microbial degradation of petroleum hydrocarbons: an environmental perspective. Microbiol Rev 45:180–209

Atlas RM (1995) Petroleum biodegradation and oil spill remediation. Mar Pollut Bull 31:178–182

Ball HA, Johnson HA, Reinhard M, Spormann AM (1996) Initial reactions in anaerobic ethylbenzene oxidation by a denitrifying bacterium, strain EB1. J Bacteriol 178: 5755–5761

Bauer JE, Capone DG (1985) Degradation and mineralization of the polycyclic aromatic hydrocarbons anthracene and naphthalene in intertidal marine sediments. Appl Environ Microbiol 50:81–90

Bauer JE, Capone DG (1988) Effects of co-occurring aromatic hydrocarbons on degradation of individual polycyclic aromatic hydrocarbons in marine sediment slurries. Appl Environ Microbiol 54:1649–1655

Bedessem ME, Swoboda-Colberg NG, Colberg PJS (1997) Naphthalene mineralization coupled to sulfate reduction in aquifer-derived enrichments. FEMS Microbiol Lett 152:213–218

Beller HR (2000) Metabolic indicators for detecting in situ anaerobic alkylbenzene degradation. Biodegradation 11:125–139

Beller HR, Ding W-H, Reinhard M (1995) Byproducts of anaerobic alkylbenzene metabolism useful as indicators of in situ bioremediation. Environ Sci Technol 29:2864–2870

Beller HR, Spormann AM (1997a) Anaerobic activation of toluene and *O*-xylene by addition to fumarate in denitrifying strain T. J Bacteriol 179:670–676

Beller HR, Spormann AM (1997b) Benzylsuccinate formation as a means of anaerobic toluene activation by sulfate-reducing strain PRTOL1. Appl Environ Microbiol 63: 3729–3731

Beller HR, Spormann AM (1999) Substrate range of benzylsuccinate synthase from *Azoarcus* sp. strain T. FEMS Microbiol Lett 178:147–153

Beller HR, Spormann AM, Sharma PK, Cole JR, Reinhard M (1996) Isolation and characterization of a novel toluene-degrading, sulfate-reducing bacterium. Appl Environ Microbiol 62:1188–1196

Biegert T, Fuchs G, Heider J (1996) Evidence that oxidation of toluene in the denitrifying bacterium *Thauera aromatica* is initiated by formation of benzylsuccinate from toluene and fumarate. Eur J Biochem 238:661–668

Birch L, Bachofen R (1988) Microbial production of hydrocarbons. Biotechnology. VCH, Weinheim

Bruce RA, Achenbach LA, Coates JD (1999) Reduction of (per)chlorate by a novel organism isolated from a paper mill waste. Environ Microbiol 1:319–331

Burland SM, Edwards EA (1999) Anaerobic benzene biodegradation linked to nitrate reduction. Appl Environ Microbiol 65:529–533

Caldwell ME, Garrett RM, Prince RC, Suflita JM (1998) Anaerobic biodegradation of long-chain n-alkanes under sulfate-reducing conditions. Environ Sci Technol 37: 2191–2195

Cerniglia CE (1992) Biodegradation of polycyclic aromatic hydrocarbons. Biodegradation 3:351–368

Cervantes FJ, Dijksma W, Duong-Dac T, Ivanova A, Lettinga G, Field JA (2001) Anaerobic mineralization of toluene by enriched sediments with quinones and humus as terminal electron acceptors. Appl Environ Microbiol 67: 4471–4478

Coates JD (2003) Bacteria that respire oxyanions of chlorine. Bergey's manual of systematic bacteriology. Springer, Berlin Heidelberg New York

Coates JD, Lovley DR (2003) Genus *Geobacter*. Bergey's manual of systematic bacteriology, Springer, Berlin Heidelberg New York

Coates JD, Anderson RT, Lovley DR (1996a) Anaerobic oxidation of polycyclic aromatic hydrocarbons under sulfate-reducing conditions. Appl Environ Microbiol 62: 1099–1101

Coates JD, Anderson RT, Woodward JC, Phillips EJP, Lovely DR (1996b) Anaerobic hydrocarbon degradation in petroleum-contaminated harbor sediments under sulfate-reducing and artificially imposed iron-reducing conditions. Environ Sci Technol 30:2784–2789

Coates JD, Phillips EJP, Lonergan DJ, Jenter H, Lovley DR (1996c) Isolation of *Geobacter* species from a variety of sedimentary environments. Appl Environ Microbiol 62:1531–1536

Coates JD, Woodward J, Allen J, Philp P, Lovley DR (1997) Anaerobic degradation of polycyclic aromatic hydrocarbons and alkanes in petroleum-contaminated marine harbor sediments. Appl Environ Microbiol 63:3589–3593

Coates JD, Michaelidou U, Bruce RA, O'Connor SM, Crespi JN, Achenbach LA (1999) The ubiquity and diversity of dissimilatory (per)chlorate-reducing bacteria. Appl Environ Microbiol 65:5234–5241

Coates JD, Bhupathiraju V, Achenbach LA, McInerney MJ, Lovley DR (2001a) *Geobacter hydrogenophilus*, *Geobacter chapellei*, and *Geobacter grbiciae* – three new strictly anaerobic dissimilatory Fe(III)-reducers. Int J Syst Evol Microbiol 51: 581–588

Coates JD, Chakraborty R, Lack JG, O'Connor SM, Cole KA, Bender KS, Achenbach LA (2001b) Anaerobic benzene oxidation coupled to nitrate reduction in pure culture by two strains of *Dechloromonas*. Nature 411:1039–1043

Coates JD, Chakraborty R, McInerney MJ (2002) Anaerobic benzene biodegradation – a new era. Res Microbiol 153:621–628

Coschigano PW (1999) Transcriptional analysis of the *tutEtutFDGH* gene cluster from the denitrifying bacterium *Thauera aromatica* strain T1. Appl Environ Microbiol 66:1147–1151

Coschigano PW, Young LY (1997) Identification and sequence analysis of two regulatory genes involved in anaerobic toluene metabolism by strain T1. Appl Environ Microbiol 63:652–660

Davis JB, Yarbrough HF (1966) Anaerobic oxidation of hydrocarbons by *Desulfovibrio desulfuricans*. Chem Geol 1:137–144

Dolfing J, Zeyer J, Binder-Eicher P, Schwarzenbach RP (1990) Isolation and characterization of a bacterium that mineralizes toluene in the absence of molecular oxygen. Arch Microbiol 134:336–341

Dyksterhouse SE, Gray JP, Herwig RP, Cano Lara J, Staley JT (1995) *Cycloclasticus pugetii* gen. nov., sp. nov., an aromatic hydrocarbon-degrading bacterium from marine sediments. Int J Syst Bacteriol 45:116–123

Edwards EA, Wills LE, Reinhard M, Grbic-Galic D (1992) Anaerobic degradation of toluene and xylene by aquifer microorganisms under sulfate-reducing conditions. Appl Environ Microbiol 58:794–800

Ehrenreich P, Behrends A, Harder J, Widdel F (2000) Anaerobic oxidation of alkanes by newly isolated denitrifying bacteria. Arch Microbiol 173:58–64

Elshahed MS, Gieg LM, McInerney MJ, Suflita JM (2001) Signature metabolites attesting to the in situ attenuation of alkylbenzenes in anaerobic environments. Environ Sci Technol 35:682–689
Evans PJ, Mang DT, Kim KS, Young LY (1991a) Anaerobic degradation of toluene by a denitrifying bacterium. Appl Environ Microbiol 57:1139–1145
Evans PJ, Mang DT, Young LY (1991b) Degradation of toluene and *m*-xylene and transformation of *o*-xylene by denitrifying enrichment cultures. Appl Environ Microbiol 57:450–454
Flyvbjerg J, Arivn E, Jensen BK, Olsen SK (1993) Microbial degradation of phenols and aromatic hydrocarbons in creosote-contaminated groundwater under nitrate-reducing conditions. J Contam Hydrol 12:133–150
Fries MR, Zhou J, Chee-Sanford J, Tiedje JM (1994) Isolation, characterization, and distribution of denitrifying toluene degraders from a variety of habitats. Appl Environ Microbiol 60:2802–2810
Galushko A, Minz D, Schink B and Widdel F (1999) Anaerobic degradation of naphthalene by a pure culture of a novel type of marine sulfate-reducing bacterium. Environ Microbiol 1:1–23
Grbic-Galic D, Vogel T (1987) Transformation of toluene and benzene by mixed methanogenic cultures. Appl Environ Microbiol 53:254–260
Griffin WM, Traxler RW (1981) Some aspects of hydrocarbon metabolism by *Pseudomonas*. Dev Ind Microbiol 22:425–435
Hambrick GA, DeLaune RD, Patrick Jr WH (1980) Effect of estuarine sediment pH and oxidation-reduction potential on microbial hydrocarbon degradation. Appl Environ Microbiol 40:365–369
Haner A, Hohener P, Zeyer J (1995) Degradation of *p*-xylene by a denitrifying enrichment culture. Appl Environ Microbiol 61:3185–3188
Harms G, Rabus R, Widdel F (1999a) Anaerobic oxidation of the aromatic plant hydrocarbon p-cymene by newly isolated denitrifying bacteria. Arch Microbiol 172:303–312
Harms G, Zengler K, Rabus R, Aeckersberg F, Minz D, Rossello-Mora R, Widdel F (1999b) Anaerobic oxidation of *o*-xylene and *m*-xylene and homologous alkylbenzenes by new types of sulfate-reducing bacteria. Appl Environ Microbiol 65: 999–1004
Hayes LA, Lovley DR (2002) Specific 16S rDNA sequences associated with naphthalene degradation under sulfate-reducing conditions in harbor sediments. Microbial Ecol 43:134–145
Hayes LA, Nevin KP, Lovley DR (1999) Role of prior exposure on anaerobic degradation of naphthalene and phenanthrene in marine harbor sediments. Org Geochem 30:937–945
Heider J, Spormann AM, Beller HR, Widdel F (1998) Anaerobic bacterial metabolism of hydrocarbons. FEMS Microbiol Rev 22:459–473
Heitkamp MA, Cerniglia CE (1987) Effects of chemical structure and exposure on the microbial degradation of polycyclic aromatic hydrocarbons in freshwater and estuarine ecosystems. Environ Toxicol Chem 6:535–546
Heitkamp MA, Freeman JP, Cerniglia CE (1987) Naphthalene biodegradation in environmental microcosms: estimates of degradation rates and characterization of metabolites. Appl Environ Microbiol 53:129–136
Herbes S (1981) Rates of microbial transformation of polycylic aromatic hydrocarbons in water and sediments in the vicinity of a coal-coking wastewater discharge. Appl Environ Microbiol 41:20–28

Hess A, Zarda B, Hahn D, Haner A, Stax D, Hohener P, Zeyer J (1997) In situ analysis of denitrifying toluene- and *m*-xylene degrading bacteria in a diesel fuel-contaminated laboratory aquifer column. Appl Environ Microbiol 65: 2136–2141

Johnson HA, Pelletier DA, Spormann AM (2001) Isolation and characterization of anaerobic ethylbenzene dehydrogenase, a novel Mo-Fe-S enzyme. J Bacteriol 183: 4536–4542

Kane SR, Beller HR, Legler TC, Anderson RT (2002) Biochemical and genetic evidence of benzylsuccinate synthase in toluene-degrading, ferric iron-reducing *Geobacter metallireducens*. Biodegradation 13:149–154

Kazumi J, Caldwell ME, Suflita JM, Lovley DR, Young LY (1997) Anaerobic degradation of benzene in diverse anoxic environments. Environ Sci Technol 31:813–818

Keith LH, Telliard WA (1979) Priority pollutants I. A perspective view. Environ Sci Technol 13:416–423

Kniemeyer O, Fischer T, Wilkes H, Glockner F, Widdel F (2003) Anaerobic degradation of ethylbenzene by a new type of marine sulfate-reducing bacterium. Appl Environ Microbiol 69:760–768

Krieger CJ (1999) Initial reactions in anaerobic oxidation of *m*-xylene by the denitrifying bacterium *Azoatcus* sp. strain T. J Bacteriol 181:6403–6410

Kropp KG, Davidova IA, Suflita JM (2000) Anaerobic oxidation of n-dodecane by an addition reaction in a sulfate-reducing bacterial enrichment culture. Appl Environ Microbiol 66:5393–5398

Kuhn EP, Zeyer J, Eicher P, Schwarzenbach RP (1988) Anaerobic degradation of alkylated benzenes in denitrifying laboratory aquifer columns. Appl Environ Microbiol 54:490–496

Langenhoff AAM, Brouwers-Ceiler DL, Engelberting JHL, Quist JJ, Wolkenfelt JGPN, Zehnder AJB, Schraa G (1997) Microbial reduction of manganese coupled to toluene oxidation. FEMS Microbiol Ecol 22:119–127

Leduc R, Samson R, Al-Bashir B, Al-Hawari J, Cseh T (1992) Biotic and abiotic disappearance of four PAH compounds from flooded soil under various redox conditions. Water Sci Technol 26:51–60

Leuthner B, Leutwein C, Schulz H, Horth P, Haehnel W, Schiltz E, Schagger H, Heider J (1998) Biochemical and genetic characterization of benzylsuccinate synthase from *Thauera aromatica*: a new glycyl radical enzyme catalysing the first step in anaerobic toluene metabolism. Mol Microbiol 28:615–628

Lonergan DJ, Lovley DR (1991) Microbial oxidation of natural and anthropogenic aromatic compounds coupled to Fe(III) reduction. In: Baker RA (ed) Organic substances and sediments in water. Lewis, Chelsea, MI, pp 327–338

Lovley DR, Lonergan DJ (1990) Anaerobic oxidation of toluene, phenol, and p-cresol by the dissimilatory iron-reducing organism, GS-15. Appl Environ Microbiol 56: 1858–1864

Lovley DR, Baedecker MJ, Lonergan DJ, Cozzarelli IM, Phillips EJP, Siegel DI (1989) Oxidation of aromatic contaminants coupled to microbial iron reduction. Nature 339:297–299

Lovley DR, Woodward JC, Chapelle FH (1994) Stimulated anoxic biodegradation of aromatic hydrocarbons using Fe(III) ligands. Nature 370:128–131

Lovley DR, Coates JD, Woodward JC, Phillips EJP (1995) Benzene oxidation coupled to sulfate reduction. Appl Environ Microbiol 61:953–958

Lovley DR, Woodward JC, Chapelle FH (1996) Rapid anaerobic benzene degradation with a variety of chelated Fe(III) forms. Appl Environ Microbiol 62:288–291

Madsen EL, Winding A, Malachowsky K, Thomas CT, Ghiorse WC (1992) Contrasts between subsurface microbial communities and their metabolic adaptation to polycyclic aromatic hydrocarbons at a forested and an urban coal-tar disposal site. Microb Ecol 24:199–213

Mancini SA, Ulrich AC, Lacrampe-Couloume G, Sleep B, Edwards EA, Lollar BS (2003) Carbon and hydrogen isotopic fractionation during anaerobic biodegradation of benzene. Appl Environ Microbiol 69:191–198

McElroy AE, Farrington JW, Teal JM (1989) Bioavailability of polycyclic aromatic hydrocarbons in the aquatic environment. In: Metabolism of polycyclic aromatic hydrocarbons in the aquatic environment. CRC Press, Boca Raton, pp 2–39

McNally DL, Mihelcic JR, Lueking DR (1998) Biodegradation of three- and four-ring polycyclic aromatic hydrocarbons under aerobic and denitrifying conditions. Environ Sci Technol 32:2633–2639

Meckenstock RU (1999) Fermentative toluene degradation in anaerobic defined syntrophic cocultures. FEMS Microbiol Lett 177:67–73

Meckenstock RU, Annweiler E, Michaelis W, Richnow HH, Schink B (2000) Anaerobic naphthalene degradation by a sulfate-reducing enrichment culture. Appl Environ Microbiol 66:2743–2747

Mihelcic JR, Luthy RG (1988a) Degradation of polycyclic aromatic hydrocarbon compounds under various redox conditions in soil-water systems. Appl Environ Microbiol 54:1182–1187

Mihelcic JR, Luthy RG (1988b) Microbial degradation of acenaphthene and naphthalene under denitrification conditions in soil-water systems. Appl Environ Microbiol 54:1188–1198

Novelli GD, Zobell CE (1944) Assimilation of petroleum hydrocarbons by sulfate-reducing bacteria. J Bacteriol 47:447–448

Pelz O, Chatzinotas A, Zarda-Hess A, Wolf-Rainer A, Zeyer J (2001) Tracing toluene-assimilating sulfate-reducing bacteria using ^{13}C-incorporation in fatty acids and whole-cell hybridization. FEMS Microbiol Ecol 38:123–131

Phelps CD, Kazumi J, Young LY (1996) Anaerobic degradation of benzene in BTX mixtures dependent on sulfate reduction. FEMS Microbiol Lett 145:433–437

Rabus R, Heider J (1998) Initial reactions of anaerobic metabolism of alkylbenzenes in denitrifying and sulfate-reducing bacteria. Arch Microbiol 170:377–384

Rabus R, Widdel F (1995) Anaerobic degradation of ethylbenzene and other aromatic hydrocarbons by new denitrifying bacteria. Arch Microbiol 163:96–103

Rabus R, Nordhaus R, Ludwig W, Widdel F (1993) Complete oxidation of toluene under strictly anoxic conditions by a new sulfate-reducing bacterium. Appl Environ Microbiol 59:1444–1451

Rabus R, Wilkes H, Behrends A, Armstroff A, Fischer T, Pierik AJ, Widdel F (2001) Anaerobic initial reaction of n-alkanes in a denitrifying bacterium: evidence for (1-methylpentyl)succinate as initial product and for involvement of an organic radical in n-hexane metabolism. J Bacteriol 183:1707–1715

Rabus R, Kube A, Beck F, Widdel F, Reinhardt R (2002) Genes involved in the anaerobic degradation of ethylbenzene in a denitrifying bacterium, strain EbN1. Arch Microbiol 178:506–516

Ridgeway HF, Safarik J, Phipps D, Carl P, Clark D (1990) Identification and catabolic activity of well-derived gasoline-degrading bacteria and a contaminated aquifer. Appl Environ Microbiol 56:3565–3575

Rockne KJ, Chee-Sanford JC, Sanford RA, Hedlund BP, Staley JT, Strand SE (2000) Anaerobic naphthalene degradation by microbial pure cultures under nitrate-reducing conditions. Appl Environ Microbiol 66:1595–1601

Rueter P, Rabus R, Wilkes H, Aeckersberg F, Rainey FA, Jannasch HW, Widdel F (1994) Anaerobic oxidation of hydrocarbons in crude oil by new types of sulphate-reducing bacteria. Nature 372:455–458

Schocher RJ, Seyfried B, Vazquez F, Zeyer J (1991) Anaerobic degradation of toluene by pure cultures of denitrifying bacteria. Arch Microbiol 157:7–12

Shiaris MP (1989) Seasonal biotransformation of naphthalene, phenanthrene, and benzo[*a*] pyrene in surficial estuarine sediments. Appl Environ Microbiol 55: 1391–1399

Sikkema J, De Bont JAM, Poolman B (1995) Mechanisms of membrane toxicity of hydrocarbons. Microbiol Rev 59:201–222

So CM, Phelps CD, Young LY (2003) Anaerobic transformation of alkanes to fatty acids by a sulfate-reducing bacterium strain Hxd3. Appl Environ Microbiol 69:3892–3900

So CM, Young LY (1999a) Initial reactions in anaerobic alkane degradation by a sulfate reducer, strain AK-01. Appl Environ Microbiol 65:5532–5540

So CM, Young LY (1999b) Isolation and characterization of a sulfate-reducing bacterium that anaerobially degrades alkanes. Appl Environ Microbiol 65:2969–26876

Spormann AM, Widdel F (2000) Metabolism of alkylbenzenes, alkanes, and other hydrocarbons in anaerobic bacteria. Biodegradation 11:85–105

Stone RW, Zobell CE (1952) Bacterial aspects of the origin of petroleum. Ind Eng Chem 44:2564–2567

Sullivan ER, Zhang X, Phelps C, Young LY (2001) Anaerobic mineralization of stable-isotope-labeled 2-methylnaphthalene. Appl Environ Microbiol 67:4353–4357

Swain HM, Somerville HJ, Cole JA (1978) Denitrification during growth of *Pseudomonas aeruginosa* on octane. J Gen Microbiol 107:103–112

Tissot BP, Welte DH (1984) Petroleum formation and occurrence, Springer, Berlin Heidelberg New York, ••

Traxler RW, Bernard JM (1969) The utilization of n-alkanes by *Pseudomonas aerogi-nosa* under conditions of anaerobiosis. 1. Preliminary observations. Int Biodetn Bull 5:21–25

Ulrich AC, Edwards EA (2003) Physiological and molecular characterization of anaerobic benzene-degrading mixed cultures. Environ Microbiol 5:92–102

Urbansky ET (1998) Perchlorate chemistry: implications for analysis and remediation. Bioremediation J 2:81–95

Vogel TM, Grbic-Galic D (1986) Incorporation of oxygen from water into toluene and benzene during anaerobic fermentative transformation. Appl Environ Microbiol 52: 200–202

Weiner J, Lovley DR (1998a) Anaerobic benzene degradation in petroleum-contaminated sediments after inoculation with a benzene-oxidizing enrichment. Appl Environ Microbiol 64:775–778

Weiner J, Lovley DR (1998b) Rapid benzene degradation in methanogenic sediments from a petroleum-contaminated aquifer. Appl Environ Microbiol 64:1937–1939

Widdel F, Bak F (1992) Gram-negative mesophilic sulfate-reducing bacteria. In: Balows A, Truper HG, Dworkin M, Harder W, Schleifer K-H (eds) The prokaryotes. Springer Berlin Heidelberg New York, pp 3353–3378

Widdel F, Rabus R (2001) Anaerobic biodegradation of saturated and aromatic hydrocarbons. Curr Op Biotechnol 12:259–276

Zengler K, Heider J, Rossello-Mora R, Widdel F (1999a) Phototrophic utilization of toluene under anoxic conditions by a new strain of *Blastochloris sulfoviridis*. Arch Microbiol 172:204–212

Zengler K, Richnow HH, Rossello-Mora R, Michaelis W, Widdel F (1999b) Methane formation from long-chain alkanes by anaerobic microorganisms. Nature 401: 266–269

Zhang X, Young LY (1997) Carboxylation as an initial reaction in the anaerobic metabolism of napthalene and phenanthrene by sulfidogenic consortia. Appl Environ Microbiol 63:4759–4764

Zhang X, Sullivan ER, Young LY (2000) Evidence for aromatic ring reduction in the biodegradation pathway of carboxylated naphthalene by a sulfate reducing consortium. Biodegradation 11:117–124
Zhou J, Fries MR, Chee-Sandford JC, Tiedje JM (1995) Phylogenetic analysis of a new group of denitrifiers capable of anaerobic growth on toluene and description of *Azoarcus tolulyticus* sp. nov. Int J Syst Bacteriol 45:500–506
Zobell CE (1945) The role of bacteria in the formation and transformation of petroleum hydrocarbons. Science 102:364–369
Zobell CE (1946) Action of microorganisms on hydrocarbons. Bacteriol Rev 10:1–49
Zobell CE (1949) Part played by bacteria in petroleum formation. Am J Bot 36:832–832
Zobell CE (1950) Assimilation of hydrocarbons by microorganisms. Adv Enzymol Relat Subj Biochem 10:443–486

5 Biotransformation, Biodegradation, and Bioremediation of Polycyclic Aromatic Hydrocarbons

Michael D. Aitken[1] and Thomas C. Long[2]

1 Introduction

Polycyclic aromatic hydrocarbons (PAHs) have been the subject of scientific study for many years, and a number of excellent reviews have been published on PAH biodegradation. The purpose of this chapter is to review those characteristics of PAH-contaminated systems and PAH-degrading microorganisms that can influence the fate of PAHs in impacted soil and sediment systems. The identification of these characteristics can be approached either from a reductionist perspective – in an effort to elucidate relevant mechanisms that can be extrapolated to more complex systems – or from an appreciation for the complexity of real PAH-contaminated environments. A perspective somewhere between these views, flexible enough to look in either direction for insights, may be required to advance our understanding of PAH biodegradation in the real world.

A prevailing reality of PAH contamination in the environment is that the biologically-mediated removal of PAHs tends to be incomplete, even if aggressive bioremediation approaches are used (US Environmental Protection Agency 2000). Why is this? We have learned from numerous well-controlled laboratory studies that virtually all of the PAHs we are concerned with are biodegradable, and that the organisms capable of transforming or degrading PAHs are essentially ubiquitous. We have also learned that the association of PAHs with nonaqueous compartments can decrease their rates of degradation. Yet, in any given contaminated system, do we really know whether limitations in bioavailability actually control the extent of degradation of the most

[1]Department of Environmental Sciences and Engineering, School of Public Health, University of North Carolina, Chapel Hill, North Carolina 27599–7431, USA, e-mail: mike_aitken@unc.edu, Tel: +1-919-9661481, Fax: +1-919-9667911
[2]Department of Environmental Sciences and Engineering, School of Public Health, University of North Carolina, Chapel Hill, North Carolina 27599–7431, USA

Soil Biology, Volume 2
Biodegradation and Bioremediation
(ed. by. A. Singh and O. P. Ward)

recalcitrant compounds? What variables do we need to control if we want to improve the extent of PAH degradation in active bioremediation processes? What processes and mechanisms should we include if we want to predict rates of natural attenuation for PAH contamination in the subsurface? There are no simple answers to these questions, but they are the ones we need to ask as a basis for developing improved approaches to bioremediation of PAH-contaminated systems.

2 Relevant Properties of PAHs

The PAHs of most concern with respect to the environment and human health range in size from two to six rings. The structures of the 16 PAHs regulated as priority pollutants by the US Environmental Protection Agency (USEPA) are shown in Fig. 1. These compounds are all hydrophobic, as illustrated by their relatively high octanol–water partition coefficients (K_{ow}) and low solubility in water (Table 1). Hydrophobicity increases with an increase in molecular size, with aqueous solubilities declining from the low mg/l range for the two- and three-ring compounds to 1 µg/l or below for the five- and six-ring compounds.

Polycyclic aromatic hydrocarbons are known to be genotoxic. The International Agency for Research on Cancer (IARC) has identified 15 PAHs, including 6 of the 16 USEPA-regulated PAHs, as “reasonably anticipated to be human carcinogens based on sufficient evidence of carcinogenicity in experimental animals” (National Toxicology Program 2001). The USEPA classifies seven PAHs as group B2 “probable human carcinogens” (US Environmental Protection Agency 1993). Chrysene is the only compound among the seven that is not among the carcinogenic PAHs designated by IARC. Toxic equivalency factors (TEF) proposed for rating the relative genotoxicity of PAHs are shown in Table 1, along with the USEPAs suggested potency factors. In both schemes, benzo[*a*]pyrene is the benchmark against which other PAHs are rated. TEFs or potency factors can be used to establish a weighted average toxicity of a given contaminated material based on measured concentrations of the individual contaminants and the relevant factor for each. This approach arguably oversimplifies the estimation of toxicity of a complex mixture (Goldstein et al. 1998; Reeves et al. 2001).

The human toxicity of PAHs does not directly impact biodegradation in contaminated systems, but it strongly influences the risk assessment that drives the need for remediation and which determines its endpoints. Human exposure to PAHs from contaminated water is more relevant for the more soluble low molecular weight (LMW) PAHs (Smith

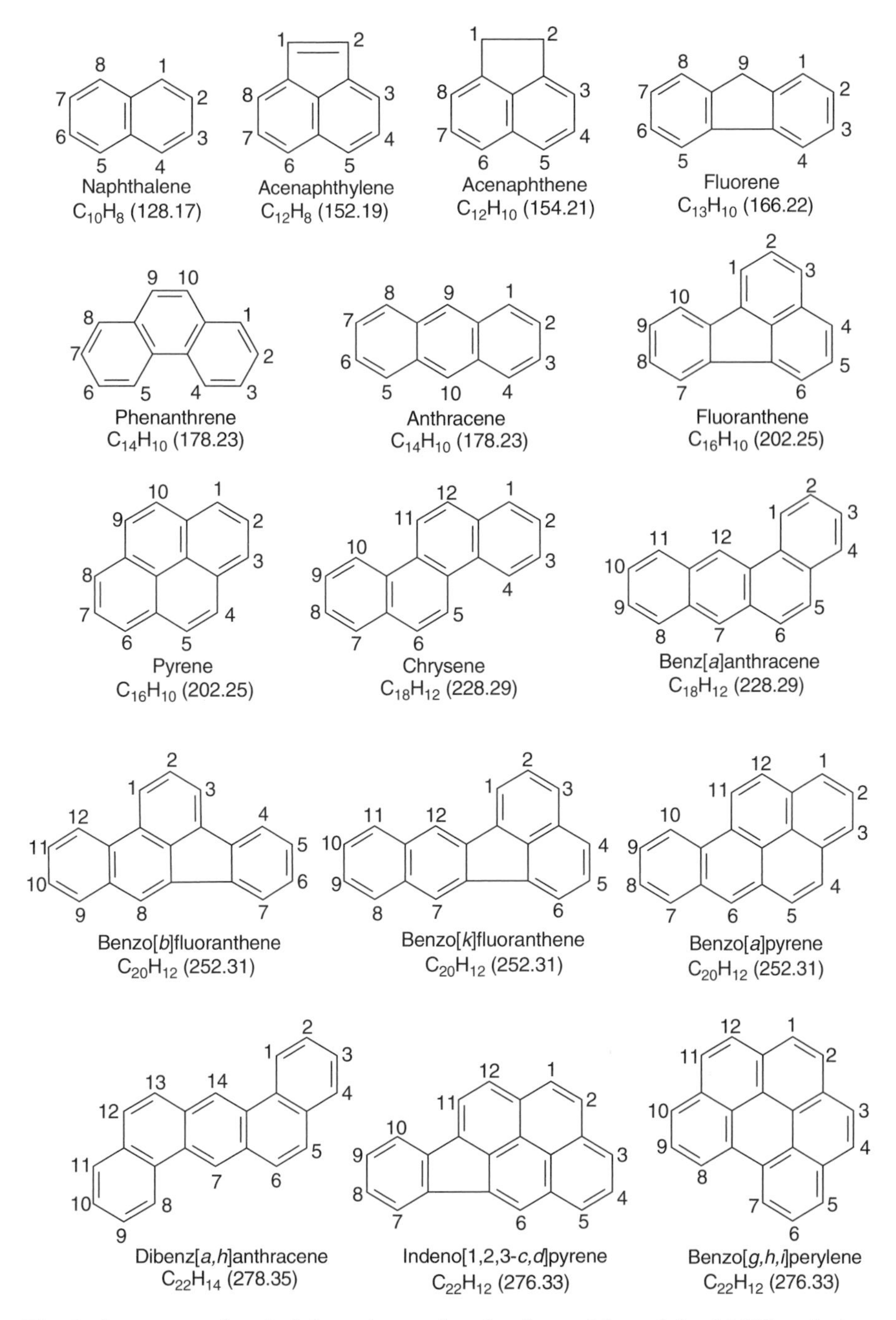

Fig. 1. Structures, chemical formulas, and molecular weights of the 16 EPA priority pollutant PAHs

Table 1. Selected properties of the 16 USEPA priority pollutant PAHs at 25 °C

Compound[a]	$\log(K_{ow})$[b]	C_{sat} (mg/l)[b,c]	TEF[d]	Carcinogenic potency[e]
Naphthalene (NAP)	3.37	31.0	0.001	–
Acenaphthylene (ACY)	4.0	16.1	0.001	–
Acenaphthene (ACE)	3.92	3.80	0.001	–
Fluorene (FLU)	4.18	1.90	0.001	–
Phenanthrene (PHN)	4.57	1.10	0.001	–
Anthracene (ANT)	4.54	0.045	0.01	–
Fluoranthene (FLA)	5.22	0.26	0.001	–
Pyrene (PYR)	5.18	0.132	0.001	–
Benz[*a*]anthracene (BaA)	5.91	0.011	0.1	0.1
Chrysene (CHR)	5.65	0.002	0.01	0.001
Benzo[*b*]fluoranthene (BbF)	5.80	0.0015	0.1	0.1
Benzo[*k*]fluoranthene (BkF)	6.0	0.0008	0.1	0.01
Benzo[*a*]pyrene (BaP)	6.04	0.0038	1	1
Benzo[*g,h,i*]perylene (BgP)	6.50	0.00026	0.01	–
Dibenz[*a,h*]anthracene (DBA)	6.75	0.0006	5	1
Indeno[1,2,3-*c,d*]pyrene (INP)	7.66	0.062	0.1	0.1

[a] Compounds listed in bold are classified by USEPA as probable human carcinogens.
[b] Data are apparent consensus values from Mackay et al. (1992). Aqueous solubilities can vary depending on the method used. Some recent measurements differ significantly from the value shown (de Maagd et al. 1998), while others support them (Reza et al. 2002).
[c] Aqueous solubility.
[d] Toxicity equivalency factor proposed by Nisbet and LaGoy (1992).
[e] USEPA (1993).

et al. 1991; Padma et al. 1998; Villholth 1999). Exposure to the carcinogenic PAHs is more likely to occur from dermal contact, ingestion of particles, inhalation of airborne dust, or bioaccumulation in the food chain (Mayer et al. 1996; Roy et al. 1998; Hussain et al. 1998; National Toxicology Program 2001; Sverdrup et al. 2002; Samsøe-Petersen et al. 2002). Cleanup goals at specific sites in the United States are typically based on the sum of the 16 USEPA-regulated PAHs, the sum of the seven carcinogenic PAHs, or both.

3
Physical and Chemical Complexity of PAH-Contaminated Systems

Soils and sediments contaminated with PAHs are physically, chemically, and biologically complex. PAH contamination is really a result of contamination with a complex chemical mixture such as coal tar, creosote, petroleum, or soot. Different source materials have different distributions of individual PAHs (Harkins et al. 1988), a feature that facilitates forensic analysis of environmental samples to identify the source(s) of contamination (Stout et al. 2001). The weathering of source materials will also change the distribution of PAHs, depleting the two- to three-ring compounds and enriching the four- to six-ring compounds (Haeseler et al. 1999a; Stroo et al. 2000; Brenner et al. 2002). Source materials also contain a variety of compounds besides PAHs. One consequence of this is that the aqueous solubility of any PAH is much lower in systems contaminated with such materials than the solubility of the pure compound in water shown in Table 1. If the mole fractions of the constituents in the complex matrix can be estimated, the aqueous solubilities of the individual compounds can be estimated using Raoult's law and the supercooled-liquid solubilities of the pure solutes (Lee et al. 1992; Lane and Loehr 1992; Peters et al. 1999; Eberhardt and Grathwohl 2002).

It is difficult to obtain comprehensive information on the contaminant matrix in which PAHs are found. Haeseler et al. (1999a) performed a rigorous analysis of the contaminant matrix in PAH-contaminated soils from five different sites formerly used to produce manufactured gas. The primary organic contaminants were distributed in similar proportions among the soils. Over 70% of the contaminant carbon was not extractable by solvent extraction, and was thought to represent a dense non-aqueous-phase liquid (DNAPL) originating from coal tar. Slightly over half of the extractable organics were classified as "resins," relatively polar compounds that are otherwise not well characterized. The 16 PAHs regulated by the USEPA accounted for approximately 10% of the extractable carbon and from 25 to 40% of the extractable aromatics, most of which are assumed to be PAHs. Ghosh et al. (2003) have recently identified, in contaminated harbor sediments, the various types of organic matter with which PAHs are associated.

Regardless of the source, it is important to remember that PAH-contaminated systems always contain a mixture of PAHs, some of which are more predominant than others. It is also important to keep in mind that it is difficult to approach the complexity of real PAH-contaminated materials through artificial contamination or "spiking" of soil or sediment in the laboratory. In particular it is known that "aging" (long-term contact) of a hydrophobic chemical in soil or sediment generally decreases its bioavailability (Hatzinger and Alexander 1995; Madsen et al. 1996; Kelsey et al. 1997; Carmichael et al. 1997; Tang et al. 1998; Guthrie and Pfaender 1998; Guthrie-Nichols et al. 2003). It may be difficult to simulate in a laboratory study the aging that occurs over decades of contamination in the field. Studies based on artificial contamination can lead to observations that have important mechanistic implications, but the significance of a particular mechanism should be expected to vary in different complex systems.

3.1 PAHs Associate with Nonaqueous Compartments

Due to their strong hydrophobicity, PAHs are generally associated with nonaqueous phases in soil and sediment. Luthy et al. (1997) summarized the relevant compartments with which hydrophobic organic chemicals (HOCs) associate as: (1) the mineral domain, which includes exposed mineral surfaces and surfaces within pore spaces; (2) natural organic matter; (3) nonaqueous-phase liquids (NAPLs); and (4) combustion residue such as soot or charcoal (black carbon). The organic compartments dominate the sorption behavior of highly hydrophobic compounds (Luthy et al. 1997) such as PAHs.

The elucidation of the sorptive behavior of the organic fraction of soil and sediment is an area of active research. Depending on the particular soil or sediment, the organic fraction can contain NAPLs, "hard" carbon such as coal and black carbon, and more flexible macromolecular substances such as humic material. The macromolecular forms of organic carbon are thought to undergo transitions between a so-called "glassy" or rigid state and a "rubbery" state that depend on temperature and on interactions with sorbed substances (Huang and Weber Jr. 1998; Xia and Pignatello 2001). The dominant mechanism of HOC association in hard carbon is believed to be adsorption to exposed surfaces or those within inflexible pores (Ghosh et al. 2000; Xia and Pignatello 2001; Accardi-Dey and Gschwend 2003). The association of HOCs with NAPLs and with the "rubbery" form of flexible macromolecules is typically described by absorption (partitioning) (Luthy et al. 1997; Xia and Pignatello 2001; Accardi-Dey and Gschwend 2003). The "glassy" form of macromolecular carbon can partition HOCs as well, but it also contains nanoscale "holes" within which HOC molecules can adsorb; the filling of the holes by competing sorbates is thought to explain the competitive sorption of chemicals that have been observed in the soil (White and Pignatello 1999).

In areas impacted directly by the source materials or by transport of source particles, PAHs are likely to be associated primarily with NAPLs (Lane and Loehr 1992; Rutherford et al. 1997; Haeseler et al. 1999a; Karimi-Lotfabad and Gray 2000; Ghosh et al. 2003) or with other source materials such as coal and combustion residue (Ghosh et al. 2000; Ghosh et al. 2003). Conversely, PAHs are likely to be associated primarily with natural organic matter in areas impacted by aqueous-phase or colloidal transport (Mackay and Gschwend 2001) of the PAHs away from the source zones. It is important to make distinctions among the principal compartments in which PAHs are likely to be found for several reasons. First, the equilibria and kinetics of PAH desorption into water will differ for different compartments (Luthy et al. 1997; White and Pignatello 1999; Talley et al. 2002; Rockne et al. 2002; Eberhardt and Grathwohl 2002; Ghosh et al. 2003; Accardi-Dey and Gschwend 2003). Secondly, the relative accessibility of the PAHs to organisms capable of degrading them will be compartment-dependent (Breedveld and Karlsen 2000; Talley et al. 2002; Ghosh et al. 2003). Thirdly, strategies to improve the biodegradation of PAHs in contaminated soils or sediments must be based on an understanding of the nature of the contamination in any given system.

4 Bioavailability and Its Impact on PAH Biodegradation

As a result of the strong association of PAHs with the organic fractions of contaminated soil and sediment, little of the PAH mass is in the aqueous phase in such systems. Most of the PAH mass is not, therefore, directly available to microorganisms. As the PAHs in the aqueous phase are depleted, continuing degradation requires mass transfer from the interior of a nonaqueous phase to a location accessible to microbial cells. In such cases rates of desorption from the nonaqueous phase(s) will drive overall rates of biodegradation. Wick et al. (2001) put it succinctly: "Limited bioavailability occurs when the capacity of the microbial biomass to consume a substrate exceeds the capacity of its environment to deliver the substrate."

Desorption of PAHs from field-contaminated soil and sediment is typically characterized by a fraction of the contaminant mass that desorbs relatively rapidly and a fraction that desorbs relatively slowly (Cornelissen et al. 1998; Stroo et al. 2000; Hawthorne and Grabanski 2000; Breedveld and Karlsen 2000; Hawthorne et al. 2001; Shor et al. 2003b). Such behavior is often quantifiable with "two-site" models. The observed desorption kinetics are consistent with the association of PAHs with different fractions of organic carbon in a given soil or sediment: organic carbon that is characterized by the strong association of HOCs or very slow rates of internal diffusion, and organic carbon that is characterized by greater mobility of the hydrophobic contaminants. The "glassy" forms of macromolecular organic matter are known to exhibit very slow rates of diffusion, leading to slow rates of contaminant desorption (Huang and Weber Jr. 1998; Xia and Pignatello 2001), and PAHs are also known to sorb tightly to surfaces of black carbon (Stroo et al. 2000; Ghosh et al. 2000; Bucheli and Gustafsson 2000; Xia and Pignatello 2001; Accardi-Dey and Gschwend 2003). The "rubbery" forms of natural organic matter and NAPLs are believed to result in more rapid contaminant desorption, corresponding to higher rates of contaminant diffusion in these materials (Xia and Pignatello 2001; Ghosh et al. 2003; Accardi-Dey and Gschwend 2003). The apparent strength of the association of PAHs with organic carbon fractions that lead to slow desorption has been illustrated by the long-term aging of laboratory-contaminated soil, in which a fraction of the added PAHs have become inaccessible even to conventional solvent extraction methods (Richnow et al. 1996; Guthrie and Pfaender 1998; Guthrie et al. 1999; Northcott and Jones 2001).

There is ample experimental evidence that rates of mass transfer out of nonaqueous phases are likely to govern the rates at which some PAHs are degraded in soil and sediment, both in laboratory-contaminated systems aged to simulate field contamination (Weissenfels et al. 1992; Hatzinger and Alexander 1995; Carmichael and Pfaender 1997; Rutherford et al. 1998; Guthrie and Pfaender 1998; Hawthorne and Grabanski 2000; Guthrie-Nichols et al. 2003) and in field-contaminated systems (Erickson et al. 1993; Williamson et al. 1997; Cornelissen et al. 1998; Talley et al. 2002; Ghosh et al. 2003; Shor et al. 2003a). In a number of cases, the extent of the degradation of a particular contaminant has correlated with an apparent rapidly-desorbing fraction (Cornelissen et al. 1998; Hawthorne and Grabanski 2000; Breedveld and Karlsen 2000; Hawthorne et al. 2001; Talley et al. 2002; Shor et al. 2003a) or was consistent with the expected strength of sorption of the organic carbon fraction (Ghosh et al. 2003). It may be possible to quantify the more readily bioavailable fraction of PAHs through mild and rapid extraction techniques (Kelsey et al. 1997; Reid et al. 2000; Hawthorne and Grabanski 2000; Cuypers et al. 2000; Hawthorne et al. 2001; Johnson and Weber Jr. 2001; Hawthorne et al. 2002; Cuypers et al. 2002).

4.1
Bioavailability May Not Be the Only Factor That Limits PAH Biodegradation

While it is clear that rates of mass transfer generally govern the rates at which PAHs are biodegraded in contaminated soil and sediment, such a generalization warrants closer scrutiny if improvements in biodegradation and bioremediation of PAHs are to be made. Over-generalizing the assumption that there is one fraction of PAHs readily available for biodegradation and another that is unavailable leads to a dichotomous perspective, a "yes/no" approach to evaluating the prospects of bioremediation. We believe that a more appropriate view is that some of the PAH mass in a given contaminated system is more immediately bioavailable and some less so (admittedly much less so in some cases). Contaminated soil and sediment can remain in contact with an aqueous phase for very long periods and usually contain microorganisms capable of degrading PAHs. What may really matter, then, is whether the rate at which a particular compound becomes available to degrader organisms is capable of sustaining meaningful rates of biological activity. Short-term bioassays cannot necessarily answer this question. In the work of Hawthorne et al. (2001), for example, up to 40% of the benzo[*a*]pyrene in a field-contaminated soil desorbed into water over a period of 120 days, a time scale consistent with the time scale for active remediation, yet little degradation of benzo[*a*]pyrene occurred during the field-scale bioremediation of the soil. Shor et al. (2003a) found a correlation between the extent of biodegradation of PAHs over a 5-day period and the estimated rapidly-desorbing fraction of each compound *except for* benz[*a*]anthracene – a compound that the inoculated degrader organism could transform but could not utilize as a sole carbon source.

Other evidence suggests that limited bioavailability alone does not dictate the rate and extent of degradation for some PAHs. Cornelissen et al. (1998) demonstrated that as much as 55% of the five- and six-ring PAHs were in a rapidly-desorbing fraction in contaminated harbor sediment, yet unlike the lower molecular weight compounds they remained undegraded after the bioremediation of the sediment. Rutherford et al. (1998) observed the selective removal of some LMW PAH in the slurry-phase treatment of creosote-contaminated soil, whereas anthracene, fluoranthene, chrysene and pyrene were degraded only after the system was inoculated with soil from another system exhibiting active degradation of these compounds. Breedveld and Karlsen (2002) found that desorption measurements overestimated the biodegradation of five- and six-ring PAHs in a contaminated topsoil. Rates of desorption significantly higher than rates of biodegradation have also been observed for five- and six-ring PAHs in aged laboratory-contaminated soils, as discussed in Chapter 2 (this Vol.; Huesemann et al. 2002, 2003). There is also evidence that increasing the rate of the mass transfer of PAHs does not necessarily lead to an increase in biodegradation: the addition of a surfactant to contaminated soil from a manufactured-gas plant (MGP) site led to significant increases in the removal of two-, three-, and four-ring PAHs, but not the five-ring compounds examined (Tiehm et al. 1997).

If there is a trend in the studies discussed above, it is that limitations in bioavailability may not explain the recalcitrance of high molecular weight (HMW) PAHs, particularly the five- and six-ring compounds. These same compounds have also been observed repeatedly to be removed less extensively than the LMW PAHs during biodegradation studies on PAH-contaminated soil and sediment (Wild and Jones 1993; R.C. Haught et al. 1995; Grosser et al. 1995; Tiehm et al. 1997; Carmichael and Pfaender 1997; Banerjee et al. 1997; Rutherford et al. 1998; Cornelissen et al. 1998; Aitken et al. 1998a; Sayles et al. 1999; Haeseler et al. 1999b; Hawthorne and Grabanski 2000; Taylor and Jones 2001; Ringelberg et al. 2001). Since the carcinogenic PAHs fall into this class, and cleanup endpoints are often based on residual concentrations of the carcinogenic PAHs, it is important to spend more effort characterizing the factors that truly limit the biodegradation of these compounds in field-contaminated systems. Factors other than bioavailability are discussed in the following sections.

5
Complexity of PAH Transformation and Degradation by Microorganisms

The following sections focus on the aerobic bacterial metabolism of PAHs. Although a number of studies have been conducted on the ability of various fungi to transform PAHs, little work has been done to quantify the relative role of fungi in PAH transformation and degradation in contaminated systems. PAH-degrading fungi are known to be present in PAH-contaminated soil and sediment (Stapleton et al. 1998; Salicis et al. 1999; Ravelet et al. 1999, 2001; Saraswathy and Hallberg 2002). In one study, however, the contribution of eukaryotes to phenanthrene transformation in coastal sediments was found to be insignificant (MacGillivray and Shiaris 1994). Attempts to inoculate field-contaminated soil with fungi known to transform PAHs have been met with mixed success (Brodkorb and Legge 1992; Davis et al. 1993;

Baud-Grasset et al. 1993; R.C. Haught et al. 1995; May et al. 1997; Andersson et al. 2000; Canet et al. 2001; Joner and Leyval 2003).

There is also relatively little known about the role of anaerobes in PAH biotransformation and biodegradation. There are a number of reports of anaerobic degradation of naphthalene and other two-ring PAHs. In some cases mineralization was observed (Coates et al. 1996; Rockne et al. 2000; Sullivan et al. 2001; Rothermich et al. 2002) and in others metabolites were identified (Zhang and Young 1997; Meckenstock et al. 2000; Annweiler et al. 2000; Gieg and Suflita 2002; Annweiler et al. 2002). Anaerobic degradation under various electron acceptor conditions has also been observed for anthracene (McNally et al. 1998; Chang et al. 2002), phenanthrene (Coates et al. 1996; Zhang and Young 1997; Rockne and Strand 1998; McNally et al. 1998; Rockne and Strand 2001; Chang et al. 2002; Eriksson et al. 2003), fluoranthene (Coates et al. 1997), and pyrene (McNally et al. 1998; Chang et al. 2002). Rothermich et al. (2002) reported significant declines in PAHs ranging in size from two to five rings in contaminated harbor sediments incubated for nearly 1 year under sulfate reducing conditions.

Most of what is known about aerobic PAH metabolism has been learned with pure cultures isolated from PAH-contaminated systems. Given that PAHs are naturally occurring compounds with widespread distribution throughout the environment, it is not surprising that PAH-degrading microorganisms are relatively easy to find and isolate from these systems. Whether the isolates represent the diversity of PAH-degrading bacteria present in PAH-impacted environments remains to be revealed by molecular approaches to studying microbial communities.

5.1 Outcomes of PAH Metabolism

Bacterial metabolism of any given PAH has the following possible outcomes: (1) growth of an organism on the PAH substrate; (2) mineralization of the PAH substrate in the absence of detectable growth; and (3) transformation of the PAH substrate to non-mineral, terminal products. Any of these outcomes might be included in the ambiguous term "degradation," which is often used to describe the removal of a particular compound from a system whether the mechanism is known or not. Either of the latter two outcomes could also be described as cometabolism. Since different metabolic capabilities are involved, we try to distinguish among the three outcomes by referring to them as growth, mineralization, and either transformation or incomplete metabolism, respectively.

Some products of incomplete metabolism are also commonly known as dead-end metabolites. We do not use this term, however, for those products that might be terminal products from one organism but which are intermediates in a known pathway for complete metabolism of the parent compound by other organisms. A dead-end metabolite, then, is one that would not enter directly into a known pathway for the complete metabolism of the parent compound.

5.1.1 Growth on PAHs

Growth on a PAH substrate as a sole carbon and energy source implies a capability to both assimilate and mineralize at least part of the carbon in the parent molecule. Growth becomes more difficult to detect for the HMW substrates, but there is now abundant evidence of bacteria able to grow on the four-ring compounds pyrene and fluoranthene (Heitkamp et al. 1988; Weissenfels et al. 1990; Mueller et al. 1990; Walter et al. 1991; Boldrin et al. 1993; Kästner et al. 1994; Fritzsche 1994; Mahro B et al. 1995; Bouchez et al. 1995; Shuttleworth and Cerniglia 1996; Schneider et al. 1996; Dean-Ross and Cerniglia 1996; Juhasz et al. 1997; Šepiè et al. 1998; Rehmann et al. 1998; Boonchan et al. 1998; Molina et al. 1999; Churchill et al. 1999; Bouchez et al. 1999; Bastiaens et al. 2000; Vila et al. 2001; Shi et al. 2001; Dean-Ross et al. 2001; Johnsen et al. 2002; Eriksson et al. 2002; Krivobok et al. 2003). There are fewer reports of bacteria which are able to utilize chrysene (Walter et al. 1991; Caldini et al. 1995; Burd and Ward 1996; Melcher et al. 2002) or benz[*a*]anthracene (Caldini et al. 1995; Juhasz et al. 1997) as growth substrates. To date, no single organism has been identified that can grow on a PAH containing five or more rings (Kanaly and Harayama 2000; Juhasz and Naidu 2000). Boonchan et al. (2000) reported that a co-culture of a bacterium and a fungus could grow on benzo[*a*]pyrene as a sole carbon source. Unless this is a more common occurrence than has been observed in laboratory studies, bacteria capable of mineralizing any five- or six-ring PAH apparently must do so during or after growth on some other substrate.

5.1.2 PAH Mineralization in the Absence of Growth

There are a number of cases in which bacteria have been reported to mineralize a particular PAH substrate without utilizing it for growth. Of the HMW compounds that do not support growth, mineralization of BaP has tended to be of most interest. Benzo[*a*]pyrene mineralization has been observed in a number of field samples and enrichment cultures, but there are relatively few reports of pure cultures able to mineralize BaP (Heitkamp and Cerniglia 1989; Grosser et al. 1991; Ye et al. 1996; Schneider et al. 1996; Aitken et al. 1998b; Khan et al. 2002; Juhasz et al. 2002). Although PAH mineralization activity is typically found in field-contaminated soil or sediment, in some cases BaP has not been mineralized when other PAHs were (Herbes and Schwall 1978; Grosser et al. 1995; Carmichael and Pfaender 1997; Aitken et al. 1998a; Ringelberg et al. 2001). This suggests either that organisms capable of mineralizing BaP were not present in the system in significant numbers, or that BaP mineralization was inhibited by a mechanism that did not affect the mineralization of other PAHs to the same extent.

Results of mineralization studies should be interpreted with caution. Because of probable differences in the bioavailability of native contaminants relative to a ^{14}C-labeled substrate spiked into a contaminated system, mineralization of the spiked substrate should only be interpreted as a *potential* for organisms in that system to mineralize the same substrate in the contaminant matrix (Sandoli et al. 1996). In addition, commercial forms of ^{14}C-labeled PAHs have a limited number of carbon atoms labeled; BaP in particular is available with ^{14}C only at the seventh and tenth positions, both of which are on the same ring (see Fig. 1). It is possible in such cases that conversion of these carbon atoms to $^{14}CO_2$ could leave a relatively large transformation product that resists further metabolism. The term "mineralization" in such cases might mislead the observer to believe that the metabolism of the compound is more extensive than it really is. For example, Schneider et al. (1996) identified a range of transformation products from the oxidation of BaP by a *Mycobacterium* strain. Several metabolites were consistent with the oxidation of the ring that contains the labeled carbon atoms in [7,10-^{14}C]BaP while leaving the other rings intact. Another metabolite was consistent with the oxidation of BaP at the 4,5 position (see Fig. 1), a transformation that would not correlate to mineralization in this case. Finally, mineralization of a given substrate in a complex system does not imply that a single organism was able to catalyze every step of the pathway leading to $^{14}CO_2$ production. Several examples of PAH mineralization requiring microbial consortia have been documented (Li et al. 1996a; Casellas et al. 1998; Boonchan et al. 2000; Kanaly et al. 2002).

It is open to question whether an organism gains benefit from an ability to mineralize a PAH substrate without being able to use it for growth. Mineralization by a single organism implies that a full complement of the necessary enzymes is available and already being expressed by the organism. Accordingly, the organism should be able to assimilate some of the carbon along the way and energy-yielding steps should exist. The apparent inability to grow on a particular substrate may, therefore, reflect rates of metabolism that are so slow that the corresponding rate of energy production is insufficient to sustain even cellular maintenance functions. It is also conceivable that the ability to mineralize compounds unable to support detectable growth in the laboratory can sustain a population of PAH degraders at very small numbers, which may be a way of life in contaminated systems in situ.

5.1.3 Incomplete Metabolism

Transformation of a PAH substrate to non-mineral products suggests that the parent compound is fortuitously acted on by the first enzyme(s) in what otherwise would be a pathway for the complete metabolism of a different substrate. This incomplete metabolism would not be expected to yield significant energy (Auger et al. 1995) nor lead to assimilation of the metabolites. In contrast, energy may actually be consumed in these processes. The dioxygenase involved in the initial oxygenation of an aromatic ring by aerobic PAH-degrading bacteria consumes reducing power during catalysis (Kiyohara and Nagao 1978), which is normally recovered in downstream metabolic steps. If ring dioxygenation is the only transformation that occurs, then energy will be consumed in the process.

Incomplete metabolism of one or more PAHs has been observed for a variety of bacteria isolated from PAH-contaminated soil and sediment, with examples for the HMW PAHs shown in Table 2. PAH transformation products have also been observed in environmental samples that were either analyzed directly or after PAHs were spiked into the samples (Table 3). Intermediates of PAH degradation pathways often accumulate extracellularly during batch incubations with organisms capable of mineralizing the parent compound. In some cases both metabolic intermediates and dead-end metabolites accumulate simultaneously (Grifoll et al. 1995).

The types of products identified from the incomplete metabolism of PAHs by aerobic bacteria are generally consistent with known metabolic steps in aerobic PAH metabolism. Such products include dihydrodiols, hydroxy acids, and *ortho*-quinones. In the latter case, *o*-quinones are not normally intermediates in PAH degradation pathways but are believed to result from the autoxidation of catechols, which are known intermediates (Davies and Evans 1964; Laurie and Lloyd-Jones 1999; Kazunga and Aitken 2000; Kazunga et al. 2001). The fate of extracellular products derived from PAHs has not been studied well. In several cases

Table 2. Studies documenting incomplete metabolism of HMW PAHs by bacteria

Compound	Organism	References[a]
Benz[*a*]anthracene	*Sphingomonas yanoikuyae*[b]	**Gibson et al. (1975); Jerina et al. (1984); Mahaffey et al. (1988)**
	Mycobacterium sp. RJGII-135	**Schneider et al. (1996)**
Benzo[*a*]pyrene	*Sphingomonas yanoikuyae*[b]	**Gibson et al. (1975)**
	Mycobacterium sp. RJGII-135	**Schneider et al. (1996)**
Chrysene	*Sphingomonas paucimobilis* EPA505[c]	Mueller et al. (1990)
Dibenz[*a,h*]anthracene	*Stenotrophomonas maltophilia* VUN 10,003	Juhasz et al. (2002)
Fluoranthene	*Bacillus cereus* R1; *Pseudomonas saccharophila* P15; *Pseudomonas stutzeri* P16; *Sphingomonas yanoikuyae* R1	**Kazunga et al. (2001)**
Pyrene	*Sphingomonas paucimobilis* EPA505[c]	Mueller et al. (1990)
	Bacillus cereus R1; *Pseudomonas saccharophila* P15; *Pseudomonas stutzeri* P16; *Sphingomonas yanoikuyae* R10	**Kazunga and Aitken (2000)**

[a]References in bold indicate studies in which specific products were identified. In general, the parent compound was not mineralized by the indicated organism, except in the case of BaP transformation by *Mycobacterium* sp. RJGII-135.
[b]Formerly described as a *Beijerinckia* sp. (Khan et al. 1996).
[c]Formerly classified as *Pseudomonas paucimobilis*.

involving consortia of two or more bacteria, one of the organisms was shown to degrade a metabolite produced by another organism (Heitkamp and Cerniglia 1989; Trzesicka-Mlynarz and Ward 1995; Li et al. 1996b; Casellas et al. 1998; Arino et al. 1998). The potential for further transformation of a terminal product from one organism is clearly enhanced in microbial communities (Heitkamp and Cerniglia 1989). However, some products of incomplete metabolism can inhibit the activity of PAH-degrading organisms (Kazunga and Aitken 2000; Kazunga et al. 2001). It is also conceivable that in complex systems some organisms might specialize in scavenging products released by other organisms (Bouchez et al. 1999).

Products of incomplete PAH metabolism can also associate with natural organic matter in soil or sediment systems, either covalently (Kacker et al. 2002) or non-covalently (Guthrie-Nichols et al. 2003). Finding ^{13}C- or ^{14}C-labeled residues associated with natural organic matter (Eschenbach et al. 1998; Nieman et al. 1999; Kästner et al. 1999)

Table 3. Studies documenting products of incomplete PAH metabolism in complex systems

Source[a]	Parent compound(s)[b]	References
Spiked uncontaminated soil	7,12-Dimethylbenz[*a*]anthracene	Park et al. (1988)
Spiked estuarine sediments	Naphthalene, phenanthrene, benzo[*a*]pyrene[c]	Shiaris (1989)
Coastal sediments	Anthracene, benz[*a*]anthracene, benz[*de*]anthracene, benzo[*a*]pyrene, dibenzofluorene, fluoranthene, fluorene, phenanthrene	Fernández et al. (1992)
Spiked coastal sediments	Phenanthrene	MacGillivray and Shiaris (1994)
Spiked contaminated soil	Naphthalene, fluoranthene	Langbehn and Steinhart (1995)
Spiked marine sediment	Pyrene	Li et al. (1996b)
Culture enriched from oil-contaminated groundwater	Anthracene, benzo[*k*]fluoranthene	Preuß et al. (1997)
Contaminated soil	Anthracene, 7,12-dimethylbenz[*a*]anthracene, fluorene	Meyer et al. (1999)
Spiked soil	Anthracene	Joner et al. (2001)
Spiked soil	Acenaphthene, anthracene, fluoranthene, fluorene, naphthalene, phenanthrene, pyrene	Meyer and Steinhart (2001)
Spiked sediment	Pyrene	Guthrie-Nichols et al. (2003)

[a] Products were identified in the native material unless the source was spiked.
[b] In unspiked materials, parent compounds were inferred from the identified product(s). In spiked materials products were chemically identified unless indicated otherwise.
[c] Accumulation of products inferred from significant accumulation of radiolabeled material in the aqueous phase after incubation with the radiolabeled parent compound.

is not necessarily evidence that the label was derived directly from one or more products of incomplete metabolism; it is also possible for the label to represent products of microbial decay after having been assimilated via metabolism of the parent compound.

The apparently common occurrence of incomplete metabolism by bacteria isolated from PAH-contaminated soil and sediment suggests that the transformation of some PAHs to non-mineral products may be inevitable in PAH-contaminated systems that are biologically active. There are two concerns with the potential accumulation of such products in the field. First, they can significantly inhibit the metabolism of other PAHs (Kazunga and Aitken 2000; Kazunga et al. 2001; Juhasz et al. 2002), and benzo[*a*]pyrene metabolism seems to be particularly susceptible to such inhibition. Secondly, the potential toxicity of these products to higher organisms must be considered (Grifoll et al. 1995). For example, several PAH-quinones are known to be genotoxic to mammals (Penning et al. 1999). Quinones were found in significant concentrations (mg/kg range) in one of the few studies designed to identify products of PAH transformation in field-contaminated soil (Meyer et al. 1999a). Anthraquinone derived from spiked anthracene was also found in mg/kg concentrations in aged spiked soil (Joner et al. 2001). Therefore, this accumulation of PAH transformation products may explain transient changes in genotoxicity that have been observed during biological treatment of PAH-contaminated soil (Belkin et al. 1994; Hughes et al. 1998).

5.2 Substrate Ranges of PAH-Degrading Bacteria

PAH-degrading bacteria are often isolated by growth on a single PAH substrate, but many of these isolates are also examined for their ability to metabolize other PAHs. Based on the available evidence from studies in which a range of PAH substrates have been tested, it is reasonable to say that virtually all bacterial isolates tested are able to grow on, mineralize, or transform a variety of PAHs. In most cases, these isolates are *not* able to metabolize one or more PAHs as well (Foght and Westlake 1988; Kästner et al. 1994; Kelley and Cerniglia 1995; Bouchez et al. 1995; Dagher et al. 1997; Aitken et al. 1998b; Surovtseva et al. 1999; Churchill et al. 1999; Daane et al. 2001). Thus, the exact PAH substrate range may be more or less unique to the given organism. Furthermore, any given organism can be expected to grow on some PAHs, mineralize others without growth, and only transform still others (Aitken et al. 1998b). Bacteria that have been isolated by growth on naphthalene have generally been found to metabolize a narrower range of PAHs than organisms isolated by growth on other compounds (Foght and Westlake 1991; Kästner et al. 1994; Daane et al. 2001).

It is not surprising that PAH-degrading bacteria have relatively broad substrate ranges, since PAH-contaminated systems always contain mixtures of PAHs. This implies, however, that even for a single organism in a PAH-contaminated system an array of different substrates may be available for metabolism; the extent to which any one or more is metabolized at a given time is essentially unknown. Add to this complex situation for a single organism the existence of multiple PAH-degraders in a contaminated system, and the result is a complex web of potentially overlapping and competing metabolic activity.

We do not yet have the ability to predict the PAH substrate range of any organism, as we do not understand the genetic predictors of complete PAH metabolism for any but the smallest of the PAHs. Such ability might, for example, help identify the range of substrates that can support the mineralization of HMW PAHs that do not support growth.

5.2.1 Substrate Interactions in Pure Cultures

PAHs are structurally similar compounds that coexist in PAH-contaminated systems. Although there can be several pathways by which a given PAH substrate is metabolized by different bacteria, the basic steps in aerobic metabolism are similar (Sutherland et al. 1995). Since most PAH-degrading bacteria can metabolize multiple PAHs, it would be efficient for an organism to have enzymes capable of transforming multiple PAHs and the intermediates derived from them. For example, in some organisms the same enzymes are involved in the transformation of phenanthrene to 1-hydroxy-2-naphthoate (1H2N) as are involved in the well-known upper pathway for the conversion of naphthalene to salicylate (Denome et al. 1993; Kiyohara et al. 1994), as illustrated in Fig. 2. There are at least three known bacterial pathways by which 1H2N can be further metabolized (Evans et al. 1965; Kiyohara and Nagao K 1978; Samanta et al. 1999; Prabhu and Phale 2003), one of which is to convert 1H2N to 1,2-dihydroxynaphthalene (large arrow in Fig. 2; Evans et al., 1965). Thus, only one extra enzyme would be required to convert the naphthalene metabolic pathway into a phenanthrene metabolic pathway. A similar overlap in the metabolism of pyrene and phenanthrene has been proposed to exist in two different species of *Mycobacterium* (Vila et al. 2001; Moody et al. 2001). In this case, dioxygenation occurs in the K region of each compound (across the 4,5-bond in pyrene and across the 9,10-bond in phenanthrene; see Fig. 1).

Various lines of evidence suggest that multiple PAH substrates are metabolized by the same enzymes for at least parts of their respective metabolic pathways: (1) direct molecular evidence that the same suite of genes codes for enzymes that metabolize multiple substrates (Denome et al. 1993; Menn et al. 1993; Yang et al. 1994; Kiyohara et al. 1994; Goyal and Zylstra 1996); (2) coinduction of the metabolism of multiple PAHs by a single inducer (Chen and Aitken 1999); and (3)

naphthalene → cis-naphthalene-1,2-dihydrodiol → 1,2-dihydroxy-naphthalene →→ salicylate

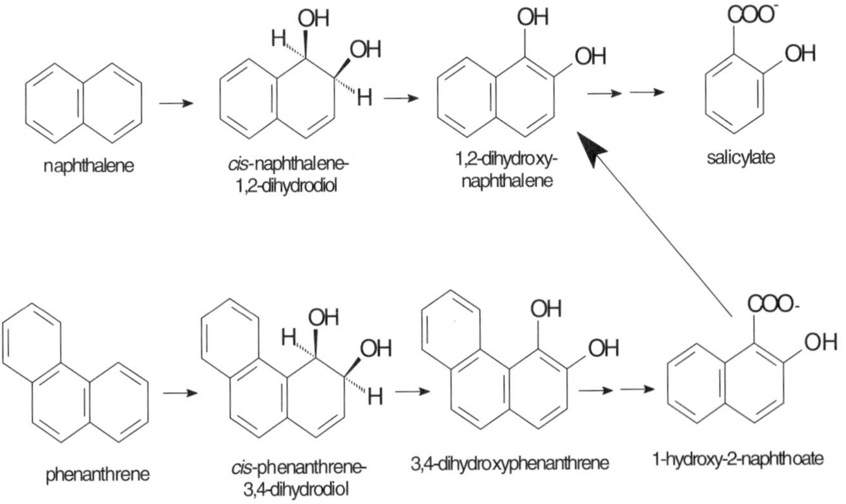

phenanthrene → cis-phenanthrene-3,4-dihydrodiol → 3,4-dihydroxyphenanthrene →→ 1-hydroxy-2-naphthoate

Fig. 2. Illustration of common steps in the upper pathway for aerobic metabolism of naphthalene and one of the pathways for aerobic metabolism of phenanthrene

broad substrate specificity of the enzymes involved, the best studied example of which is the initial ring dioxygenase (Selifonov et al. 1996; Foght and Westlake 1996; Resnick et al. 1996; Di Gennaro et al. 1997; Khan et al. 2001).

There are significant implications of the existence of pathways that use at least some of the same enzymes to metabolize multiple substrates, all of which are likely to be present in the system at the same time. It has been shown that the rate of degradation of one PAH is susceptible to competitive inhibition by another PAH (Stringfellow and Aitken 1995; Guha et al. 1999; Dean-Ross et al. 2002), consistent with competition for one or more enzymes in at least the initial part of the pathway. Apparent competitive effects have been observed in other studies as well (Millette et al. 1995; Kelley and Cerniglia 1995; Shuttleworth and Cerniglia 1996; Millette et al. 1998; Yuan et al. 2000; Lotfabad and Gray 2002). The extent to which one PAH influences another would depend on the relative rates of catalysis for each PAH substrate and its metabolic intermediates. Simple competition might not be manifested in situations in which one or more of the substrates are a growth substrate (Guha et al. 1999). Competition might also be masked in mixed cultures (Lotfabad and Gray 2002), in which there are opportunities for more than one organism to be acting on the different substrates.

The presence of an energy-yielding substrate can also stimulate, rather than inhibit, the rate of removal of a substrate that does not support growth (Bouchez et al. 1995; Beckles et al. 1998; Yuan et al. 2000; Dean-Ross et al. 2002). These observations underscore the potential for significant energy consumption resulting from incomplete metabolism (Sect. 5.1.3).

The use of common enzymes or pathway elements to metabolize multiple PAHs also implies that the induction of those enzymes should coinduce the metabolism of multiple PAHs (Molina et al. 1999; Chen and Aitken 1999). This has implications for degradation of the HMW PAHs that do not serve as growth substrates, because their metabolism must be induced some other way. Salicylate induces naphthalene metabolism (Yen and Serdar 1988), phenanthrene metabolism in some bacteria (Chen and Aitken 1999; Meyer et al. 1999b), and benz[*a*]anthracene transformation in at least one organism (Mahaffey et al. 1988). However, the inducers for the full range of PAH metabolic pathways are not well understood.

5.3 Diversity of PAH Metabolism

Just as it is not surprising that PAH-degrading bacteria are relatively easy to find in PAH-contaminated systems, it is also not surprising that these bacteria represent a wide range of genera. Representatives of the α-, β- and γ-*Proteobacteria* are known to degrade PAHs (Mueller et al. 1997; Meyer et al. 1999b; Daane et al. 2001) as are many Gram-positive bacteria, particularly from the genus *Mycobacterium* (Dean-Ross and Cerniglia 1996; Meyer et al. 1999b; Daane et al. 2001). Recent work indicates that the diversity of PAH-degrading bacteria can be attributed in part to horizontal gene transfer (Wilson et al. 2003) and to recombination events that led to mosaic pathways containing recognizable components of other known pathways for PAH metabolism (Bosch et al. 1999). While there is obviously phylogenetic diversity among PAH degraders, we must question how this relates to diversity in PAH metabolism.

Although the general steps in the aerobic metabolism of aromatic compounds are similar, there is emerging evidence for substantial diversity in the exact steps by which PAHs are metabolized. Indeed, a number of "novel" pathways or metabolites have been reported recently for several different PAHs (Vila et al. 2001; Moody et al. 2001; Kazunga et al. 2001; Dean-Ross et al. 2001; Daane et al. 2001; van Herwijnen et al. 2003; Prabhu and Phale 2003). The unique range of substrates that a given organism can grow on, mineralize, or metabolize incompletely (Sect. 5.2) may be a manifestation of such diversity in PAH metabolism. Another explanation may be that some organisms have genes that code for more than one pathway of PAH metabolism (Vila et al. 2001; Khan et al. 2001; Moody et al. 2001; Ferrero et al. 2002) or more than one ring dioxygenase (Krivobok et al. 2003).

Some of the recent reports of novel pathways have been based on the identification of intermediates in the "lower" parts of pathways that deviate from a previously observed pathway for complete metabolism of the parent compound. However, the diversity of PAH metabolism may also begin with the initial steps of metabolism, particularly in the initial dioxygenation reaction. For example, some dioxygenases catalyze the dioxygenation of phenanthrene at the 3,4-bond (Sutherland et al. 1995) and others at the 9,10-bond (Vila et al. 2001; Moody et al. 2001). Clearly, different dioxygenases can exhibit regioselectivity for a given substrate. Even if different dioxygenases act at the same site on a particular PAH, there may be differences in the range of PAHs that are substrates for a given dioxygenase. Although ring dioxygenases are known to have broad substrate specificity (Resnick et al., 1996; also see Chap. 8), none have been examined for their ability to catalyze the oxygenation of the range of PAH substrates of most concern in PAH-contaminated systems.

The first well-characterized ring dioxygenase able to act on a PAH substrate was the multi-component dioxygenase involved in naphthalene catabolism by *Pseudomonas putida* NCIB 9816 (Ensley et al. 1982), and the first well-characterized genes for naphthalene catabolism were from the NAH7 plasmid of *Pseudomonas putida* G7 (Yen and Gunsalus 1982). Accordingly, investigators studying PAH metabolism and genetics have used the *P. putida* dioxygenase (NahA) and the NAH7 genes for naphthalene catabolism (*nah* genes) as benchmarks. Many PAH-degrading bacteria do not, however, possess genes homologous to the *nah* genes (Foght and Westlake 1991; Rosselló-Mora et al. 1994; Goyal and Zylstra 1996; Foght and Westlake 1996; Geiselbrecht et al. 1998; Berardesco et al. 1998; Saito et al. 1999; Laurie and Lloyd-Jones 1999; Churchill et al. 1999; Meyer et al. 1999b; Daane et al. 2001). While some of these genes have sequences similar to those of the *nah* genes, others have low sequence similarity (Goyal and Zylstra 1996; Foght and Westlake 1996; Laurie and Lloyd-Jones 1999; Saito et al. 2000; Moser and Stahl 2001; Daane et al. 2001; Widada et al. 2002). Foght and Westlake (1991) suggested that probes based on the *nah* genes might be suitable for the detection of naphthalene-utilizing organisms but not for more general PAH-degraders, which has been borne out in one recent study (Laurie and Lloyd-Jones 2000). The term "PAH dioxygenase" was coined in recognition that ring dioxygenases act on a range of PAHs other than naphthalene (Takizawa et al. 1994). Others have used the PAH substrate that an organism was grown on to identify its ring dioxygenase, such as "phenanthrene dioxygenase" (Goyal and Zylstra 1996; Saito et al. 2000).

The genetic diversity of PAH dioxygenases makes it difficult to develop consensus gene probes that target a given PAH-degrading phenotype based on its initial ring dioxygenase (Moser and Stahl 2001). Similar diversity of the genes for ring-cleavage dioxygenases involved in PAH metabolism have been noted (Meyer et al. 1999b). However, differences in regioselectivity and in substrate ranges of ring dioxygenases should correspond to detectable genetic differences as well, which could eventually lead to the development of predictive gene probes. For example, pyrene-degrading *Mycobacterium* spp. have been reported to contain ring dioxygenases involved in pyrene transformation that are unique to Gram-positive bacteria (Khan et al. 2001; Krivobok et al. 2003; Brezna et al. 2003).

Overall, the genetic diversity of PAH metabolism suggests that a range of probes will need to be developed to identify the full PAH-degradation potential of a microbial community in any PAH-contaminated system (Foght and Westlake 1991; Moser and Stahl 2001). One approach to identifying the full range of microorganisms represented in a functional guild that is known to be highly diverse is to develop microarrays containing probes that can cover the known representatives in the guild (Loy et al. 2002; Taroncher-Oldenburg et al. 2003). An emerging technique to identify organisms capable of growing on a particular PAH substrate is the use of ^{13}C-labeled substrates followed by density-gradient ultracentrifugation to separate the ^{13}C-labeled nucleic acids, which can then be amplified and subjected to conventional methods of molecular microbial community analysis (Padmanabhan et al. 2003). One limitation of this approach is the availability of suitably labeled substrates.

It should also be kept in mind that the specificity of one enzyme for a particular range of PAHs will not be indicative of how any of those PAHs is metabolized beyond that enzyme. As noted in Section 5.1.3, the accumulation of products of incomplete metabolism suggests that there is a point in a pathway at which the product of one enzyme is not recognized as a substrate for the next enzyme in that pathway. Given our general ignorance of the molecular genetics of all but a few relevant enzymes involved in PAH metabolism, we may be a long way from being able to use molecular tools to predict whether a given compound can be metabolized completely in a given system. Put another way, what will it take to develop molecular tools to predict (and eventually quantify) the capacity of a system to mineralize, for example, benzo[*a*]pyrene?

6
Prospects for Bioremediation of PAH-Contaminated Systems

Bioremediation involves the use of biological systems, typically relying on microbial processes, to remove target contaminants in soil, sediment or groundwater to acceptable levels. While bioremediation has been used successfully to clean up PAH-contaminated sites, biological treatment often leaves relatively high residual levels of HMW PAHs (Sect. 4.1). Such findings are usually explained by limitations in bioavailability, whether the mechanism(s) controlling the degradation of a particular PAH have been evaluated or not.

We have suggested that factors other than bioavailability may limit the biodegradation of the five- and six-ring PAHs. These factors can include: (1) low numbers of organisms capable of degrading the more recalcitrant PAHs; (2) inherently slow enzyme activity for a given compound, so that removal is controlled by the biodegradation rate rather than the mass transfer rate even at relatively high concentrations of the relevant organism(s); (3) slow rates of degradation exacerbated by competition among substrates, particularly the more water-soluble PAHs; (4) depletion of growth substrates, leading to a decline in activity on the compounds that do not serve as growth substrates; and (5) the accumulation of inhibitory products. In addition to these factors, limitations in oxygen (Madsen et al. 1996), other electron acceptors or other nutrients might also explain limitations in PAH biodegradation in situ. The incomplete removal of the HMW PAHs is still observed, however, in engineered systems designed to provide mixing, sufficient oxygen, and macronutrients for microbial activity.

If improvements are to be made to the extent to which any of the target PAHs are removed in a biological treatment process, then strategies to do so must be based on knowledge of the factor(s) limiting biodegradation in the first place.

6.1
Mass Transfer Limitations

If mass transfer rates limit the rate of degradation then methods of increasing the rate of mass transfer should be tried. These methods may have varying impacts on PAHs associated with different hydrophobic compartments in soil and sediment (Sects. 3 and 4). Mixing can enhance contact between NAPLs and water, and may also increase the surface area for mass transfer by breaking up soil aggregates or NAPL blobs, but may have a limited impact on the availability of PAHs associated with "hard" carbon.

The use of surfactants has also been suggested by a number of investigators to increase the bioavailability of hydrophobic chemicals (Rouse et al. 1994). If there are no secondary effects of surfactants such as toxicity, then the influence of surfactants on mass transfer rates should correspond to predictable effects on microbial activity (Grimberg et al. 1996). The addition of a surfactant to field-contaminated soil has led to results ranging from improvements in PAH removal (Tiehm et al. 1997) to no effect or inhibition of PAH removal (Deschênes et al. 1996). The possibility that the surfactants themselves can also be biodegraded should be taken into account (Deschênes et al. 1996; Tiehm et al. 1997). To date, the comparative effects of surfactants on rates of PAH desorption from different compartments in field-contaminated soil have not been studied.

We also have to be aware that trying to solve one problem can create another. In several laboratory studies, increases in rates of PAH mass transfer have led to the accumulation of products of incomplete metabolism, particularly quinones, that in turn led to a decreased rate of transformation of the parent compound (Auger et al. 1995; Ghoshal and Luthy 1998; Willumsen and Arvin 1999). Such findings suggest that increases in the rate of parent compound availability can lead to faster rates of metabolism in the upper part of a pathway, with an eventual "bottleneck" that causes accumulation of an intermediate. In the case of quinone accumulation, it is likely that the precursor catechol accumulates and that the ring-cleavage dioxygenase is the rate-limiting enzyme in the pathway.

6.2 Bioagumentation

If biodegradation of one or more PAHs is limited by the concentration of organisms that can degrade them, then inoculating the system with known degrader organisms (bioaugmentation) may help. Bioaugmentation of field-contaminated soil with PAH-degrading microorganisms has led to increases in the extent of degradation of several HMW PAHs relative to the uninoculated soils (Rutherford et al. 1998; Juhasz et al. 2000). Bioaugmentation is not likely to help, however, if other factors such as bioavailability primarily influence the biodegradability of a particular contaminant in a given system. One way to diagnose low concentrations of specific degrader organisms is from an apparent lack of mineralization activity against the compound of interest.

6.3 Removal Kinetics

Our current knowledge of specific rates of PAH metabolism (rates per unit biomass) are inadequate, particularly for the HMW PAHs. Knightes and Peters (2003) quantified first-order removal rate coefficients for PAHs ranging in size from two to four rings in a mixed culture of PAH degraders. The rates were of the same magnitude for most of the substrates, suggesting that inherent removal kinetics may be similar for various PAHs at equivalent aqueous-phase concentrations. However, data provided in Chen and Aitken (1999) suggest a first-order removal rate coefficient for benzo[*a*]pyrene transformation by *Pseudomonas saccharophila* P15 that is two orders of magnitude lower than those reported by Knightes and Peters. Rate studies need to be conducted with many more cultures and communities before we can begin to see patterns in inherent removal kinetics for different PAHs. Ideally, removal kinetics will eventually be linked to other indicators such as the presence of a particular PAH dioxygenase or other critical enzymes, or whether the compound can be metabolized completely. Kinetic information is important for modeling, which in turn is an important component of design and prediction for bioremediation systems.

6.4 Supplemental Carbon Sources

Most of the carcinogenic PAHs are not known to be growth substrates for any organism, suggesting that organisms capable of degrading these compounds must obtain energy from other sources if they are to sustain long-term activity against these more recalcitrant PAHs. Obvious sources of energy are the LMW PAHs that do sustain microbial growth. This leads to a paradox of sorts if the same compounds that sustain growth and provide energy also competitively inhibit the metabolism of the HMW PAHs (Sect. 5.2.1). Since the LMW compounds are consistently removed to greater extents than the HMW compounds (Sect. 4.1), it is reasonable to expect that depletion of energy sources reduces the collective ability of the microbial community to remove the HMW PAHs. However, if this is the case in a given system, then it is also reasonable to expect that such activity can be enhanced through the addition of supplemental carbon sources.

There is some evidence that degradation of HMW PAHs can be enhanced by the addition of non-specific carbon sources (Keck et al. 1989). Natural organic matter has been observed to stimulate the degradation of pyrene (Holman et al. 2002; Pecher et al. 2002) and to enhance the degradation of HMW PAHs in creosote-contaminated soil (Bengtsson and Zerhouni 2003). Other investigators have proposed the use of water-soluble carbon sources more relevant to PAH metabolism, such as phthalate and salicylate (Ogunseitan et al. 1991; Ogunseitan and Olson 1993; Tittle et al. 1995; Poeton et al. 1999; Chen and Aitken 1999). Salicylate added to soil has been shown to enhance and sustain the guild of naphthalene degraders for periods of time that may be consistent with periods over which enhanced degradation of PAHs can occur (Ogunseitan et al. 1991). Salicylate addition to soil has also been used to sustain populations of bacteria inoculated into the soil (Colbert et al. 1993a, b, c; Ji and Wilson, 2003).

Carbon source supplementation is probably superfluous in a system that contains adequate concentrations of *available* LMW PAHs or other biodegradable carbon. In any approach to stimulating PAH degradation relying on supplemental carbon sources, we must also consider longer-term effects of those carbon sources on the microbial community. For example, the continuous addition of the carbon sources could easily select organisms that do not degrade PAHs. For these reasons, an appropriate strategy for adding carbon sources would be to do so intermittently in a batch process or in the second stage of a two-stage treatment process.

6.5 Other Considerations

Our suggestions that products of incomplete metabolism may accumulate in PAH-contaminated systems, and that these products may inhibit the degradation of the HMW PAHs in particular, are speculative. Nevertheless, there is substantial evidence that such products are formed by pure cultures and microbial communities capable of metabolizing PAHs, and they have been found in contaminated environments (Sect. 5.3). Clearly, more research is needed to determine the relevance of these observations at contaminated sites. It also should be considered that more than one strategy might be necessary to optimize PAH biodegradation in the field.

All of the approaches that have been considered for enhancing PAH removal are more difficult to implement in situ than in above-ground processes used to treat excavated soil or dredged sediment. In the case of natural attenuation, we are less interested in interventions that will increase PAH removal than we are in being able to predict their fates over time scales of interest. Given the complexity of PAH-contaminated systems and the complexity of microbial activity on PAHs, much more work needs to be done before we have adequate conceptual and quantitative models to make such predictions with confidence.

7
Conclusions

Polycyclic aromatic hydrocarbons are inherently biodegradable compounds. As molecular size increases, these compounds transition from readily degraded carbon and energy sources to poorly available cosubstrates that do not appear to support significant growth of microorganisms. Nevertheless, the higher molecular weight PAHs are of most concern with respect to human health.

One purpose of this chapter was to present a case that further research into PAH biodegradation should be aimed at elucidating factors that govern the degradation of the carcinogenic PAHs, particularly the five- and six-ring compounds, in real contaminated systems. Complexity is a fact of life in these systems. Mechanistic insights are still needed, of course, and many of these will continue to derive from well-controlled laboratory studies. However, more laboratory work needs to be done with contaminated materials as a means of testing theories and evaluating the potential significance of specific mechanisms in the field.

Molecular research on PAH-degrading microorganisms and microbial communities should be aimed at elucidating the determinants of metabolism for specific PAHs – again the carcinogenic PAHs are of most interest – in contaminated systems. Eventually, those tools need to become quantitative to assess the potential microbial activity against any PAH of interest in a given system.

We need better information on rates of relevant processes in PAH-contaminated systems, particularly intrinsic rates of microbial metabolism, and we need a better understanding of substrate interactions – first in pure cultures and then in more complex systems. We need to develop conceptual models that account for the major processes and mechanisms influencing PAH degradation as a prerequisite to developing quantitative models with reasonable predictive capabilities. We also need the quantitative models, in particular to understand the potential risks associated with contaminated materials that are not subjected to active remediation.

Acknowledgments. The lead author's research on PAH biodegradation was supported by the National Institute of Environmental Health Sciences under grant number 5 P42 ES05948.

References

Accardi-Dey A, Gschwend PM (2003) Reinterpreting literature sorption data considering both absorption into organic carbon and adsorption onto black carbon. Environ Sci Technol 37:99–106

Aitken MD, Nitz DC, Roy DV, Kazunga C (1998a) Slurry-phase Bioremediation of Contaminated Soil from a Former Manufactured-Gas Plant Site, Report No 320. University of North Carolina Water Resources Research Institute, Raleigh, NC

Aitken MD, Stringfellow WT, Nagel RD, Kazunga C, Chen S-H (1998b) Characteristics of phenanthrene-degrading bacteria isolated from soils contaminated with polycyclic aromatic hydrocarbons. Can J Microbiol 44:743–752

Andersson BE, Welinder L, Olsson PA, Olsson S, Henrysson T (2000) Growth of inoculated white-rot fungi and their interactions with the bacterial community in soil contaminated with polycyclic aromatic hydrocarbons, as measured by phospholipid fatty acids. Bioresour Technol 73:29–36

Annweiler E, Materna A, Safinowski M, Kappler A, Richnow HH, Michaelis W, Meckenstock RU (2000) Anaerobic degradation of 2-methylnaphthalene by a sulfate-reducing enrichment culture. Appl Environ Microbiol 66:5329–5333

Annweiler E, Michaelis W, Meckenstock RU (2002) Identical ring cleavage products during anaerobic degradation of naphthalene, 2-methylnaphthalene, and tetralin indicate a new metabolic pathway. Appl Environ Microbiol 68:852–858

Arino S, Marchal R, Vandecasteele JP (1998) Involvement of a rhamnolipid-producing strain of *Pseudomonas aeruginosa* in the degradation of polycyclic aromatic hydrocarbons by a bacterial community. J Appl Microbiol 84:769–776

Auger RL, Jacobson AM, Domach MM (1995) Effect of nonionic surfactant addition on bacterial metabolism of naphthalene: Assessment of toxicity and overflow metabolism potential. J Hazard Mater 43:263–272

Banerjee DK, Gray MR, Dudas MJ, Pickard MA (1997) Protocol to enhance the extent of biodegradation of contamination in soil. In: Alleman BC, Leeson A (eds) In Situ and On-site Bioremediation. Battelle Press, Columbus, OH, pp 163–168

Bastiaens L, Springael D, Wattiau P, Harms H, deWachter R, Verachtert H, Diels L (2000) Isolation of adherent polycyclic aromatic hydrocarbon (PAH)-degrading bacteria using PAH-sorbing carriers. Appl Environ Microbiol 66:1834–1843

Baud-Grasset S, Baud-Grasset F, Bifulco J, Meier JR, Ma T-H (1993) Reduction of genotoxicity of a creosote-contaminated soil after fungal treatment determined by the Tredescantia-micronucleus test. Mutat Res 303:77–82

Beckles DM, Ward CH, Hughes JB (1998) Effect of mixtures of polycyclic aromatic hydrocarbons and sediments on fluoranthene biodegradation patterns. Environ Toxicol Chem 17:1246–1251

Belkin S, Steiber M, Tiehm A, Frimmel F, Abeliovich A, Werner P, Ulitzur S (1994) Toxicity and genotoxicity enhancement during polycyclic aromatic hydrocarbons' degradation. Environ Toxicol Water Qual 9:303–309

Bengtsson G, Zerhouni P (2003) Effects of carbon substrate enrichment and DOC concentration on biodegradation of PAHs in soil. J Appl Microbiol 94: 608–617

Berardesco G, Dyhrman S, Gallagher E, Shiaris MP (1998) Spatial and temporal variation of phenanthrene-degrading bacteria in intertidal sediments. Appl Environ Microbiol 64:2560–2565

Boldrin B, Tiehm A, Fritzsche C (1993) Degradation of phenanthrene, fluorene, fluoranthene, and pyrene by a *Mycobacterium* sp. Appl Environ Microbiol 59:1927–1930

Boonchan S, Britz ML, Stanley GA (1998) Surfactant-enhanced biodegradation of high molecular weight polycyclic aromatic hydrocarbons by *Stenotrophomonas maltophilia*. Biotechnol Bioeng 59:482–494

Boonchan S, Britz ML, Stanley GA (2000) Degradation and mineralization of high-molecular-weight polycyclic aromatic hydrocarbons by defined fungal-bacterial cocultures. Appl Environ Microbiol 66:1007–1019

Bosch R, García-Valdés E, Moore ERB (1999) Genetic characterization and evolutionary implications of a chromosomally encoded naphthalene-degradation upper pathway from *Pseudomonas stutzeri* AN10. Gene 236:149–157

Bouchez M, Blanchet D, Vandecasteele JP (1995) Degradation of polycyclic aromatic hydrocarbons by pure strains and by defined strain associations: Inhibition phenomena and cometabolism. Appl Microbiol Biotechnol 43:156–164

Bouchez M, Blanchet D, Bardin V, Haeseler F, Vandecasteele JP (1999) Efficiency of defined strains and of soil consortia in the biodegradation of polycyclic aromatic hydrocarbon (PAH) mixtures. Biodegradation 10:429–435

Breedveld GD, Karlsen DA (2000) Estimating the availability of polycyclic aromatic hydrocarbons for bioremediation of creosote contaminated soils. Appl Microbiol Biotechnol 54:255–261

Brenner RC, Magar VS, Ickes JA, Abbott JE, Stout SA, Crecelius EA, Bingler LS (2002) Characterization and FATE of PAH-contaminated sediments at the Wyckoff/Eagle Harbor Superfund site. Environ Sci Technol 36:2605–2613

Brezna B, Khan AA, Cerniglia CE (2003) Molecular characterization of dioxygenases from polycyclic aromatic hydrocarbon-degrading *Mycobacterium* spp. FEMS Microbiol Lett 223:177–183

Brodkorb TS, Legge RL (1992) Enhanced biodegradation of phenanthrene in oil tar-contaminated soils supplemented with *Phanerochaete chrysosporium*. Appl Environ Microbiol 58:3117–3121

Bucheli TD, Gustafsson O (2000) Quantification of the soot-water distribution coefficient of PAHs provides mechanistic basis for enhanced sorption observations. Environ Sci Technol 34:5144–5151

Burd G, Ward OP (1996) Involvement of a surface-active high molecular weight factor in degradation of polycyclic aromatic hydrocarbons by *Pseudomonas marginalis*. Can J Microbiol 42:791–797

Caldini G, Cenci G, Manenti R, Morozzi G (1995) The ability of an environmental isolate of *Pseudomonas fluorescens* to utilize chrysene and other four-ring polynuclear aromatic hydrocarbons. Appl Microbiol Biotechnol 44:225–229

Canet R, Birnstingl JG, Malcolm DG, Lopez-Real JM, Beck AJ (2001) Biodegradation of polycyclic aromatic hydrocarbons (PAHs) by native microflora and combinations of white-rot fungi in a coal-tar contaminated soil. Bioresour Technol 76: 113–117

Carmichael LM, Pfaender FK (1997) Polynuclear aromatic hydrocarbon metabolism in soils: relationship to soil characteristics and preexposure. Environ Toxicol Chem 16:666–675

Carmichael LM, Christman RF, Pfaender FK (1997) Desorption and mineralization kinetics of phenanthrene and chrysene in contaminated soils. Environ Sci Technol 31:126–132

Casellas M, Grifoll M, Sabaté J, Solanas AM (1998) Isolation and characterization of a 9-fluorenone-degrading bacterial strain and its role in synergistic degradation of fluorene by a consortium. Can J Microbiol 44:734–742

Chang BV, Shiung LC, Yuan SY (2002) Anaerobic biodegradation of polycyclic aromatic hydrocarbon in soil. Chemosphere 48:717–724

Chen S-H, Aitken MD (1999) Salicylate stimulates the degradation of high-molecular weight polycyclic aromatic hydrocarbons by *Pseudomonas saccharophila* P15. Environ Sci Technol 33:435–439

Churchill SA, Harper JP, Churchill PF (1999) Isolation and characterization of a *Mycobacterium* species capable of degrading three- and four-ring aromatic and aliphatic hydrocarbons. Appl Environ Microbiol 65:549–552

Coates JD, Anderson RT, Lovley DR (1996) Oxidation of polycyclic aromatic hydrocarbons under sulfate-reducing conditions. Appl Environ Microbiol 62:1099–1101

Coates JD, Woodward J, Allen J, Philp P, Lovley DR (1997) Anaerobic degradation of polycyclic aromatic hydrocarbons and alkanes in petroleum-contaminated marine harbor sediments. Appl Environ Microbiol 63:3589–3593

Colbert SF, Hendson M, Ferri M, Schroth MN (1993a) Enhanced growth and activity of a biocontrol bacterium genetically engineered to utilize salicylate. Appl Environ Microbiol 59:2071–2076

Colbert SF, Schroth MN, Weinhold AR, Hendson M (1993b) Enhancement of population densities of *Pseudomonas putida* PpG7 in agricultural ecosystems by selective feeding with the carbon source salicylate. Appl Environ Microbiol 59: 2064–2070

Colbert S, Isakeit T, Ferri M, Weinhold AR, Hendson M, Schroth M (1993c) Use of an exotic carbon source to selectively increase metabolic activity and growth of *Pseudomonas putida* in soil. Appl Environ Microbiol 59:2056–2063

Cornelissen G, Rigterink H, Ferdinandy MMA, Van Noort PCM (1998) Rapidly desorbing fractions of PAHs in contaminated sediments as a predictor of the extent of bioremediation. Environ Sci Technol 32:966–970

Cuypers C, Grotenhuis T, Joziasse J, Rulkens W (2000) Rapid persulfate oxidation predicts PAH bioavailability in soils and sediments. Environ Sci Technol 34:2057–2063

Cuypers C, Pancras T, Grotenhuis T, Rulkens W (2002) The estimation of PAH bioavailability in contaminated sediments using hydroxypropyl-β-cyclodextrin and Triton X-100 extraction techniques. Water Res 36:1235–1245

Daane LL, Harjono I, Zylstra GJ, Haggblom MM (2001) Isolation and characterization of polycyclic aromatic hydrocarbon-degrading bacteria associated with the rhizosphere of salt marsh plants. Appl Environ Microbiol 67:2683–2691

Dagher F, Déziel E, Lirette P, Paquette G, Bisaillon J-G, Villemur R (1997) Comparative study of five polycyclic aromatic hydrocarbon degrading bacterial strains isolated from contaminated soils. Can J Microbiol 43:368–377

Davies JI, Evans WC (1964) Oxidative metabolism of naphthalene by soil pseudomonads. Biochem J 91:251–261

Davis MW, Glaser JA, Evans JW, Lamar RT (1993) Field evaluation of the lignin-degrading fungus *Phanerochaete sordida* to treat creosote-contaminated soil. Environ Sci Technol 27:2572–2576

de Maagd PGJ, ten Hulscher ThEM, van den Heuvel H, Opperhuizen A, Sijm DTHM (1998) Physicochemical properties of polycyclic aromatic hydrocarbons: aqueous solubilities, *n*-octanol/water partition coefficients, and Henry's law constants. Environ Toxicol Chem 17:251–257

Dean-Ross D, Cerniglia CE (1996) Degradation of pyrene by *Mycobacterium flavescens*. Appl Microbiol Biotechnol 46:307–312

Dean-Ross D, Moody JD, Freeman JP, Doerge DR, Cerniglia CE (2001) Metabolism of anthracene by a *Rhodococcus* species. FEMS Microbiol Lett 204:205–211

Dean-Ross D, Moody J, Cerniglia CE (2002) Utilization of mixtures of polycyclic aromatic hydrocarbons by bacteria isolated from contaminated sediment. FEMS Microbiol Ecol 41:1–7

Denome SA, Stanley DC, Olson ES, Young KD (1993) Metabolism of dibenzothiophene and naphthalene in *Pseudomonas* strains: complete DNA sequence of an upper naphthalene catabolic pathway. J Bacteriol 175:6890–6901
Deschênes L, Lafrance P, Villeneuve J-P, Samson R (1996) Adding sodium dodecyl sulfate and *Pseudomonas aeruginosa* UG2 biosurfactants inhibits polycyclic aromatic hydrocarbon biodegradation in a weathered creosote-contaminated soil. Appl Microbiol Biotechnol 46:638–646
Di Gennaro P, Sello G, Bianchi D, D'Amico P (1997) Specificity of substrate recognition by *Pseudomonas fluorescens* N3 dioxygenase. J Biol Chem 272:30254–30260
Eberhardt C, Grathwohl P (2002) Time scales of organic contaminant dissolution from complex source zones: coal tar pools vs. blobs. J Contam Hydrol 59:45–66
Ensley BD, Gibson DT, Laborde AL (1982) Oxidation of naphthalene by a multicomponent enzyme system from *Pseudomonas* sp. strain NCIB 9816. J Bacteriol 149: 948–954
Erickson DC, Loehr RC, Neuhauser EF (1993) PAH loss during bioremediation of manufactured gas plant site soils. Water Res 27:911–919
Eriksson M, Dalhammar G, Mohn WW (2002) Bacterial growth and biofilm production on pyrene. FEMS Microbiol Ecol 40:21–27
Eriksson M, Sodersten E, Yu Z, Dalhammar G, Mohn WW (2003) Degradation of polycyclic aromatic hydrocarbons at low temperature under aerobic and nitrate-reducing conditions in enrichment cultures from northern soils. Appl Environ Microbiol 69:275–284
Eschenbach A, Wienberg R, Mahro B (1998) Fate and stability of nonextractable residues of [^{14}C]PAH in contaminated soils under environmental stress conditions. Environ Sci Technol 32:2585–2590
Evans WC, Fernley HN, Griffiths E (1965) Oxidative metabolism of phenanthrene and anthracene by soil pseudomonads. Biochem J 95:819–831
Fernández P, Grifoll M, Solanas AM, Bayona JM, Albaigés J (1992) Bioassay-directed chemical analysis of genotoxic components in coastal sediments. Environ Sci Technol 26:817–829
Ferrero M, Llobet-Brossa E, Lalucat J, Garcia-Valdes E, Rossello-Mora R, Bosch R (2002) Coexistence of two distinct copies of naphthalene degradation genes in *Pseudomonas* strains isolated from the western Mediterranean region. Appl Environ Microbiol 68:957–962
Foght JM, Westlake DWS (1988) Degradation of polycyclic aromatic hydrocarbons and aromatic heterocycles by a *Pseudomonas* species. Can J Microbiol 34:1135–1141
Foght JM, Westlake DWS (1991) Cross hybridization of plasmid and genomic DNA from aromatic and polycyclic aromatic hydrocarbon degrading bacteria. Can J Microbiol 37:924–932
Foght JM, Westlake DWS (1996) Transposon and spontaneous deletion mutants of plasmid-borne genes encoding polycyclic aromatic hydrocarbon degradation by a strain of *Pseudomonas fluorescens*. Biodegradation 7:353–366
Fritzsche C (1994) Degradation of pyrene at low defined oxygen concentrations by a *Mycobacterium* sp. Appl Environ Microbiol 60:1687–1689
Geiselbrecht AD, Hedlund BP, Tichi MA, Staley JT (1998) Isolation of marine polycyclic aromatic hydrocarbon (PAH)-degrading *Cycloclasticus* strains from the Gulf of Mexico and comparison of their PAH degradation ability with that of Puget Sound *Cycloclasticus* strains. Appl Environ Microbiol 64:4703–4710
Ghosh U, Gillette JS, Luthy RG, Zare RN (2000) Microscale location, characterization, and association of polycyclic aromatic hydrocarbons on harbor sediment particles. Environ Sci Technol 34:1729–1736

Ghosh U, Zimmerman JR, Luthy RG (2003) PCB and PAH speciation among particle types in contaminated harbor sediments and effects on PAH bioavailability. Environ Sci Technol 37:2209–2217

Ghoshal S, Luthy RG (1998) Biodegradation kinetics of naphthalene in nonaqueous phase liquid-water mixed batch systems: comparison of model predictions and experimental results. Biotechnol. Bioeng. 57:356–366

Gibson D, Mahadevan V, Jerina D, Yagi H, Yeh HJC (1975) Oxidation of the carcinogens benzo[*a*]pyrene and benzo[*a*]anthracene to dihydrodiols by a bacterium. Science 189:295–297

Gieg LM, Suflita JM (2002) Detection of anaerobic metabolites of saturated and aromatic hydrocarbons in petroleum-contaminated aquifers. Environ Sci Technol 36: 3755–3762

Goldstein LS, Weyand EH, Safe S, Steinberg M, Culp SJ, Gaylor DW, Beland FA, Rodriguez LV (1998) Tumors and DNA adducts in mice exposed to benzo[*a*]pyrene and coal tars: implications for risk assessment. Environ Health Perspect 106 (Supplement 6):1325–1330

Goyal AK, Zylstra GJ (1996) Molecular cloning of novel genes for polycyclic aromatic hydrocarbon degradation from *Comamonas testosteroni* GZ39. Appl Environ Microbiol 62:230–236

Grifoll M, Selifonov SA, Gatlin CV, Chapman PJ (1995) Actions of a versatile fluorene-degrading bacterial isolate on polycyclic aromatic compounds. Appl Environ Microbiol 61:3711–3723

Grimberg SJ, Stringfellow WT, Aitken MD (1996) Quantifying the biodegradation of phenanthrene by *Pseudomonas stutzeri* P16 in the presence of a nonionic surfactant. Appl Environ Microbiol 62:2387–2392

Grosser RJ, Warshawsky D, Vestal JR (1991) Indigenous and enhanced mineralization of pyrene, benzo[*a*]pyrene, and carbazole in soils. Appl Environ Microbiol 57: 3462–3469

Grosser RJ, Warshawsky D, Vestal JR (1995) Mineralization of polycyclic and *N*-heterocyclic aromatic compounds in hydrocarbon-contaminated soils. Environ Toxicol Chem 14:375–382

Guha S, Peters CA, Jaffé PR (1999) Multisubstrate biodegradation kinetics of naphthalene, phenanthrene, and pyrene mixtures. Biotechnol Bioeng 65:491–499

Guthrie EA, Pfaender FK (1998) Reduced pyrene bioavailability in microbially active soils. Environ Sci Technol 32:501–508

Guthrie EA, Bortiatynski JM, Van Heemst JDH, Richman JE, Hardy KS, Kovach EM, Hatcher PG (1999) Determination of [^{13}C]pyrene sequestration in sediment microcosms using flash pyrolysis-GC-MS and ^{13}C NMR. Environ Sci Technol 33:119–125

Guthrie-Nichols E, Grasham A, Kazunga C, Sangaiah R, Gold A, Bortiatynski J, Salloum M, Hatcher P (2003) The effect of aging on pyrene transformation in sediments. Environ Toxicol Chem 22:40–49

Haeseler F, Blanchet D, Druelle V, Werner P, Vandecasteele J-P (1999a) Analytical characterization of contaminated soils from former manufactured gas plants. Environ Sci Technol 33:825–830

Haeseler F, Blanchet D, Druelle V, Werner P, Vandecasteele J-P (1999b) Ecotoxicological assessment of soils of former manufactured gas plant sites: bioremediation potential and pollutant mobility. Environ Sci Technol 33:4379–4384

Harkins SM, Truesdale RS, Hill R, Hoffman P, Winters S (1988) U.S. Production of Manufactured Gases: Assessment of Past Disposal Practices, EPA/600/2–88/012. US Environmental Protection Agency, Cincinnati, OH

Hatzinger PB, Alexander M (1995) Effect of aging of chemicals in soil on their biodegradability and extractability. Environ Sci Technol 29:537–545

Haught RC, Neogy R, Vonderhaar SS, Krishnan ER, Safferman SI, Ryan J (1995) Land treatment alternatives for bioremediating wood preserving wastes. Hazard Waste Hazard Mater 12:329–344
Hawthorne SB, Grabanski CB (2000) Correlating selective supercritical fluid extraction with bioremediation behavior of PAHs in a field treatment plot. Environ Sci Technol 34:4103–4110
Hawthorne SB, Poppendieck DG, Grabanski CB, Loehr RC (2001) PAH release during water desorption, supercritical carbon dioxide extraction, and field bioremediation. Environ Sci Technol 35:4577–4583
Hawthorne SB, Poppendieck DG, Grabanski CB, Loehr RC (2002) Comparing PAH availability from manufactured gas plant soils and sediments with chemical and biological tests. 1. PAH release during water desorption and supercritical carbon dioxide extraction. Environ Sci Technol 36:4795–4803
Heitkamp MA, Cerniglia CE (1989) Polycyclic aromatic hydrocarbon degradation by a *Mycobacterium* sp. in microcosms containing sediment and water from a pristine ecosystem. Appl Environ Microbiol 55:1968–1973
Heitkamp MA, Franklin W, Cerniglia CE (1988) Microbial metabolism of polycyclic aromatic hydrocarbons: isolation and characterization of a pyrene-degrading bacterium. Appl Environ Microbiol 54:2549–2555
Herbes SE, Schwall LR (1978) Microbial transformation of polycyclic aromatic hydrocarbons in pristine and petroleum-contaminated sediments. Appl Environ Microbiol 35:306–316
Holman H-YN, Nieman K, Sorensen DL, Miller CD, Martin MC, Borch T, McKinney WR, Sims RC (2002) Catalysis of PAH biodegradation by humic acid shown in synchrotron infrared studies. Environ Sci Technol 36:1276–1280
Huang W, Weber Jr WJ (1998) A distributed reactivity model for sorption by soils and sediments. 11. Slow concentration-dependent sorption rates. Environ Sci Technol 32:3549–3555
Huesemann MH, Hausmann TS, Fortman TJ (2002) Microbial factors rather than bioavailability limit the rate and extent of PAH biodegradation in aged crude oil contaminated model soils. Bioremed J 6:321–336
Huesemann MH, Hausmann TS, Fortman TJ (2003) Assessment of bioavailability limitations during slurry biodegradation of petroleum hydrocarbons in aged soils. Environ Toxicol Chem 22:2853–2860
Hughes TJ, Claxton LD, Brooks L, Warren S, Brenner R, Kremer F (1998) Genotoxicity of bioremediated soils from the Reilly Tar Site, St. Louis Park, Minnesota. Environ Health Perspect 106, (Supp6):1427–1433
Hussain M, Rae J, Gilman A, Kauss P (1998) Lifetime health risk assessment from exposure of recreational users to polycyclic aromatic hydrocarbons. Arch Environ Contam Toxicol 35:527–531
Jerina DM, van Bladeren PJ, Yagi H, Gibson DT, Mahadeven V, Neese AS, Koreeda M, Sharma ND, Boyd D (1984) Synthesis and absolute configuration of *cis*-1,2-, 8,9- and 10,11-dihydrodiol metabolites of benz[*a*]anthracene formed by a strain of *Beijerinckia*. J Org Chem 49:1075–1082
Ji P, Wilson M (2003) Enhancement of population size of a biological control agent and efficacy in control of bacterial speck of tomato through salicylate and ammonium sulfate amendments. Appl Environ Microbiol 69:1290–1294
Johnsen AR, Bendixen K, Karlson U (2002) Detection of microbial growth on polycyclic aromatic hydrocarbons in microtiter plates by using the respiration indicator WST-1. Appl Environ Microbiol 68:2683–2689
Johnson MD, Weber WJ Jr (2001) Rapid prediction of long-term rates of contaminant desorption from soils and sediments. Environ Sci Technol 35:427–433

Joner EJ, Leyval C (2003) Rhizosphere gradients of polycyclic aromatic hydrocarbon (PAH) dissipation in two industrial soils and the impact of arbuscular mycorrhiza. Environ Sci Technol 37:2371–2375

Joner EJ, Johansen A, Loibner AP, dela Cruz MA, Szolar OHJ, Portal J-M, Leyval C (2001) Rhizosphere effects on microbial community structure and dissipation and toxicity of polycyclic aromatic hydrocarbons (PAHs) in spiked soil. Environ Sci Technol 35:2773–2777

Juhasz AL, Naidu R (2000) Bioremediation of high molecular weight polycyclic aromatic hydrocarbons: a review of the microbial degradation of benzo[*a*]pyrene. Int Biodeterior Biodeg 45:57–88

Juhasz AL, Britz ML, Stanley GA (1997) Degradation of fluoranthene, pyrene, benz[*a*]anthracene and dibenz[*a,h*]anthracene by *Burkholderia cepacia*. J Appl Microbiol 83:189–198

Juhasz AL, Stanley GA, Britz ML (2000) Degradation of high molecular weight PAHs in contaminated soil by a bacterial consortium: effects on Microtox and mutagenicity assays. Bioremed J 4:271–283

Juhasz AL, Stanley GA, Britz ML (2002) Metabolite repression inhibits degradation of benzo[*a*]pyrene and dibenz[*a,h*]anthracene by *Stenotrophomonas maltophilia* VUN 10,003. J Ind Microbiol Biotechnol 8:88–96

Kacker T, Haupt ETK, Garms C, Francke W, Steinhart H (2002) Structural characterisation of humic acid-bound PAH residues in soil by ^{13}C-CPMAS-NMR-spectroscopy: evidence of covalent bonds. Chemosphere 48:117–131

Kanaly RA, Harayama S (2000) Biodegradation of high-molecular-weight polycyclic aromatic hydrocarbons by bacteria. J Bacteriol 182:2059–2067

Kanaly RA, Harayama S, Watanabe K (2002) *Rhodanobacter* sp. strain BPC1 in a benzo[*a*]pyrene-mineralizing bacterial consortium. Appl Environ Microbiol 68: 5826–5833

Karimi-Lotfabad S, Gray MR (2000) Characterization of contaminated soils using confocal laser scanning microscopy and cryogenic-scanning electron microscopy. Environ Sci Technol 34:3408–3414

Kästner M, Breuer-Jammali M, Mahro B (1994) Enumeration and characterization of the soil microflora from hydrocarbon-contaminated soil sites able to mineralize polycyclic aromatic hydrocarbons (PAH). Appl Microbiol Biotechnol 41:267–273

Kästner M, Streibich S, Beyrer M, Richnow HH, Fritsche W (1999) Formation of bound residues during microbial degradation of [^{14}C]anthracene in soil. Appl Environ Microbiol 65:1834–1842

Kazunga C, Aitken MD (2000) Products from the incomplete metabolism of pyrene by polycyclic aromatic hydrocarbon-degrading bacteria. Appl Environ Microbiol 66: 1917–1922

Kazunga C, Aitken MD, Gold A, Sangaiah R (2001) Fluoranthene-2,3- and -1,5-diones are novel products from the bacterial transformation of fluoranthene. Environ Sci Technol 35:917–922

Keck J, Sims R, Coover M, Park K, Symons B (1989) Evidence for cooxidation of polynuclear aromatic hydrocarbons in soil. Water Res 23:1467–1476

Kelley I, Cerniglia CE (1995) Degradation of a mixture of high-molecular-weight polycyclic aromatic hydrocarbons by a *Mycobacterium* strain PYR-1. J Soil Contam 4:77–91

Kelsey JW, Kottler BD, Alexander M (1997) Selective chemical extractants to predict bioavailability of soil-aged organic chemicals. Environ Sci Technol 31:214–217

Khan AA, Wang R-F, Cao W-W, Franklin W, Cerniglia CE (1996) Reclassification of a polycyclic aromatic hydrocarbon-metabolizing bacterium, *Beijerinckia* sp. strain

B1, as *Sphingomonas yanoikuyae* by fatty acid analysis, protein pattern analysis, DNA-DNA hybridization, and 16S ribosomal DNA sequencing. Int J Syst Bacteriol 46:466–499

Khan AA, Wang R-F, Cao W-W, Doerge DR, Wennerstrom D, Cerniglia CE (2001) Molecular cloning, nucleotide sequence, and expression of genes encoding a polycyclic aromatic ring dioxygenase from *Mycobacterium* sp. strain PYR-1. Appl Environ Microbiol 67:3577–3585

Khan AA, Kim S-J, Paine DD, Cerniglia CE (2002) Classification of a polycyclic aromatic hydrocarbon-metabolizing bacterium, *Mycobacterium* sp. strain PYR-1, as *Mycobacterium vanbaalenii* sp. nov. Int J Syst Evol Microbiol 52:1997–2002

Kiyohara H, Nagao K (1978) The catabolism of phenanthrene and naphthalene by bacteria. J Gen Microbiol 105:69–75

Kiyohara H, Torigoe S, Kaida N, Asaki T, Iida T, Hayashi H, Takizawa N (1994) Cloning and characterization of a chromosomal gene cluster, *pah*, that encodes the upper pathway for phenanthrene and naphthalene utilization by *Pseudomonas putida* OUS82. J Bacteriol 176:2439–2443

Knightes CD, Peters CA (2003) Aqueous phase biodegradation kinetics of 10 PAH compounds. Environ Eng Sci 20:207–218

Krivobok S, Kuony S, Meyer C, Louwagie M, Willison JC, Jouanneau Y (2003) Identification of pyrene-induced proteins in *Mycobacterium* sp. strain 6PY1: evidence for two ring-hydroxylating dioxygenases. J Bacteriol 185:3828–3841

Lane WF, Loehr RC (1992) Estimating the equilibrium aqueous concentrations of polynuclear aromatic hydrocarbons in complex mixtures. Environ Sci Technol 26: 983–990

Langbehn A, Steinhart H (1995) Biodegradation studies of hydrocarbons in soils by analyzing metabolites formed. Chemosphere 30:855–868

Laurie AD, Lloyd-Jones G (1999) The *phn* genes of *Burkholderia* sp. strain RP007 constitute a divergent gene cluster for polycyclic aromatic hydrocarbon catabolism. J Bacteriol 181:531–540

Laurie AD, Lloyd-Jones G (2000) Quantification of *phnAc* and *nahAc* in contaminated New Zealand soils by competitive PCR. Appl Environ Microbiol 66:1814–1817

Lee LS, Rao PSC, Okuda I (1992) Equilibrium partitioning of polycyclic aromatic hydrocarbons from coal tar into water. Environ Sci Technol 26:2110–2115

Li X-F, Cullen WR, Reimer KJ, Le X-C (1996a) Microbial degradation of pyrene and characterization of a metabolite. Sci Total Environ 177:17–29

Li X-F, Le X-C, Simpson CD, Cullen WR, Reimer KJ (1996b) Bacterial transformation of pyrene in a marine environment. Environ Sci Technol 30:1115–1119

Lotfabad SK, Gray MR (2002) Kinetics of biodegradation of mixtures of polycyclic aromatic hydrocarbons. Appl Microbiol Biotechnol 60:361–366

Loy A, Lehner A, Lee N, Adamczyk J, Meier H, Ernst J, Schleifer K-H, Wagner M (2002) Oligonucleotide microarray for 16S rRNA gene-based detection of all recognized lineages of sulfate-reducing prokaryotes in the environment. Appl Environ Microbiol 68:5064–5081

Luthy RG, Aiken GR, Brusseau ML, Cunningham SD, Gschwend PM, Pignatello JJ, Reinhard M, Traina SJ, Weber Jr WJ, Westall JC (1997) Sequestration of hydrophobic organic contaminants by geosorbents. Environ Sci Technol 31:3341–3347

MacGillivray AR, Shiaris MP (1994) Relative role of eukaryotic and prokaryotic microorganisms in phenanthrene transformation in coastal sediments. Appl Environ Microbiol 60:1154–1159

Mackay AA, Gschwend PM (2001) Enhanced concentrations of PAHs in groundwater at a coal tar site. Environ Sci Technol 35:1320–1328

Mackay D, Shia YW, Ma KC (1992) Illustrated Handbook of Physical and Environmental Fate for Organic Chemicals, Vol. 2: Polynuclear Aromatic Hydrocarbons, Polychlorinated Dioxins, and Dibenzofurans. Lewis, Chelsea, MI

Madsen EL, Mann CL, Bilotta S (1996) Oxygen limitations and aging as explanation for the persistence of naphthalene in coal-tar contaminated surface sediments. Environ Toxicol Chem 15:1876–1882

Mahaffey WR, Gibson DT, Cerniglia CE (1988) Bacterial oxidation of chemical carcinogens: formation of polycyclic aromatic acids from benz[*a*]anthracene. Appl Environ Microbiol 54:2415–2423

Mahro B, Rode K, Kasche V (1995) Non-selective precultivation of bacteria able to degrade different polycyclic aromatic hydrocarbons (PAH). Acta Biotechnologia 15: 337–345

May R, Schröder P, Sandermann Jr H (1997) Ex-situ process for treating PAH-contaminated soil with *Phanerochaete chrysosporium*. Environ Sci Technol 31:2626–2633

Mayer LM, Chen Z, Findlay RH, Fang J, Sampson S, Self RFL, Jumars PA, Quetel C, Donard OFX (1996) Bioavailability of sedimentary contaminants subject to deposit-feeder digestion. Environ Sci Technol 30:2641–2645

McNally DL, Mihelcic JR, Lueking DR (1998) Biodegradation of three- and four-ring polycyclic aromatic hydrocarbons under aerobic and denitrifying conditions. Environ Sci Technol 32:2633–2639

Meckenstock RU, Annweiler E, Michaelis W, Richnow HH, Schink B (2000) Anaerobic naphthalene degradation by a sulfate-reducing enrichment culture. Appl Environ Microbiol 66:2743–2747

Melcher RJ, Apitz SE, Hemmingsen BB (2002) Impact of irradiation and polycyclic aromatic hydrocarbon spiking on microbial populations in marine sediment for future aging and biodegradability studies. Appl Environ Microbiol 68:2858–2868

Menn F-M, Applegate BM, Sayler GS (1993) NAH plasmid-mediated catabolism of anthracene and phenanthrene to naphthoic acids. Appl Environ Microbiol 59: 1938–1942

Meyer S, Cartellieri S, Steinhart H (1999a) Simultaneous determination of PAHs, hetero-PAHs (N, S, O), and their degradation products in creosote-contaminated soils. Method development, validation, and application to hazardous waste sites. Analyt Chem 71:4023–4029

Meyer S, Steinhart H (2001) Fate of PAHs and hetero-PAHs during biodegradation in a model soil/compost-system: formation of extractable metabolites. Water Air Soil Pollut 132:215–231

Meyer S, Moser R, Neef A, Stahl U, Kampfer P (1999b) Differential detection of key enzymes of polyaromatic-hydrocarbon-degrading bacteria using PCR and gene probes. Microbiology 145:1731–1741

Millette D, Barker JF, Comeau Y, Butler BJ, Frind EO, Clement B, Samson R (1995) Substrate interaction during aerobic biodegradation of creosote-related compounds: a factorial batch experiment. Environ Sci Technol 29:1944–1952

Millette D, Butler BJ, Frind EO, Comeau Y, Samon R (1998) Substrate interaction during aerobic biodegradation of creosote-related compounds in columns of sandy aquifer material. J Contam Hydrol 29:165–183

Molina M, Araujo R, Hodson RE (1999) Cross-induction of pyrene and phenanthrene in a *Mycobacterium* sp isolated from polycyclic aromatic hydrocarbon contaminated river sediments. Can J Microbiol 45:520–529

Moody JD, Freeman JP, Doerge DR, Cerniglia CE (2001) Degradation of phenanthrene and anthracene by cell suspensions of *Mycobacterium* sp. strain PYR-1. Appl Environ Microbiol 67:1476–1483

Moser R, Stahl U (2001) Insights into the genetic diversity of initial dioxygenases from PAH-degrading bacteria. Appl Microbiol Biotechnol 55:609–618

Mueller JG, Chapman PJ, Blattmann BO, Pritchard PH (1990) Isolation and characterization of a fluoranthene-utilizing strain of *Pseudomonas paucimobilis.* Appl Environ Microbiol 56:1079–1086

Mueller JG, Devereux R, Santavy DL et al. (1997) Phylogenetic and physiological comparisons of PAH-degrading bacteria from geographically diverse soils. Antonie Van Leeuwenhoek Int J Gen Mol Microbiol 71:329–343

National Toxicology Program (2001) Ninth Report on Carcinogens. US Department of Health and Human Services, Washington, DC

Nieman JKC, Sims RC, Sims JL, Sorensen DL, McLean JE, Rice JA (1999) [^{14}C]Pyrene bound residue evaluation using MIBK fractionation method for creosote-contaminated soil. Environ Sci Technol 33:776–781

Nisbet ICT, LaGoy PK (1992) Toxic equivalency factors (TEFs) for polycyclic aromatic hydrocarbons (PAHs). Regulatory Toxicol Pharmacol 16:290–300

Northcott GL, Jones KC (2001) Partitioning, extractability, and formation of nonextractable PAH residues in soil 1. Compound differences in aging and sequestration. Environ Sci Technol 35:1103–1110

Ogunseitan A, Delgado I, Tsai Y-L, Olson B (1991) Effect of 2-hydroxybenzoate on the maintenance of naphthalene-degrading pseudomonads in seeded and unseeded soil. Appl Environ Microbiol 57:2873–2879

Ogunseitan O, Olson B (1993) Effect of 2-hydroxybenzoate on the rate of naphthalene mineralization in soil. Appl Microbiol Biotechnol 38:799–807

Padma TV, Hale RC, Roberts Jr MH (1998) Toxicity of water-soluble fractions derived from whole creosote and creosote-contaminated sediments. Environ Toxicol Chem 17:1606–1610

Padmanabhan P, Padmanabhan S, DeRito C, Gray A, Gannon D, Snape JR, Tsai CS, Park W, Jeon C, Madsen EL (2003) Respiration of ^{13}C-labeled substrates added to soil in the field and subsequent 16S rRNA gene analysis of ^{13}C-labeled soil DNA. Appl Environ Microbiol 69:1614–1622

Park KS, Sims RC, Doucette WJ, Matthews JE (1988) Biological transformation and detoxification of 7,12-dimethylbenz(a)anthracene in soil systems. J Water Pollut Control Fed 60:1822–1825

Pecher K, Haderlein SB, Schwarzenbach RP (2002) Reduction of polyhalogenated methanes by surface-bound Fe(II) in aqueous suspensions of iron oxides. Environ Sci Technol 36:1734–1741

Penning TM, Burczynski ME, Hung C-F, McCoull KD, Palackal NT, Tsuruda LS (1999) Dihydrodiol dehydrogenases and polycyclic aromatic hydrocarbon activation: generation of reactive and redox active *o*-quinones. Chem Res Toxicol 12:1–18

Peters CA, Brown DG, Knightes CD (1999) Long-term composition dynamics of PAH-containing NAPLs and implications for risk assessment. Environ Sci Technol 33: 4499–4507

Poeton TS, Stensel HD, Strand SE (1999) Biodegradation of polyaromatic hydrocarbons by marine bacteria: effect of solid phase on degradation kinetics. Water Res 33: 868–880

Prabhu Y, Phale PS (2003) Biodegradation of phenanthrene by *Pseudomonas* sp. strain PP2: novel metabolic pathway, role of biosurfactant and cell surface hydrophobicity in hydrocarbon assimilation. Appl Microbiol Biotechnol 61:342–351

Preuß S, Wittneben D, Lorber KE (1997) Microbiological degradation of polycyclic aromatic hydrocarbons: anthracene, benzo(*k*)fluoranthene, dibenzothiophene, benzo(*h*)quinoline and 2-nitronaphthalene. Toxicol Environ Chem 58:179–195

Ravelet C, Krivobok S, Sage L, Steiman R (1999) Biodegradation of pyrene by sediment fungi. Chemosphere 40:557–563

Ravelet C, Grosset C, Krivobok S, Montuelle B, Alary J (2001) Pyrene degradation by two fungi in a freshwater sediment and evaluation of fungal biomass by ergosterol content. Appl Microbiol Biotechnol 56:803–808

Reeves WR, Barhoumi R, Burghardt RC, Lemke SL, Mayura K, Mcdonald TJ, Phillips TD, Donnelly KC (2001) Evaluation of methods for predicting the toxicity of polycyclic aromatic hydrocarbon mixtures. Environ Sci Technol 35: 1630–1636

Rehmann K, Noll HP, Steinberg CEW, Kettrup AA (1998) Pyrene degradation by *Mycobacterium* sp. strain KR2. Chemosphere 36:2977–2992

Reid BJ, Stokes JD, Jones KC, Semple KT (2000) Nonexhaustive cyclodextrin-based technique for the evaluation of PAH bioavailability. Environ Sci Technol 34: 3174–3179

Resnick SM, Lee K, Gibson DT (1996) Diverse reactions catalyzed by naphthalene dioxygenase from *Pseudomonas* sp strain NCIB 9816. J Ind Microbiol 17:438–457

Reza J, Trejo A, Vera-Avila LE (2002) Determination of the temperature dependence of water solubilities of polycyclic aromatic hydrocarbons by a generator column-on-line solid-phase extraction-liquid chromatographic method. Chemosphere 47: 933–945

Richnow HH, Seifert R, Kästner M, Mahro B, Horsfield B, Tiedgen U, Bohm S, Michaelis W (1996) Rapid screening of PAH-residues in bioremediated soils. Chemosphere 31:3991–3999

Ringelberg DB, Talley JW, Perkins EJ, Tucker SG, Luthy RG, Bouwer EJ, Fredrickson HL (2001) Succession of phenotypic, genotypic, and metabolic community characteristics during in vitro bioslurry treatment of polycyclic aromatic hydrocarbon-contaminated sediments. Appl Environ Microbiol 67:1542–1550

Rockne KJ, Strand SE (1998) Biodegradation of bicyclic and polycyclic aromatic hydrocarbons in anaerobic enrichments. Environ Sci Technol 32:3962–3967

Rockne KJ, Chee-Sanford JC, Sanford RA, Hedlund BP, Staley JT, Strand SE (2000) Anaerobic naphthalene degradation by microbial pure cultures under nitrate-reducing conditions. Appl Environ Microbiol 66:1595–1601

Rockne KJ, Strand SE (2001) Anaerobic biodegradation of naphthalene, phenanthrene, and biphenyl by a denitrifying enrichment culture. Water Res 35:291–299

Rockne KJ, Shor LM, Young LY, Taghon GL, Kosson DS (2002) Distributed sequestration and release of PAHs in weathered sediment: the role of sediment structure and organic carbon properties. Environ Sci Technol 36:2636–2644

Rosselló-Mora RA, Lalucat J, García-Valdés E (1994) Comparative biochemical and genetic analysis of naphthalene degradation among *Pseudomonas stutzeri* strains. Appl Environ Microbiol 60:966–972

Rothermich MM, Hayes LA, Lovley DR (2002) Anaerobic, sulfate-dependent degradation of polycyclic aromatic hydrocarbons in petroleum-contaminated harbor sediment. Environ Sci Technol 36:4811–4817

Rouse JD, Sabatini DA, Suflita JM, Harwell JH (1994) Influence of surfactants on microbial degradation of organic compounds. Crit Rev Environ Sci Technol 24:325–370

Roy TA, Krueger AJ, Taylor BB, Mauro DM, Goldstein LS (1998) Studies estimating the dermal bioavailability of polynuclear aromatic hydrocarbons from manufactured gas plant tar-contaminated soils. Environ Sci Technol 32:3113–3117

Rutherford PM, Gray MR, Dudas MJ (1997) Desorption of [^{14}C]naphthalene from bioremediated and nonbioremediated soils contaminated with creosote compounds. Environ Sci Technol 31:2515–2519

Rutherford PM, Banerjee DK, Luther SM, Gray MR, Dudas MJ, McGill WB, Pickard MA, Salloum MJ (1998) Slurry-phase bioremediation of creosote and petroleum-contaminated soils. Environ Technol 19:683–696
Saito A, Iwabuchi T, Harayama S (1999) Characterization of genes for enzymes involved in the phenanthrene degradation in *Nocardioides* sp. KP7. Chemosphere 38:1331–1337
Saito A, Iwabuchi T, Harayama S (2000) A novel phenanthrene dioxygenase from *Nocardioides* sp. strain KP7: expression in *Escherichia coli*. J Bacteriol 182:2134–2141
Salicis F, Krivobok S, Jack M, Benoit-Guyod JL (1999) Biodegradation of fluoranthene by soil fungi. Chemosphere 38:3031–3039
Samanta SK, Chakraborti AK, Jain RK (1999) Degradation of phenanthrene by different bacteria: evidence for novel transformation sequences involving the formation of 1-naphthol. Appl Microbiol Biotechnol 53:98–107
Samsøe-Petersen L, Larsen EH, Larsen PB, Bruun P (2002) Uptake of trace elements and PAHs by fruit and vegetables from contaminated soils. Environ Sci Technol 36:3057–3063
Sandoli RL, Ghiorse WC, Madsen EL (1996) Regulation of microbial phenanthrene mineralization in sediment samples by sorbent-sorbate contact time, inocula and gamma irradiation-induced sterilization artifacts. Environ Toxicol Chem 15:1901–1907
Saraswathy A, Hallberg R (2002) Degradation of pyrene by indigenous fungi from a former gasworks site. FEMS Microbiol Lett 210:227–232
Sayles GD, Acheson CM, Kupferle MJ, Shan Y, Zhou Q, Meier JR, Chang L, Brenner RC (1999) Land treatment of PAH-contaminated soil: performance measured by chemical and toxicity assays. Environ Sci Technol 33:4310–4317
Schneider J, Grosser R, Jayasimhulu K, Xue W, Warshawsky D (1996) Degradation of pyrene, benz[*a*]anthracene, and benzo[*a*]pyrene by *Mycobacterium* sp strain RJGII-135, isolated from a former coal gasification site. Appl Environ Microbiol 62:13–19
Selifonov SA, Grifoll M, Eaton RW, Chapman PJ (1996) Oxidation of naphthenoaromatic and methyl-substituted aromatic compounds by naphthalene 1,2-dioxygenase. Appl Environ Microbiol 62:507–514
Šepiè E, Bricelj M, Leskovšek H (1998) Degradation of fluroanthene by *Pasteurella* sp. IFA and *Mycobacterium* sp. PYR-1: isolation and identification of metabolites. J Appl Microbiol 85:746–754
Shi T, Fredrickson JK, Balkwill DL (2001) Biodegradation of polycyclic aromatic hydrocarbons by *Sphingomonas* strains isolated from the terrestrial subsurface. J Ind Microbiol Biotechnol 26:283–289
Shiaris MP (1989) Seasonal biotransformation of naphthalene, phenanthrene, and benzo[*a*]pyrene in surficial estuarine sediments. Appl Environ Microbiol 55:1391–1399
Shor LM, Rockne KJ, Kosson DS (2003a) Intra-aggregate mass transport-limited bioavailability of polycyclic aromatic hydrocarbons to *Mycobacterium* strain PC01. Environ Sci Technol 37:1545–1552
Shor LM, Taghon GL, Kosson DS (2003b) Desorption kinetics for field-aged polycyclic aromatic hydrocarbons from sediments. Environ Sci Technol 37:1535–1544
Shuttleworth KL, Cerniglia CE (1996) Bacterial degradation of low concentrations of phenanthrene and inhibition by naphthalene. Microb Ecol 31:305–317
Smith JD, Bagg J, Wrigley I (1991) Extractable polycyclic hydrocarbons in waters from rivers in south-eastern Australia. Water Res 25:1145–1150
Stapleton RD, Savage DC, Sayler GS, Stacey G (1998) Biodegradation of aromatic hydrocarbons in an extremely acidic environment. Appl Environ Microbiol 64: 4180–4184

Stout S, Magar V, Uhler A, Ickes J, Abbott J, Brenner R (2001) Characteristics of naturally occurring and anthropogenic PAHs in urban sediments – Wycoff/Eagle Harbor Superfund site. Environ Forensics 2:287–300

Stringfellow WT, Aitken MD (1995) Competitive metabolism of naphthalene, methylnaphthalenes, and fluorene by phenanthrene-degrading pseudomonads. Appl Environ Microbiol 61:357–362

Stroo HF, Jensen R, Loehr RC, Nakles DV, Fairbrother A, Liban CB (2000) Environmentally acceptable endpoints for PAHs at a manufactured gas plant site. Environ Sci Technol 34:3831–3836

Sullivan ER, Zhang X, Phelps C, Young LY (2001) Anaerobic mineralization of stable-isotope-labeled 2-methylnaphthalene. Appl Environ Microbiol 67:4353–4357

Surovtseva EG, Ivoilov VS, Belyaev SS (1999) Physiological and biochemical properties of *Beijerinckia mobilis* 1F Phn(+) capable of degrading polycyclic aromatic hydrocarbons. Microbiology 68:746–750

Sutherland JB, Rafii F, Khan AA, Cerniglia CE (1995) Mechanisms of polycyclic aromatic hydrocarbon degradation. In: Young LY, Cerniglia CE (eds) Microbial Transformation and Degradation of Toxic Organic Chemicals. Wiley-Liss, New York, pp 269–306

Sverdrup LE, Nielsen T, Krogh PH (2002) Soil ecotoxicity of polycyclic aromatic hydrocarbons in relation to soil sorption, lipophilicity, and water solubility. Environ Sci Technol 36:2429–2435

Takizawa N, Kaida N, Torigoe S, Moritani T, Sawada T, Satoh S, Kiyohara H (1994) Identification and characterization of genes encoding polycyclic aromatic hydrocarbon dioxygenase and polycyclic aromatic hydrocarbon dihydrodiol dehydrogenase in *Pseudomonas putida* OUS82. J Bacteriol 176:2444–2449

Talley JW, Ghosh U, Tucker SG, Furey JS, Luthy RG (2002) Particle-scale understanding of the bioavailability of PAHs in sediment. Environ Sci Technol 36: 477–483

Tang J, Carroquino MJ, Robertson BK, Alexander M (1998) Combined effect of sequestration and bioremediation in reducing the bioavailability of polycyclic aromatic hydrocarbons in soil. Environ Sci Technol 32:3586–3590

Taroncher-Oldenburg G, Griner EM, Francis CA, Ward BB (2003) Oligonucleotide microarray for the study of functional gene diversity in the nitrogen cycle in the environment. Appl Environ Microbiol 69:1159–1171

Taylor LT, Jones DM (2001) Bioremediation of coal tar PAH in soils using biodiesel. Chemosphere 44:1131–1136

Tiehm A, Stieber M, Werner P, Frimmel FH (1997) Surfactant-enhanced mobilization and biodegradation of polycyclic aromatic hydrocarbons in manufactured gas plant soil. Environ Sci Technol 31:2570–2576

Tittle PC, Liu Y-T, Strand SE, Stensel HD (1995) Use of alternative growth substrates to enhance PAH degradation. In: Hinchee RE, Anderson DB, Hoeppel RE (eds) Bioremediation of Recalcitrant Organics. Battelle Press, Columbus,OH, pp 1–7

Trzesicka-Mlynarz D, Ward OP (1995) Degradation of polycyclic aromatic hydrocarbons (PAHs) by a mixed culture and its component pure cultures, obtained from PAH-contaminated soil. Can J Microbiol 41:470–476

US Environmental Protection Agency (1993) Provisional Guidance for Quantitative Risk Assessment of Polycyclic Aromatic Hydrocarbons, EPA/600/R-93/089. US Environmental Protection Agency, Washington, DC

US Environmental Protection Agency (2000) A Resource for MGP Site Characterization and Remediation. Expedited Site Characterization and Source Remediation at

Former Manufactured Gas Plant Sites, EPA-542-R-00-005. US Environmental Protection Agency, Washington, DC

van Herwijnen R, Springael D, Slot P, Govers HAJ, Parsons JR (2003) Degradation of anthracene by *Mycobacterium* sp. strain LB501T proceeds via a novel pathway, through *o*-phthalic acid. Appl Environ Microbiol 69:186–190

Vila J, Lopez Z, Sabate J, Minguillon C, Solanas AM, Grifoll M (2001) Identification of a novel metabolite in the degradation of pyrene by *Mycobacterium* sp. strain AP1: actions of the isolate on two- and three-ring polycyclic aromatic hydrocarbons. Appl Environ Microbiol 67:5497–5505

Villholth KG (1999) Colloid characterization and colloidal phase partitioning of polycyclic aromatic hydrocarbons in two creosote-contaminated aquifers in Denmark. Environ Sci Technol 33:691–699

Walter U, Beyer M, Klein J, Rehm H-J (1991) Degradation of pyrene by *Rhodococcus* sp. UW1. Appl Microbiol Biotechnol 34:671–676

Weissenfels WD, Beyer M, Klein J (1990) Degradation of phenanthrene, fluorene and fluoranthene by pure bacterial cultures. Appl Microbiol Biotechnol 32:479–484

Weissenfels WD, Klewer H-J, Langhoff J (1992) Adsorption of polycyclic aromatic hydrocarbons (PAHs) by soil particles: Influence on biodegradability and biotoxicity. Appl Microbiol Biotechnol 36:689–696

White JC, Pignatello JJ (1999) Influence of bisolute competition on the desorption kinetics of polycyclic aromatic hydrocarbons in soil. Environ Sci Technol 33:4292–4298

Wick LY, Colangelo T, Harms H (2001) Kinetics of mass transfer-limited bacterial growth on solid PAHs. Environ Sci Technol 35:354–361

Widada J, Nojiri H, Kasuga K, Yoshida T, Habe H, Omori T (2002) Molecular detection and diversity of polycyclic aromatic hydrocarbon-degrading bacteria isolated from geographically diverse sites. Appl Microbiol Biotechnol 58:202–209

Wild SR, Jones KC (1993) Biological and abiotic losses of polynuclear aromatic hydrocarbons (PAHs) from soils freshly amended with sewage sludge. Environ Toxicol Chem 12:5–12

Williamson DG, Loehr RC, Kimura Y (1997) Measuring release and biodegradation kinetics of aged hydrocarbons from soils. In: Alleman BC, Leeson A (eds) In Situ and On-site Bioremediation. Battelle Press, Columbus, pp 605–610

Willumsen PA, Arvin E (1999) Kinetics of degradation of surfactant-solubilized fluoranthene by a *Sphingomonas paucimobilis*. Environ Sci Technol 33:2571–2578

Wilson MS, Herrick JB, Jeon CO, Hinman DE, Madsen EL (2003) Horizontal transfer of phnAc dioxygenase genes within one of two phenotypically and genotypically distinctive naphthalene-degrading guilds from adjacent soil environments. Appl Environ Microbiol 69:2172–2181

Xia GS, Pignatello JJ (2001) Detailed sorption isotherms of polar and apolar compounds in a high-organic soil. Environ Sci Technol 35:84–94

Yang Y, Chen RF, Shiaris MP (1994) Metabolism of naphthalene, fluorene, and phenanthrene: preliminary characterization of a cloned gene cluster from *Pseudomonas putida* NCIB 9816. J Bacteriol 176:2158–2164

Ye D, Siddiqi MA, Maccubbin AE, Kumar S, Sikka HC (1996) Degradation of polynuclear aromatic hydrocarbons by *Sphingomonas paucimobilis*. Environ Sci Technol 30:136–142

Yen KM, Gunsalus IC (1982) Plasmid gene organization: Naphthalene/salicylate oxidation. Proc Nat Acad Sci USA 79:874–878

Yen K-M, Serdar CM (1988) Genetics of naphthalene catabolism in pseudomonads. CRC Crit Rev Microbiol 15:247–268

Yuan SY, Wei SH, Chang BV (2000) Biodegradation of polycyclic aromatic hydrocarbons by a mixed culture. Chemosphere 41:1463–1468

Zhang X, Young LY (1997) Carboxylation as an initial reaction in the anaerobic metabolism of naphthalene and phenanthrene by sulfidogenic consortia. Appl Environ Microbiol 63:4759–4764

6 Biodegradation and Bioremediation of Halogenated Organic Compounds

William W. Mohn[1]

1 Introduction: The Problem with Halo-organic Compounds

Halo-organic compounds are among the most problematic pollutants. The relatively great electronegativity of halogens often confers chemical stability to halo-organic compounds, thereby making these compounds recalcitrant to biodegradation. Halogen substituents can increase the hydrophobicity of organic compounds, increasing their tendency to bioaccumulate in food chains as well as to sorb to soil. Finally, halogen substituents can contribute to harmful biological effects of organic compounds, increasing their toxicity, mutagenicity and other detrimental capacities. This chapter will focus on types of halo-organic compounds that have become important soil pollutants and that have the potential, demonstrated or theoretical, to be bioremediated. Generally, the potential for bioremediation requires that a halo-organic compound can be biodegraded, partly or completely destroyed by metabolism. Despite the metabolic challenges posed by halo-organic compounds, microorganisms have demonstrated a remarkable capacity to biodegrade such compounds.

Many halo-organic compounds are xenobiotic and were created by humans in order for them to have some of the above characteristics that make them pollutants. These compounds were produced for use as pesticides, solvents, heat-transfer agents, fire retardants and other purposes. The short history of xenobiotic compounds in the environment may further contribute to their recalcitrance. There is circumstantial evidence that microorganisms have only recently evolved mechanisms to biodegrade certain xenobiotic compounds, such as pentachlorophenol (PCP), discussed below. It follows from this situation that those biodegradation mechanisms may not be highly efficient, possibly being

[1]Department of Microbiology and Immunology, University of British Columbia, 300–6174 University Boulevard, Vancouver, British Columbia V6T 1Z3, Canada, e-mail: wmohn@interchange.ubc.ca, Tel: +1-604-8224285, Fax: +1-604-8226041

Soil Biology, Volume 2
Biodegradation and Bioremediation
(ed. by. A. Singh and O. P. Ward)

incomplete or yielding dead-end intermediates that might even be harmful to the responsible organism and the environment. This appears to be the case for the fortuitous anaerobic degradation of the solvents perchloroethene (PCE) and trichloroethene (TCE) by many microorganisms. On the other hand, there is also evidence that a very broad range of halo-organic compounds were created by natural processes, such as pyrolysis of vegetation and biosynthesis (Neidleman and Geigert 1986; Gribble 1994). For the latter chemical structures, one would expect there to be efficient biodegradation mechanisms that are the result of long-term evolution. Furthermore, these mechanisms might fortuitously also degrade xenobiotic compounds in addition to their natural substrates.

2 Biodegradation

2.1 Some Generalizations About Halo-organic Biodegradation

Within the scope of this chapter, it is impossible to comprehensively cover the remarkably diverse mechanisms of the biodegradation of halo-organic compounds. Therefore, the focus will be on principles that contribute to a broad understanding of those mechanisms. This chapter will also focus on microbial biodegradation, while acknowledging that plant and even animal metabolism may also have useful applications to bioremediation. To illustrate the above principles, a small number of examples will be presented that have particular relevance to bioremediation. These examples are ones that are relatively well understood and serve as paradigms. It is important to appreciate that our knowledge of microbial metabolism is far from complete, with novel processes still being reported. Within the vast metabolic diversity of microorganisms there are likely to be undiscovered processes, which could be used to rid the environment of halo-organic pollutants.

Generalizations provide the benefit of making complex phenomena more readily understandable, while obscuring details that may be interesting or critical. Keeping this in mind, some generalization is probably a useful starting point for considering halo-organic biodegradation. One instructive distinction is between degradation processes "specifically evolved" for halo-organic substrates versus "co-metabolic" or "fortuitous" processes that are the result of enzymes that evolved for transformation of non-halogenated substrates. This distinction is largely based on speculation as to why degradation processes occur (i.e.

how they evolved), however, these reasons are not always clear-cut. Another instructive distinction among the specifically evolved processes is between oxidative processes in which halo-organic compounds are used as sources of reductant and carbon versus reductive processes in which the compounds are used as respiratory electron acceptors (dehalorespiration).

2.2 Co-metabolic Processes

In some cases, degradation of halo-organic compounds appears to be a fortuitous result of the low substrate specificity of a mechanism for degradation of a non-halogenated analogue. Oxygenases, which often have broad substrate specificity, typically play key roles in these fortuitous processes. Usually, there is a limit to the halogen substitution tolerated by oxygenases. Thus, polychlorinated biphenyls (PCBs) with chlorine substituents at particular positions are very recalcitrant. Similarly, while trichloroethene (TCE) can be co-metabolized, perchloroethene (PCE), with all available positions chlorinated, is largely recalcitrant to co-metabolism. The above co-metabolism by oxygenases is often referred to as co-oxidation. Anaerobic co-metabolic processes also exist.

Co-metabolic processes are typically unproductive, failing to support growth and often incompletely degrading the substrate. In fact, oxygenases involved often require reductant, making co-oxidation an energetic burden to the responsible organism and making an additional substrate necessary to sustain the process. Also, the expression of the critical enzymes may be regulated, and the fortuitous substrates may not serve as inducers. Thus, for co-metabolic processes to be exploited for bioremediation, it may be necessary to supply the degradative organisms with appropriate substrates to serve as electron donors and inducers.

2.2.1 Aerobic Polychlorinated Biphenyl Biodegradation

Due to their low dielectric constants and high heat capacity, polychlorinated biphenyls (PCBs) were widely used for dielectric fluids, plasticizers, heat transfer fluids, fire retardants and other applications. There are over 200 PCB congeners, and complex mixtures of these congeners were used for the above applications. When the persistence and toxicity of PCBs were recognized in the 1970s, production of PCBs was banned in North America and other regions. However, many sites where

PCBs were dumped or PCB-containing equipment was stored remain contaminated today.

A certain fraction of aerobic bacteria able to grow on biphenyl are also able to co-oxidize PCBs. For the most part, PCBs do not support growth of these organisms. In rare cases, bacteria have been reported to grow on mono- and di-chlorobiphenyls. From the available evidence, it appears that PCBs are most commonly co-oxidized by a process involving attack by a 2,3-biphenyl dioxygenase (BPDO), preferably on the least substituted aromatic ring (Fig. 1A). The ability of organisms to degrade individual PCB congeners varies considerably due to a

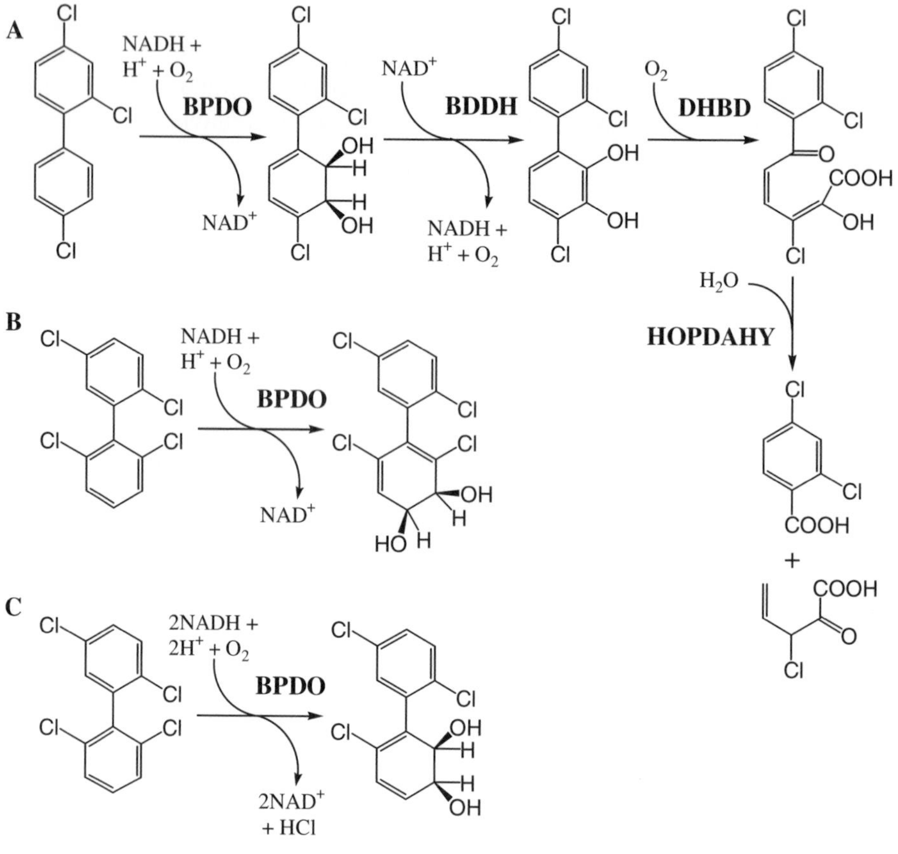

Fig. 1. Mechanisms for co-oxidation of PCBs. **A** 2,3-BPDH attack on the less chlorinated ring followed by ring cleavage and hydrolysis. **B** 3,4-BPDH attack on a congener hindering 2,3-attack. **C** Oxidative dechlorination by 2,3-BPHD. Enzymes include BPDH, biphenyl dioxygenase; BDDH, biphenyl dihydrodiol dehydrogenase; DHBD, dihydroxybiphenyl dioxygenase; HOPDAHY, 2-hydroxy-6-oxo-6-phenyl-hexa-2,4-dienoate hydrolase. (Adapted from Komancova et al. 2003)

number of factors. A primary factor appears to largely be substrate specificity of BPDO, with homologous enzymes differing substantially in their tolerance of chorine substituents. The limiting factor does not appear to be the number of chlorine substituents, per se, but rather the presence of chlorine in key positions (Arnett et al. 2000). It stands to reason that a congener with no unsubstituted, adjacent *ortho* and *meta* positions (2,3-positions), is resistant to 2,3-BPDO attack. However, chlorine substituents at the *para* positions also contribute to recalcitrance, and the overall influence of chlorine substituents on BPDO activity appears complex. Some variants of BPDO with exceptionally broad specificity for PCBs appear to additionally have 3,4-BPDO activity (Haddock et al. 1995). The attack at the 3,4-positions circumvents hindrance by *ortho* chlorine substituents (Fig. 1B). Another mechanism to overcome hindrance by *ortho* or *meta* chlorine substituents is attack at the 2,3-positions with a BPDPO capable of displacement of chlorine with a hydroxyl group, oxidative dechlorination (Fig. 1C). Komancova et al. (2003) showed some evidence for this mechanism, but it is not yet clear whether this is a primary route for degradation and to what extent it contributes to the degradation of PCBs.

Further degradation of PCBs is possible, with ring cleavage of some congeners by an extradiol dioxygenase (Fig. 1A). Aerobic bacteria degrading PCBs typically accumulate chlorobenzoates and other dead-end metabolites. With complex PCB mixtures, the dead-end metabolites have not been well characterized, and the actual extent of PCB degradation is not well understood. Thus, in bioremediation applications, co-oxidation will destroy PCBs, but undefined chlorinated products will remain. The potential exists for additional organisms to degrade and more extensively dechlorinate dead-end metabolites from PCB degraders. In particular, most characterized PCB degraders do not degrade chlorobenzoates, but distinct chlorobenzoate-degrading organisms do exist.

While much of the focus of investigating PCB co-oxidation has been on BPDO, the tolerance of chlorine substituents by other enzymes can also be an important factor. Accumulating metabolites may be inhibitory to further PCB degradation or generally toxic to the degrading organism. An example of inhibition involves 2,3-dihydroxybiphenyl 1,2-dioxygenase (DHBD), which catalyzes aromatic ring cleavage. *Ortho*-chlorinated PCB metabolites bind to DHBD but are not transformed (Dai et al. 2002). These metabolites strongly inhibit DHBD, promote its suicide inactivation and interfere with the degradation of other compounds. Additional enzymes may exist with the function of removing toxic metabolites. Gilmartin et al. (2003) demonstrated that a gene clustered with PCB-degradation genes in *Burkholderia* sp. LB400 encodes a glutathione S-transferase that can dechlorinate 4-chlorobenzoate, a metabolite from some PCBs. Thus, this enzyme might have a role either in removing a toxic metabolite or permitting productive degradation of the chlorobenzoate metabolite to yield reductant and carbon.

2.2.2
Aerobic Trichloroethene Biodegradation

Chlorinated ethenes, ethanes and methanes were used extensively for degreasing machinery and textile processing. Improper disposal has caused these solvents, in particular trichloroethene (TCE), to be among the most abundant environmental contaminants. These solvents are relatively mobile, and when they contaminate soil, they tend to be transported to underlying aquifers and to continue moving with the groundwater flow.

Aerobic co-oxidation of TCE is another example of co-metabolism that is very relevant to bioremediation. This co-oxidation is catalyzed by various monooxygenases and dioxygenases. Importantly, these oxygenases can often attack other halogenated alkyl solvents, including dichloroethene, and vinyl chloride, chloroethanes and chloromethanes. The completely chlorinated PCE and carbon tetrachloride are very recalcitrant to co-oxidation, however, oxidation of PCE by a monooxygenase has been recently reported (Ryoo et al. 2000). Oxygenases attacking these compounds are diverse and evolved for the degradation of a range of substrates, including toluene, phenol, methane, propane, propylene, 2,4-dichlorophenoxyacetate and isopropylbenzene (Arp 1995). The oxygenases degrade the chlorinated solvents during the oxidation of their normal substrates, which are required as inducers for the expression of the oxygenases and as electron donors for the process.

Several major products of TCE co-oxidation have been identified, but the factors controlling the product yields and ratios are not well understood. TCE degradation by methane monooxygenase has been studied in detail (Wackett 1995), while other co-oxidations of TCE are not well understood. Methane monooxygenase primarily forms an unstable epoxide from TCE (Fig. 2), however, other products are also possible. The epoxide decomposes, often losing chlorine substituents, forming several products, including carbon monoxide, glyoxylic acid, formic acid and dichloroacetic acid. This decomposition is largely spontaneous, but enzymatic reactions may also be involved in the formation of these and other products. Some products of TCE co-oxidation have the potential to be inhibitory. Thus, organisms capable

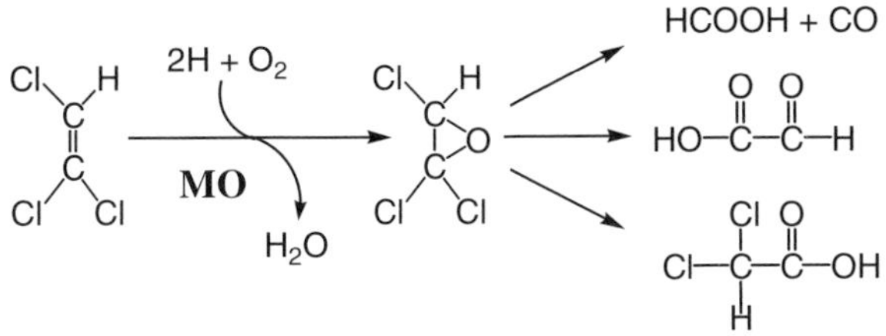

Fig. 2. Co-oxidation of TCE. MO. *First step* catalyzed by various monooxygenases. *Second step* involves spontaneous reactions forming several products

of detoxifying such products would be better able to tolerate the co-oxidation process.

2.3 Specifically Evolved Aerobic Processes

Productive degradation processes, ones supporting growth, have clear advantages for bioremediation applications. A growing biocatalyst can be maintained or increased in a properly designed system; while, a non-growing one can be maintained only with effort and expense. Also, as illustrated in the examples below, certain compounds, which probably have too many halogen substituents to be efficiently attacked by oxygenases, can be degraded by productive pathways. To accomplish this, various dechlorination mechanisms are employed prior to attack by an oxygenase.

2.3.1 Aerobic Pentachlorophenol Biodegradation and Patchwork Evolution

Pentachlorophenol (PCP) was used for wood preservation and many other pesticide applications. PCP is very toxic to microorganisms. However, as PCP is also toxic to humans and very persistent, its production was banned in many countries for over 20 years. Despite this, soils heavily contaminated with PCP exist at former wood treatment facilities and other sites.

A number of aerobic bacteria can grow on and mineralize PCP using a metabolic process with great potential for bioremediation. Copley et al. (2000) have provided evidence for a "patchwork" mechanism of the evolution of the PCP degradation pathway of *Sphingobium chlorophenolicus* (formerly *Sphingomonas chlorophenolica*) in which enzymes were recruited from two unrelated catabolic pathways. The first enzyme

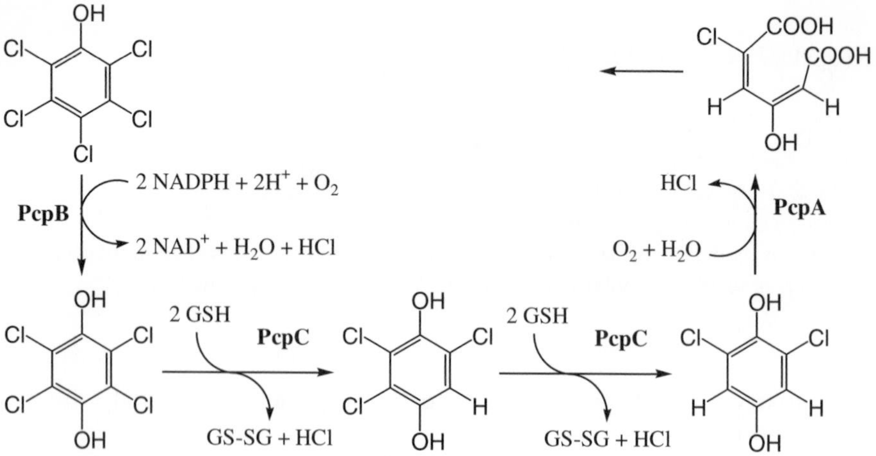

Fig. 3. Pathway for biodegradation of PCP by *Sphingomonas chlorophenolica*. Enzymes include *PcpB*, a flavin monooxygenase capable of oxidative dehalogenation; *PcpC*, a glutathione (*GSH*)-dependent reductive dechlorinase; and *PcpA*, an extradiol ring-cleavage dioxygenase that also has halidohydrolase activity. The ring-cleavage product can be further degraded and presumably mineralized. (Adapted from Ohtsubo et al. 1999; Copley 2000)

in the pathway (Fig. 3), PCP hydroxylase, is a flavin monooxygenase that may have originated from a pathway for degradation of less-chlorinated phenols. This hydroxylase was likely to have been modified to yield broader substrate specificity (i.e. to tolerate additional chlorine substituents) at the cost of lower catalytic efficiency. The second enzyme, tetrachlorohydroquinone dehalogenase, is a glutathione-dependent reductive dechlorinase that catalyses two sequential dechlorination steps. This dechlorinase appears to have originated from maleylacetoacetate isomerase, which has a very different function in a pathway for tyrosine degradation. The third enzyme, dichlorohydroquinone dioxygenase, and subsequent enzymes, may be from the above pathway for degradation of less-chlorinated phenols without modification. A number of observations support this evolutionary scenario and suggest that the PCP pathway has recently evolved, including the low catalytic efficiency of PCP hydroxylase and the constitutive expression of the reductive dechlorinase despite the induction of the other enzymes of the pathway by PCP. Also consistent with the above mechanism is the location of the genes for the hydroxylase and dioxygenase at one locus and the gene for the dechlorinase at a distinct locus on the *S. chlorophenolica* chromosome (Cai and Xun 2002).

2.3.2 Aerobic Hexachlorocyclohexane Biodegradation

Hexachlorocyclohexane (HCH) exists as several isomers. The gamma isomer, also called lindane, was used intensively as an insecticide, and is now banned in many countries due to its toxicity and persistence. Several other isomers, particularly alpha and delta, are by-products from lindane production. These three isomers and traces of other ones are soil contaminants at sites where lindane was produced and the by-products disposed.

The best understood pathway for HCH biodegradation is that of *Sphingomonas paucimobilis* UT26 (Nagata et al. 1999). This pathway degrades γ-HCH, and also appears to degrade α-, β- and δ-HCH (Johri et al. 1998). It involves three distinct mechanisms to remove five of six chlorine substituents from HCH prior to the cleavage of the ring (Fig. 4). The first two steps involve dehydrochlorination, and the second two, hydrolytic dechlorination. Following oxidation to aromatize the ring, another chlorine substituent is removed by reductive dechlorination. The reductive dechlorinase is also able to remove the sixth chlorine substituent, but in vivo, it appears likely that the mono-chlorinated ring is cleaved and dechlorinated by an extradiol dioxygenase.

Evolution of the above HCH pathway may have been by patchwork assembly as for the PCP pathway, with common origins for some of the enzymes in both pathways. The two pathways appear to have homolo-

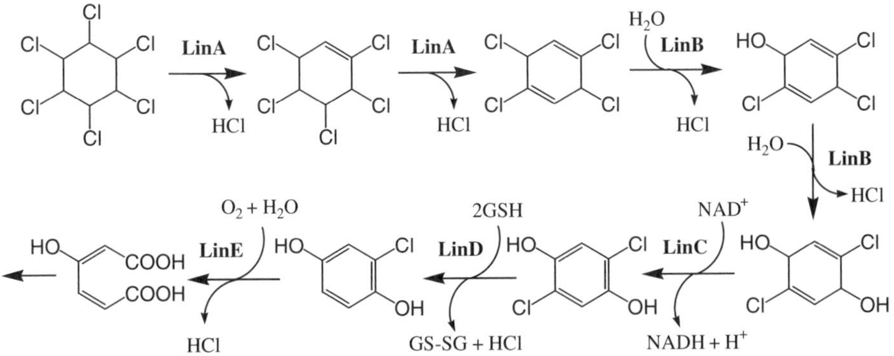

Fig. 4. Pathway for biodegradation of γ-HCH (lindane) by *Sphingomonas paucimobilis* UT26. Enzymes include *LinA*, a dehydrochlorinase; *LinB*, a halidohydrolase; *LinC*, a dehydrogenase; *LinD*, a glutathione (*GSH*)-dependent reductive dechlorinase; and *LinE*, an extradiol ring-cleavage dioxygenase that also has halidohydrolase activity. The ring-cleavage product can be mineralized. (Adapted from Nagata et al. 1999)

gous reductive dechlorinases and extradiol dioxygenases. The reductive dehalogenases of both pathways are glutathione-dependent and share sequence similarity to one another as well as to their proposed ancestor, maleylacetoacetate isomerase. Furthermore, the two pathways have homologous extradiol dioxygenases, PcpA and LinE (Ohtsubo et al. 1999). Van Hylckama Vlieg et al. (2000) proposed that these dioxygenases share a reaction mechanism similar to that of CbzE, an extradiol dioxygenase involved in chloro-benzene metabolism by *Pseudomonas putida* GJ31. The proposed mechanism involves ring cleavage by dioxygen insertion plus simultaneous hydrolytic dechlorination. This proposed mechanism has the benefit of avoiding the formation of acyl chloride products, which are known to inactivate extradiol dioxygenases that form them.

2.4 Dehalorespiration

Anaerobic processes are often very effective for degradation of halo-organic compounds. Indeed, for some of the most highly-halogenated compounds (e.g. highly chlorinated PCBs), we do not (yet) know of aerobic degradation processes, but efficient anaerobic processes readily occur. Thus, anaerobic processes are the only option for the bioremediation of such compounds. The primary basis for these anaerobic processes is the reductive removal of the halogen substituents, which is thermodynamically very favorable. A great deal of progress has recently been made in understanding some of the enzyme catalysts of reductive dehalogenation, substantially increasing our understanding of the underlying mechanisms (reviewed in Holliger et al. 1998). Reductive dehalogenases are covered in Chapter 7 (this Vol.).

Unlike many aerobic processes for halo-organic biodegradation, anaerobic processes usually begin with the removal of all halogen substituents, via reductive dehalogenation, prior to the attack of the organic parent compound. As mentioned above, reductive dechlorination can be a fortuitous process. However, the focus here is on what appear to be cases of specifically evolved reductive dehalogenation processes. These specifically evolved processes seem to have more potential for bioremediation than co-metabolic processes, partly because rates of the former are generally greater than those of the latter. Reductive dehalogenation appears to be specifically evolved in cases where it supports anaerobic respiration (dehalorespiration), with the halo-organic compound serving as a terminal electron acceptor. Dehalorespiration has been demonstrated with chlorobenzoates and chloroethenes, and there is strong circumstantial evidence indicating that it can occur with PCBs, PCP and other halo-aromatic compounds. Another major advantage of dehalorespiration for bioremediation is that there is selective pressure for the degrading organisms, which can grow on the pollutant if an electron donor is available. Also, anaerobic processes are more practical for in situ treatment in environments where oxygen is limiting.

2.4.1 Probable Dehalorespiration with Polychlorinated Biphenyls

Many PCB mixtures, including those that are most persistent in the environment, are composed of relatively highly chlorinated PCB congeners. Such mixtures can have an average number of chlorines per biphenyl of five or more. These highly chlorinated PCBs can be substantially biodegraded only via reductive dehalogenation. This degradation activity is known to occur only in undefined enriched anaerobic consortia. The fact that the dechlorinating activity can be enriched in these consortia and that the cultures can be maintained through serial transfers, strongly suggest, but does not conclusively prove that growth is supported by dehalorespiration.

These consortia partly dechlorinate PCBs by reductive dechlorination, yielding less chlorinated PCBs as products. This partial degradative process reduces the mass of PCBs by removing chlorine substituents but does not reduce the molar concentration of PCBs. This process degrades some of the most toxic congeners, the co-planar PCBs. There is no convincing evidence that such anaerobic consortia can degrade the biphenyl parent compound. Consortia exhibit the specificity for the removal of chlorine substituents from particular positions of biphenyl, and specificity can vary considerably between different consortia (reviewed in Bedard and Quensen 1995). As a very general rule, chlorine substituents are increasingly less able to remove from the *meta*, *para* and *ortho* positions, respectively.

In general, PCB-dechlorinating consortia remain poorly understood. No pure culture capable of degrading PCBs has been isolated. The consortia typically require the addition of anaerobic sediment. The sediment can be autoclaved, so it presumably contributes little to the biological complexity of a consortium. But, the sediment is chemically complex and undefined. Organisms associated with PCB dechlorination have been identified on the basis of rRNA gene sequences (Pulliam Holoman et al. 1998; Wu et al. 2002). However, confirmation that these phylotypes are responsible for dechlorination is lacking, and further study of these organisms is severely limited by the fact that they are not isolated. Important characteristics of PCB dechlorinating consortia are their low rates of growth and dechlorination. The consortia typically require incubation for a month before dechlorination begins and several additional weeks of incubation before dechlorination is complete. These low rates further limit the investigation of the organisms and dechlorination mechanism(s), and these low rates are a definite limitation to exploiting this process for bioremediation.

2.4.2
Dehalorespiration with Perchloroethene

Several pure cultures of diverse phylogeny have been shown to grow via dehalorespiration with PCE as an electron acceptor (Holliger et al. 1998). Most of these organisms only partially dechlorinate PCE to 1,2-*cis*-dichloroethene (Fig. 5). *Dehalococcoides ethenogenes* is the only pure culture that can use PCE for dehalorespiration and completely dechlorinate the compound to ethene (Maymó-Gatell et al. 1997). Clearly, this complete dechlorination is advantageous for bioremediation, since it avoids harmful products. Interestingly, *D. ethenogenes* is fastidious and is not known to use any other electron accepters than chloroethenes and 1,2-dichloroethane, nor any other electron donor than hydrogen. Some other organisms that dehalorespire with PCE are similarly fastidious. Thus, the niche of these organisms in a natural environment is a puzzle.

Several reductive dehalogenases from organisms able to dehalorespire with PCE have been purified (Holliger et al. 1998). The investigation of these enzymes is currently making great contributions to our understanding of dehalorespiration. These enzymes have corinoid plus iron-sulfur clusters as co-factors. All or most of the enzymes

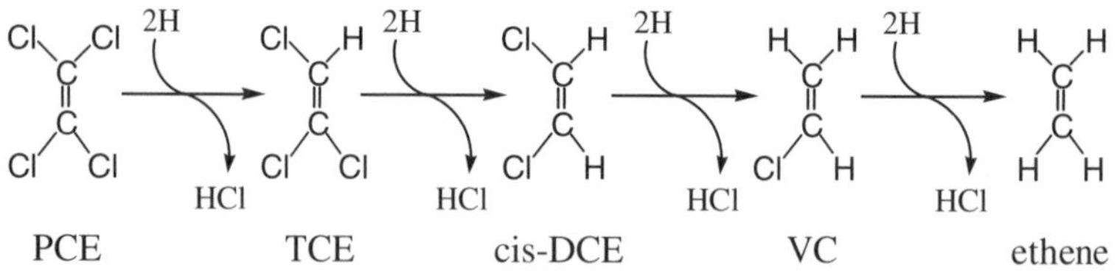

PCE → TCE → cis-DCE → VC → ethene (each step: + 2H, − HCl)

Fig. 5. Pathway for complete reductive dechlorination of PCE by *Dehalococcoides ethenogenes.* Other organisms are capable of partial dechlorination of PCE, often yielding *cis*-DCE as a product. *PCE* Perchlorethene; *TCE* trichloroethene; *cis-DCE*, *cis*-1,2-dichloroethene; *VC* vinyl chloride

that have been localized are membrane-associated and facing the cytoplasm, which is likely to be essential for their roles in dehalorespiration. The organisms that partly declorinate PCE to cis-DCE appear to have a single enzyme capable of both dechlorination steps (Fig. 5). While, *D. ethenogenes* appears to have two dehalogenases, PCE reductive dehalogenase, which catalyzes the first step, and TCE reductive dehalogenase, which catalyzes the remaining three steps to yield ethene (Magnuson et al. 1998).

2.5 Ligninase Systems

Lignin is biodegraded by a special type of aerobic enzyme system found in certain fungi. These ligninase systems are highly non-specific and often will degrade halo-organic compounds. Ligninase degradation of pollutants has been recently reviewed (Pointing 2001) and will not be covered in detail here. In general, this type of fungal system degrades lignin by free-radical reactions that attack a broad range of substrates, particularly aromatic ones. Lignin-degrading fungi have been shown to degrade a variety of organo-chlorine compounds, including dieldrin, heptachlor, chlordane, lindane, and mirex (Kennedy et al. 1990), DDT, dicofol, and methoxychlor (Bumpus and Aust 1987), 2,4,5-T and 2,4-D (Yadav and Reddy 1993), atrazine (Hickey et al. 1994), PCP (Mileski et al. 1988; Lin et al. 1990) as well as PCBs (Vyas et al. 1994; Dietrich et al. 1995; Beaudette et al. 2000). This degradation capacity is promising, but typically such experiments demonstrate less than 25% mineralization of the halo-organic compounds, with the notable exception of PCP. Substantial fractions of halo-organic compounds are often undegraded by lignolytic systems, and by-products are often formed. In many cases, the by-products have not been identified, and there is the possibility that the by-products themselves are harmful. In many studies of the biodegradation of halo-organic compounds by lignolytic fungi, the role of extracellular ligninases remains unclear, and it is possible that additional enzymatic systems contribute to the degradation observed. Thus, lignolytic systems have great potential to degrade halo-organic compounds, but at present, such systems have generally not been demonstrated to degrade such compounds sufficiently for bioremediation applications.

3 Bioremediation

3.1 The Current State of Halo-organic Bioremediation

The potential for bioremediation of a large number of halo-organic soil contaminants has been demonstrated. This evidence is primarily in the form of laboratory studies demonstrating the degradation of a particular compound in some type of model soil system. There is a great difference between these studies and the practical application of bioremediation in the field on a large scale. A number of issues must be addressed in progressing to practical applications, a few of which include the bioavailability of weathered contaminants, the presence of adequate biocatalysts, the heterogeneity of soil systems and appropriate physicochemical conditions for biocatalysts. Some of these issues are covered in other chapters of this book. The relatively great recalcitrance of halo-organic compounds makes it more likely that the biodegradation of these compounds, compared to that of other compounds will be limited by a lack of an adequate biocatalyst. This is particularly true for cases where co-metabolism is required. Thus, bioaugmentation may have a more important role in the bioremediation of halo-organic compounds than in the bioremediation of other compounds.

The economics of halo-organic compound bioremediation differ in some cases from those of other pollutants, resulting in different possibilities for biotreatment systems. Many halo-organic compounds are high-priority pollutants, both from public and regulatory perspectives. Often, the alternative non-biological remediation technologies for these compounds (e.g., incineration under highly-controlled conditions) are expensive and controversial. Finally, the volumes of soil contaminated with these compounds are sometimes smaller than those with other pollutants. All of these factors may justify the use of more elaborate technology for the cleanup of halo-organic pollutants than for the cleanup of many other pollutants. This situation makes bioremediation of halo-organic compounds a possibility in many circumstances despite the relative recalcitrance of these compounds to biodegradation.

There are relatively few field studies demonstrating bioremediation of halo-organic soil contaminants that have been published in peer-reviewed journals. A large proportion of such studies have been reported in conference proceedings or commercial literature. Thus, due to the limited availability of essential information, it is difficult to

Table 1. Selected field studies demonstrating bioremediation of halo-organic compounds

Compound	Treatment system	References
Chloroethenes	Aquifer, in situ	Beeman et al. (1994); Buchanan et al. (1995); Major et al. (1995); Spuij et al. (1997)
Chlroroethanes	Aquifer, in situ	Boyer et al. (1988); Fathepure et al. (1995)
Chloromethanes	Aquifer, in situ	Semprini et al. (1992); Hooker et al. (1998)
PCP	Compost	Valo and Salkinoja-Salonen (1986); Laine and Jørgensen (1997)

critically evaluate many field studies. However, there have clearly been some demonstrated successes in field-scale bioremediation of certain halo-organic compounds (Table 1), a few of which are described below.

3.2 Composting of PCP-Contaminated Soil

The use of composting systems to treat halo-organic soil contaminants is relatively new, and only a few controlled studies have been done. Bioremediation of PCP-contaminated soil with composting technologies is the best-demonstrated example. Composting is clearly a practical treatment for PCP-contaminated soil and is also being assessed for the treatment of other soil contaminants. In a field experiment, PCBs with three or fewer chlorine substituents were reportedly degraded in a compost system (Michel et al. 2001). However, for this system to be practical, it would need to degrade the more-chlorinated PCBs normally found as contaminants. In a laboratory-scale composting system with grass trimmings, extensive mineralization of 2,4-D was demonstrated (Michel et al. 1995). Thus, composting appears to be a promising soil decontamination strategy and is likely to receive further attention.

Effective bioremediation of PCP and other chlorophenols was demonstrated in field scale compost systems (Table 1). Initial concentrations greater than 100 mg of PCP per kg of soil (ppm) were reduced to approximately 10 ppm. Separate laboratory-scale experiments showed extensive (>50%) mineralization of PCP in compost systems. However, associated dioxins were not removed. In a laboratory-scale manure compost system, PCP was also very efficiently biodegraded (Jaspers et al. 2002). However, evidence indicated that in this system, in contrast to the above composts, the primary mechanism of PCP degradation was initiated by reductive dechlorination. This activity may be a result of the microflora derived from the manure.

Composting is an approach that provides a relatively well-controlled microbial environment. Contaminated soil is typically amended with various materials that, in principle, may serve a number of functions (reviewed in Semple et al. 2001). However, few studies have actually demonstrated these functions. Soil amendments may supply degradable organic material, which could act as a co-substrate for co-oxidative processes. Another function of soil amendments is to increase the porosity of soil, thus, improving the aeration of compost systems. Soil amendments also may adsorb water, increasing the capacity for water while reducing the negative effect of water on aeration. Soil amendments may also sorb pollutants, reducing their toxic effects on microorganisms while permitting their slow uptake by microorganisms. When added to an aerobic soil system, bark chips reduced the toxic effect of PCP (Apajalahti and Salkinoja-Salonen 1984). Compost systems are usually well mixed, which improves the bioavailability of pollutants and so may result in lower residual concentrations of pollutants.

A true compost system will reach a high temperature (commonly 60 °C) as a result of high metabolic activity. Added organic matter is likely to be necessary for such activity, as halo-organic compounds are likely to be present in concentrations too low, and likely to be degraded too slowly, to generate substantial heat. High temperature may directly increase biodegradation rates and might make treatment possible at ambient temperatures that would otherwise be too low for substantial biodegradation activity. Heat may also have the benefit of increasing the bioavailability of pollutants due to thermal desorption from soil. However, biodegradation of halo-organic compounds at a high temperature is not well studied. It is possible that temperature could inhibit degradation processes by excluding essential degradative organisms. Alternatively, high temperatures might select degradative organisms or degradation processes unlike those which are currently known.

3.3 In Situ Treatment of Chlorinated Ethenes

Several studies have demonstrated in situ degradation of PCE and TCE in shallow aquifers (Table 1). Field studies have demonstrated effective in situ bioremediation of PCE (Beeman et al. 1994; Buchanan et al. 1995). A sequential anaerobic-aerobic system was also found to be effective (Spuij et al. 1997). Thus, in situ bioremediation of chloroethenes, as well as chloroethanes and chloromethanes, is definitely a practical option for site cleanup.

This mobility of chlorinated solvents is a challenge for bioremediation. Their depth and spreading in aquifers make in situ bioremediation a very attractive option. Pump-and-treat systems, with biological or physical removal of the solvents, can effectively purify groundwater. However, this approach is often futile, as a large pool of chlorinated solvents sorbed to aquifer solids typically re-contaminates treated groundwater. A more effective bioremediation strategy involves biostimulation or bioaugmentation to colonize the aquifer solids with a population capable of biodegrading the solvents in situ. A series of injection and recovery wells are often employed to deliver amendments to the aquifer and promote desired water movement. A good understanding of the aquifer structure and flow patterns is required for this to be successful.

Different strategies exist for in situ biostimulation, based on the different processes for chloroethene biodegradation. Those processes include (1) complete reductive dechlorination, (2) partial dechlorination followed by oxidation and (3) oxidation. As indicated above, PCE cannot be extensively degraded by oxidation alone. These processes are also generally applicable to chloroethanes and chloromethanes. Anaerobic processes are often preferable, because organic contaminants typically deplete oxygen in aquifers, and delivering oxygen to this environment can be difficult. A concern with anaerobic treatment alone is the potential accumulation of toxic and carcinogenic metabolites, particularly vinyl chloride.

A large number of factors affect in situ biodegradation of chloroethenes (reviewed in Lee et al. 1998), but the primary engineering concern is normally the provision of electron donors or acceptors required for the desired degradation process and required to provide an appropriate redox state. For anaerobic treatment, a variety of electron donors, including hydrogen, methanol, ethanol, acetate, lactate, butyrate and benzoate have been added to reduce systems and to drive reductive dehalogenation. For aerobic treatment, oxygen must typically be added, either directly or indirectly in the form of more-soluble peroxide. Alternating anaerobic-aerobic conditions appears to be a promising strategy, which can accomplish degradation of PCE while avoiding the accumulation of harmful less-chlorinated ethanes (Spuij et al. 1997). Additional nutrients such as N, P and vitamins may be necessary when trying to promote growth of degrading organisms in aquifer environments, which are typically nutrient-poor.

While PCE-dechlorinating organisms appear common in aquifers, particularly anaerobic ones high in organic content, they may not be in sufficient abundance for desired degradation activity. Furthermore, the presence of organisms that completely dechlorinate PCE may limit that process as a treatment option. Inoculation with a PCE-dechlorinating culture that included a *Dehalococcoides* sp. was demonstrated in a large artificial aquifer (Adamson et al. 2003). However, effectively seeding real aquifers with an inoculum is technically very challenging.

Natural attenuation of chloroethenes is another possibility of obvious interest to those responsible for contaminated sites. Several studies have provided evidence for substantial rates of natural biodegradation of chloroethenes in contaminated aquifers (Major et al. 1995; Witt et al. 2002). An increasingly acceptable option for handling chloroethene-contaminated sites is to monitor with the objective of showing natural biodegradation and a lack of immediate risk to the environment or human health.

3.4 The Promise of PCB Bioremediation

Despite the large body of research on PCB biodegradation, there does not yet exist a well-demonstrated system for PCB bioremediation. Evidence for the potential for PCB bioremediation is largely based on laboratory studies. These studies demonstrate the necessary biocatalysts for PCB degradation but do not demonstrate practical means to employ those biocatalysts to soil on a large scale. Many attempts to measure PCB biodegradation in soil have been confounded by a volatile loss of PCBs stimulated by manipulation of the soil and by variability in PCB recovery during analysis.

Some laboratory studies have demonstrated co-oxidation of PCBs in soil (Focht and Brunner 1985; Hill et al. 1989; Barriault and Sylvestre 1993). From these studies, the need for biphenyl as a co-substrate is clear. More recent work has indicated that certain other compounds might replace biphenyl as a co-substrate that induces PCB co-oxidation by some bacteria (Donnelly et al. 1994; Billingsley et al. 1997; Gilbert and Crowley 1997). The above studies also underscore the importance of PCB bioavailability as a limiting factor. Surfactants can improve bioavailability, if their use is carefully optimized. However, it seems likely that adequate PCB degradation may not be possible with in situ treatment and that some type of mixed system (e.g., soil slurry reactor) may be necessary. Finally, many PCB mixtures contain highly-chlorinated congeners that are not effectively degraded by aerobic bacteria.

The investigation of anaerobic PCB dechlorination has mainly focused on contaminated anaerobic sediments, but a few studies have demonstrated successful treatment of contaminated soil (Tiedje et al. 1993; Kuipers et al. 2003). This treatment was accomplished using biological activity to make the slurried soil anaerobic. For aerobic soils, in situ anaerobic treatment is probably not an option due to the need to isolate the system from oxygen and to use relatively high temperatures (20 °C) to achieve acceptable dechlorination rates. Even at such a temperature, the long time required for anaerobic treatment (usually months) is a major challenge for applications. Finally, anaerobic treatment alone is probably not adequate for PCB bioremediation, because dechlorination will always leave residual PCBs as products.

Due to various limitations, neither aerobic nor anaerobic treatment alone is likely to be adequate for the bioremediation of PCBs, at least for the relatively highly-chlorinated PCB mixtures of greatest concern. Sequential anaerobic-aerobic treatment at a laboratory scale has been shown to substantially degrade PCBs (Master et al. 2001). This strategy appears feasible to use on a large scale, but it is relatively complicated and expensive. This strategy requires excavation of soil, mixing of a slurry, long-term (months) maintenance at moderate temperature (room temperature) and inoculation with both anaerobic and aerobic microorganisms. For this strategy to be practical, alternative treatment options would have to be more expensive.

An alternative strategy for PCB bioremediation is the engineering of aerobic bacteria to improve their ability to degrade highly-chlorinated congeners. Progress has been made in modifying biphenyl dioxygenase using a variety of strategies, resulting in variants with broadened PCB substrate ranges. The strategies for modification include site-directed mutagenesis (Erickson and Mondello 1993), chimeric genes (Suenaga et al. 1999), DNA shuffling (Kumamaru et al. 1998; Brühlmann and Chen 1999) and hybrid enzymes from exchanging subunits (Hurtubise et al. 1998; Chebrou et al. 1999). While these advances are promising, a number of challenges remain before engineered strains can be effective for PCB bioremediation. Effective engineered strains are likely to require complex modification of several enzymes and the optimization of regulation of multiple enzymes in order to achieve efficient pathways with balanced enzyme activities and to avoid accumulation of inhibitory metabolites. Furthermore, effective engineered strains will need to survive and be metabolically active in a soil environment likely to contain competitors and predators. The work to meet these challenges and achieve effective engineered strains is only in its very early stages.

3.5 The Potential of Exploiting Mobile Genetic Elements

Many degradation pathways for halo-organic compounds are associated with mobile genetic elements, including pathways for chlorinated benzoate, catechol, biphenyl, propene and propionate. It is hypothesized that these mobile genetic elements have played a role in the evolution of the pathways via horizontal (interspecies) gene transfer. A few studies have demonstrated the interspecies transfer of biodegradative capabilities for halo-organic compounds (Brokamp and Schmidt 1991; Fulthorpe and Wyndham 1991). These observations have prompted the proposal to exploit mobile genetic elements for bioremediation. The basic strategy is to deliver a limiting metabolic capacity to a microbial community via a mobile genetic element. A major advantage might occur if the genes encoding the metabolic capacity transferred from a delivery strain to natural members of the community, which would presumably be more fit and better adapted to the particular environment than the delivery strain would be. Top et al. (2002) have reviewed laboratory studies testing the feasibility of this strategy, which show very limited evidence for success. Interspecies transfer of the capacity to degrade biphenyl and 2,4-D was demonstrated along with an improved capacity to degrade these compounds in soil communities. However, unrealistically high substrate concentrations were required to select transconjugants. It appears that this strategy of using mobile genetic elements might be practical in certain very particular circumstances where biodegradation potential is the limiting factor and strong selective pressure for degradative genes can be applied.

References

Adamson DT, McDade JM, Hughes JB (2003) Inoculation of DNAPL source zone to initiate reductive dechlorination of PCE. Environ Sci Technol 37:2525–2533

Apajalahti JA, Salkinoja-Salonen MS (1984) Absorption pf pentachlorophenol (PCP) by bark chips and its role in microbial PCP degradation. Microb Ecol 10:359–367

Arnett CM, Parales JV, Haddock JD (2000) Influence of chlorine substituents on rates of oxidation of chlorinated biphenyls by the biphenyl dioxygenase of *Burkholderia* sp. Strain LB400. Appl Environ Microbiol 66:2928–2933

Arp DJ (1995) Understanding the diversity of trichloroethene co-oxidations. Curr Opin Biotechnol 6:352–358

Barriault D, Sylvestre M (1993) Factors affecting PCB degradation by an implanted bacterial strain in soil microcosms. Can J Microbiol 39:594–602

Beaudette LA, Ward OP, Pickard MA, Fedorak PM (2000) Low surfactant concentration increases fungal mineralization of a polychlorinated biphenyl congener but has no effect on overall metabolism. Lett Appl Microbiol 30:155–160

Bedard DL, Quensen III JF (1995) Microbial reductive dechlorination of polychlorinated biphenyls. In: Young LY, Cerniglia CE (eds) Microbial transformation and degradation of toxic organic chemicals. Wiley-Liss, New York, pp 127–216

Beeman RE, Shoemaker SH, Howell JE, Salazar EA, Buttram JR (1994) A field evaluation of in situ microbial reductive dehalogenation by the biotransformation of chlorinated ethenes. In: Hinchee RE, Leeson A, Semprini L, Ong SK (eds) Bioremediation of chlorinated and polycyclic aromatic hydrocarbon compounds. Boca Raton, pp 14–27

Billingsley KA, Backus SM, Juneson C, Ward OP (1997) Comparison of the degradation patterns of polychlorinated biphenyl congeners in Aroclors by *Pseudomonas* strain LB400 after growth on various carbon sources. Can J Microbiol 43:1172–1179

Boyer JD, Ahlert AC, Kosson DS (1988) Pilot plant demonstration of in-situ biodegradation of 1,1,1-trichloroethane. J Water Pollut Control Fed 60:1843–1849

Brokamp A, Schmidt FRJ (1991) Survival of Alcaligenes xylosoxidans degrading 2,2-dichloropropionate and horizontal transfer of its halidohydrolase gene in a soil microcosm. Curr Microbiol 22:299–306

Brühlmann F, Chen W (1999) Tuning biphenyl dioxygenase for extended substrate specificity. Biotechnol Bioeng 63:544–551

Buchanan RJ Jr, Ellis DE, Odom JM, Mazierski PF, Lee MDSRA (1995) Intrinsic and accelerated anaerobic biodegradation of perchloroethylene in groundwater. In: Hinchee RE, Wilson, JT, Downey DC (eds) Intrinsic bioremediation. Batelle, Columbus, OH, pp 245–252

Bumpus JA, Aust SD (1987) Biodegradation of DDT [1,1,1-trichloro-2,2-bis(4-chlorophenyl)ethane] by the white-rot fungus *Phanerochaete chrysosporium*. Appl Environ Microbiol 53:2001–2008

Cai M, Xun LY (2002) Organization and regulation of pentachlorophenol-degrading genes in *Sphingobium chlorophenolicum* ATCC 39723. J Bacteriol 184:4672–4680

Chebrou H, Hurtubise Y, Barriault D, Sylvestre M (1999) Heterologous expression and characterization of the purified oxygenase component of *Rhodococus globerulus* P6 biphenyl dioxygenase and of chimeras derived from it. J Bacteriol 181:4805–4811

Copley SD (2000) Evolution of a metabolic pathway for degradation of a toxic xenobiotic: the patchwork approach. Trends Biochem Sci 25:261–265

Dai SD, Vaillancourt FH, Maaroufi H, Drouin HM, Neau DB, Snieckus V, Bolin JT, Eltis LD (2002) Identification and analysis of a bottleneck in PCB biodegradation. Nat Struct Biol 9:934–939

Dietrich D, Hickey WJ, Lamar RT (1995) Degradation of 4,4′-dichlorobiphenyl, 3,3′, 4,4′-tetrachlorobiphenyl and 2,2′, 4,4′, 5,5′-hexachlorobiphenyl by the white-rot fungus *Phanerochaete chrysosporium*. Appl Environ Microbiol 61:3904–3909

Donnelly PK, Hedge RS, Fletcher JS (1994) Growth of PCB-degrading bacteria on compounds from photosynthetic plants. Chemosphere 28:981–988

Erickson BD, Mondello FJ (1993) Enhanced biodegradation of polychlorinated biphenyls after site-directed mutagenesis of a biphenyl dioxygenase gene. Appl Environ Microbiol 59:3858–3862

Fathepure BZ, Youngers GA, Richter DL, Downs CE (1995) In situ bioremediation of chlorinated hydrocarbons under field aerobic-anaerobic environments. In: Hinchee RE, Wilson, JT, Downey DC (eds) Intrinsic bioremediation. Batelle, Columbus, OH, pp 169–186

Focht DD, Brunner W (1985) Kinetics of biphenyl and polychlorinated biphenyl metabolism in soil. Appl Environ Microbiol 50:1058–1063

Fulthorpe RR, Wyndham RC (1991) Transfer and expression of the catabolic plasmid pBRC60 in wild bacterial recipients in a freshwater ecosystem. Appl Environ Microbiol 57:1546–1553

Gilbert ES, Crowley DE (1997) Plant compounds that induce polychlorinated biphenyl biodegradation by *Arthrobacter* sp. strain B1B. Appl Environ Microbiol 63: 1933–1938

Gilmartin N, Ryan D, Sherlock O, Dowling D (2003) BphK shows dechlorination activity against 4-chlorobenzoate, an end product of bph-promoted degradation of PCBs. FEMS Microbiol Lett 222:251–255

Gribble GW (1994) The natural production of chlorinated compounds. Environ Sci Technol 28:310–319

Haddock JD, Horton JR, Gibson DT (1995) Dihydroxylation and dechlorination of chlorinated biphenyls by purified biphenyl 2,3-dioxygenase from *Pseudomonas* sp. strain LB400. J Bacteriol 177:20–26

Hickey WJ, Fuster DJ, Lamar RT (1994) Transformation of atrazine in soil by *Phanerochaete chrysosporium*. Soil Biol Biochem 26:1665–1671

Hill DL, Phelps TL, Palumbo AV, White DC (1989) Bioremediation of polychlorinated biphenyls. Degradation capabilities in field lysimeters. Appl Biochem Biotechnol 20–21:233–243

Holliger C, Wohlfarth G, Diekert G (1998) Reductive dechlorination in the energy metabolism of anaerobic bacteria. FEMS Microbiol Rev 22:383–398

Hooker BS, Skeen RS, Truex MJ, Johnson CD, Peyton BM (1998) In situ bioremediation of carbon tetrachloride: field test results. Bioremed J 1:181–182

Hurtubise Y, Barriault D, Sylvestre M (1998) Involvement of the terminal oxygenase beta subunit in the biphenyl dioxygenase reactivity pattern toward chlorobiphenyls. J Bacteriol 180:5828–5835

Jaspers CJ, Ewbank G, McCarthy AJ, Penninckx MJ (2002) Successive rapid reductive dehalogenation and mineralization of pentachlorophenol by the indigenous microflora of farmyard manure compost. J Appl Microbiol 92:127–133

Johri AK, Dua M, Tuteja D, Saxena R, Saxena DM, Lal R (1998) Degradation of alpha, beta, gamma and delta-hexachlorocyclohexanes by *Sphingomonas paucimobilis*. Biotechnol Lett 120:885–889

Kennedy DW, Aust SD, Bumpus JA (1990) Comparative biodegradation of alkyl halide insecticides by the white-rot fungus *Phanerochaete chrysosporium* (BKM-F-1767). Appl Environ Microbiol 56:2347–2353

Komancova M, Jurcova I, Kochankova L, Burkhard J (2003) Metabolic pathways of polychlorinated biphenyls degradation by *Pseudomonas* sp. 2. Chemosphere 50:537–543

Kuipers B, Cullen WR, Mohn WW (2003) Reductive dechlorination of weathered Aroclor 1260 during anaerobic biotreatment of Arctic soils. Can J Microbiol 49:9–14

Kumamaru T, Suenaga H, Mitsuoka M, Watanabe T, Furukawa K (1998) Enhanced degradation of polychlorinated biphenyls by directed evolution of biphneyl dioxygenase. Nat Biotechnol 16:663–666

Laine MM, Jørgensen KS (1997) Effective and safe composting of chlorophenol-contaminated soil in pilot scale. Environ Sci Technol 31:371–378

Lee MD, Odom JM, Buchanan RJ (1998) New perspectives on microbial dehalogenation of chlorinated solvents: Insights from the field. Annu Rev Microbiol 52:423–452

Lin JE, Wang HY, Hickey RF (1990) Degradation kinetics of pentachlorophenol by *Phanerochaete chrysosporium*. In: Hinchee RE, Wilson JT, Downey DC (eds) Intrinsic Bioremediation, Batelle, Columbus, OH, pp 1125–1134

Magnuson JK, Stern RV, Gossett JM, Zinder SH, Burris DR (1998) Reductive dechlorination of tetrachloroethene to ethene by a two-component enzyme pathway. Appl Environ Microbiol 64:1270–1275

Major D, Cox E, Edwards E, Hare P (1995) Intrinsic dechlorination of trichloroethene to ethene in a bedrock aquifer. In: Hinchee RE, Wilson, JT, Downey DC (eds) Intrinsic Bioremediation, Batelle, Columbus, OH, pp 197–203

Master ER, Lai W-M, Kuipers B, Cullen WR, Mohn WW (2001) Seqeuntial anaerobic-aerobic treatment of soil contaminated with weathered Aroclor 1260. Environ Sci Technol 36:100–103

Maymó-Gatell X, ChienY-T, Gossett JM, Zinder SH (1997) Isolation of a bacterium that reductively dechlorinates tetrachloroethene to ethene. Science 276:1567–1571

Michel FC Jr, Reddy CA, Forney LJ (1995) Microbial degradation and humification of the lawn care pesticide 2,4-dichlorophenoxyacetic acid during composting of yard trimmings. Appl Environ Microbiol 61:2566–2571

Michel FC, Quensen J, Reddy CA (2001) Bioremediation of a PCB-contaminated soil via composting. Comp Sci Util 9:274–284

Mileski GJ, Bumpus JA, Jurek MA, Aust SD (1988) Biodegradation of pentachlorophenol by the white-rot fungus *Phanerochaete chrysosporium*. Appl Environ Microbiol 54:2885–2889

Nagata Y, Miyauchi K, Takagi M (1999) Complete analysis of genes and enzymes for γ-hexachlorocyclohexane degradation in *Sphingomonas paucimobilis* UT26. J Ind Microbiol Biotech 23:380–390

Neidleman SL, Geigert J (1986) Biohalogenation. Principles, basic roles and applications. Ellis Horwood, Chichester

Ohtsubo Y, Miyauchi K, Kanda K, Hatta T, Kiyohara H, Senda T, Nagata M, Mitsui Y, Takagi M (1999) PcpA, which is involved in the degradation of pentachlorophenol in *Sphingomonas chlorophenolica* ATCC 39723, is a novel type of ring-cleavage dioxygenase. FEBS Lett 459:395–398

Pointing SB (2001) Feasibility of bioremediation by white-rot fungi. Appl Microbiol Biotechnol 57:20–33

Pulliam Holoman TR, Elberson MA, Cutter LA, May HD, Sowers KR (1998). Characterization of a defined 2,3,5,6-tetrachlorobiphenyl-ortho-dechlorinating microbial community by comparative sequence analysis of genes coding for 16S rRNA. Appl Environ Microbiol 64:3359–3367

Ryoo D, Shim H, Canada K, Barbieri P, Wood TK (2000) Aerobic degradation of tetrachloroethylene by toluene-o-xylene monooxygenase of *Pseudomonas stutzeri* OX1. Nat Biotechnol 18:775–778

Semple KT, Reid BJ, Fermor TR (2001) Impact of composting strategies on the treatment of soils contaminated with organic pollutants. Environ Pollut 112:269–283

Semprini L, Hopkins GD, Roberts PV, McCarty PL (1992) In situ transformation of carbon tetrachloride and other halogenated compounds resulting from biostimulation under anoxic conditions. Environ Sci Technol 26:2454–2461

Spuij F, Alphenaar A, de Wit H, Lubbers R, van der Brink K (1997) Full-scale application of in situ bioremediation of PCE-contaminated soil. 4th International In Situ and On Site Bioremediation Symposium, New Orleans, LA. Battelle, Columbus, OH

Suenaga H, Nishi A, Watanabe T, Sakai M, Furukawa K (1999) Engineering a hybrid pseudomonad to acquire 3,4-dioxygenase activity for polychlorinated biphenyls. J Biosci Bioeng 87:430–435

Tiedje JM, Quensen III JF, Chee-Sanford J, Schimel JP, Boyd SA (1993) Microbial reductive dechlorination of PCBs. Biodegradation 4:231–240

Top EM, Springael D, Boon N (2002) Catabolic mobile genetic elements and their potential use in bioaugmentation of polluted soils and waters. FEMS Microbiol Ecol 42: 199–208

Valo R, Salkinoja-Salonen MS (1986) Bioreclamation of chlorophenol-contaminated soil by composting. Appl Microbiol Biotechnol 25:68–75 van Hylckama Vlieg JET, Poelarends GJ, Mars AE, Janssen DB (2000) Detoxification of reactive intermediates during microbial metabolism of halogenated compounds. Curr Opin Microbiol 3:257–262

Vyas BRM, Sasek V, Matucha M, Bubner M (1994) Degradation of 3,3′,4,4′-tetrachlorobiphenyl by selected white-rot fungi. Chemosphere 28:1127–1134

Wackett LP (1995) Bacterial co-metabolism of halogenated organic compounds. In: Young LY, Cerniglia CE (eds) Microbial transformation and degradation of toxic organic chemicals. Wiley-Liss, New York, pp 217–241

Witt ME, Klecka GM, Lutz EJ, Ei TA, Grosso NR, Chapelle FH (2002) Natural attenuation of chlorinated solvents at Area 6, Dover Air Force Base: groundwater biogeochemistry. J Contam Hydrol 57:61–80

Wu Q, Watts JEM, Sowers KR, May HD (2002) Identification of a bacterium that specifically catalyzes the reductive dechlorination of polychlorinated biphenyls with doubly flanked chlorines. Appl Environ Microbiol 68:807–812

Yadav JS, Reddy CA (1993) Mineralization of 2,4-dichlorophenoxyacetic acid (2,4-D) and mixtures of 2,4-D and 2,4,5-trichlorophenoxyacetic acid by *Phanerochaete chrysosporium*. Appl Environ Microbiol 59:2904–2908

7 Biodegradation of N-Containing Xenobiotics

Jing Ye,[1] Ajay Singh,[3] and Owen P. Ward[2]

1 Introduction

Nitrobenzene, nitrotoluenes, nitrophenols, nitrobenzoates and nitrate esters constitute a major class of widely distributed environmental contaminants. Theses nitrogen-containing organic compounds are frequently used as pesticides, explosives, dyes, and in the manufacture of polymers and pharmaceuticals and their industrial manufacture and application have generated a serious disposal problem. These chemicals, their by-products and metabolites can be highly toxic, mutagenic and carcinogenic, thereby threatening the environment and human health (Yinon 1990; Gong et al. 2001). Microorganisms play an important role in transforming these contaminants through a wide diversity of genes and evolved mechanisms to degrade these synthetic organic structures (Spain 1995; Freeman and Sutherland 1998; Hawari 2000; Ralebitso et al. 2002). Genomics research has assisted microbiologists to further understand the role of microorganisms in biodegradation and in the development of new bioremediation solutions (Wackett and Hershberger 2001; Wackett et al. 2002). However, combinations of nitroaromatic compounds present in the contaminated environments complicate the bioremediation efforts.

Research on the biodegradation of nitroaromatic compounds, undertaken over the past few years, has yielded a wealth of information on the microbiological and molecular aspects of the process. This has led to the discovery of new metabolic pathways and to the identification and characterization of some of the genes and enzymes responsible for the

[1] Department of Biology, University of Waterloo, Waterloo, Ontario, Canada N2L 3G1
[2] Department of Biology, University of Waterloo, Waterloo, Ontario, Canada N2L 3G1, e-mail: opward@sciborg.uwaterloo.ca, Tel.: +1-519-8884567 ext. 2427, Fax: +1-519-7460614
[3] Petrozyme Technologies Inc., 7496 Wellington Road 34, R.R.#3, Guelph, Ontario N1H 6H9, Canada

Soil Biology, Volume 2
Biodegradation and Bioremediation
(ed. by. A. Singh and O. P. Ward)

key transformation reactions (Wackett and Hershberger 2001; Ralebitso et al. 2002). This new knowledge and advances in pathway engineering have helped further understanding of the nature of nitroaromatics biodegradation. This chapter provides an overview of recent research on the biodegradation of nitrogen-containing xenobiotics.

2 Overview of Key N-Containing Xenobiotics

The key N-containing xenobiotic species of interest in bioremediation are the nitroaromatics, the nitrate esters and compounds containing nitrogen-ring heterocycles.

Nitroaromatic Compounds. Due to higher solubility and xenobiotic nature, most of the nitroaromatic compounds are on the United States Environmental Protection Agency (EPA) priority pollutant list (EPA 2003). Simpler nitroaromatic compounds such as nitrobenzene, nitrophenol, nitrotoluene and nitrobenzoate, their hazardous metabolic intermediates or dead-end products, are major environmental contaminants, and their fate is of serious concern. Understanding the metabolism of these compounds provides insight into the mechanisms for biodegradation of more complex nitroaromatics (Zylstra et al. 2000). Nitrobenzene is the simplest nitroaromatic compound, which is mainly used for the manufacture of aniline and as a starting material for the synthesis of some dyes, drugs, pesticides, and synthetic rubber and lubricating oil. Nitrotoluenes, 2- and 4-nitrotoluenes, 2,4- and 2,6-dinitrotoluenes (DNT) are precursors of 2,4,6-trinitrotoluene (TNT) and are important by-products of the explosive producing industry. 2,4- and 2,6-DNT are listed as EPA priority pollutants (EPA 2003). Nitrophenols are among the most versatile industrial organic compounds with applications as ingredients in pesticides, pharmaceuticals, pigments, dyes, and rubber chemicals. 2,4,6-Trinitrophenol (TNP, picric acid), an explosive and major by-product of large-scale nitration of benzene, is a common anthropogenic compound and environmental contaminant (Russ et al. 2000).

Nitrate Esters. Most commonly used nitrate esters in the ammunition and pharmaceutical industries include glycerol trinitrate (GTN), pentaerythritol tetranitrate (PETN), nitrocellulose (NC), isosorbide dinitrate (ISDN), ethylene glycol dinitrate (EGDN) and the nitrate ester chemical treatment products, epoxides glycidol and glycidyl nitrate (White and Snape 1993). Their metabolites are soluble at high concentrations and exhibit high toxicity.

Nitrogen-Ring Heterocycles. The *s*-triazines constitute a group of heterocyclic compounds, characterized by a symmetrical hexameric nitrogen-containing ring (Wackett and Hershberger 2001). These compounds are xenobiotics and exhibit toxicity, mutagenicity and carcinogenicity. Atrazine (2-chloro-4(ethylamino)-6-(isopropylamino)-*s*-triazine) is a widely employed *s*-triazine herbicide. RDX (hexahydro-1,3,5-trinitro-1, 3,5-triazine) is one of the most widely used explosive compounds because of its relative stability and great explosive power. RDX is commonly used in detonators, primers, mines, rocket boosters and plastic explosive applications (Yinon 1990). Since it does not adsorb strongly to soil (Sheremata et al. 2001; Price et al. 1998; Pennington and Brannon 2002), it can migrate into the groundwater and offsite.

The structures of the most important N-containing xenobiotics are provided in Fig. 1.

3 Nitroaromatics

Nitroaromatics are degraded by various pathways using different biochemical mechanisms; the specific functional nitro-group in these compounds plays a key role. The aromatic π electron nucleophilic mechanism with the additional nitro (—NO_2) electron-withdrawing property protects nitroaromatics from initial attack by oxygenases, but is favourable for reductive attack (Rieger and Knackmuss 1995). On the other hand, anaerobic reductive attack produces the nitroamine (—NH_2), an electron-donating group which represents a barrier to further attack by anaerobes. Thus, nitroaromatics often either persist or become amine end products in the environment.

The electron-donating amino-groups resulting from nitroaromatic reductive attack by aerobic bacteria activate the aromatic nucleus for further electrophilic attack by a dioxygenase. Since the pathway involving reduction of the nitro-group to the amine by aerobic bacteria appeared to be an ineffective degradation system until recently, there was little evidence showing complete reduction of nitroaromatics by a single aerobe via the amine prior to ring cleavage (Nishino et al. 2000a). However, it has been demonstrated that a *Pseudomonas* sp. JX165 strain mineralizes nitrobenzene via both 2-aminophenol and aniline, followed by oxidative catabolism under fully aerobic conditions (Wang et al. 2001). Johnson and Spain (2003) used molecular genetic analysis to show how catabolic pathways for 2,4-dinitrotoluene and nitrobenzene evolved in response to the introduction of these synthetic chemicals into the biosphere.

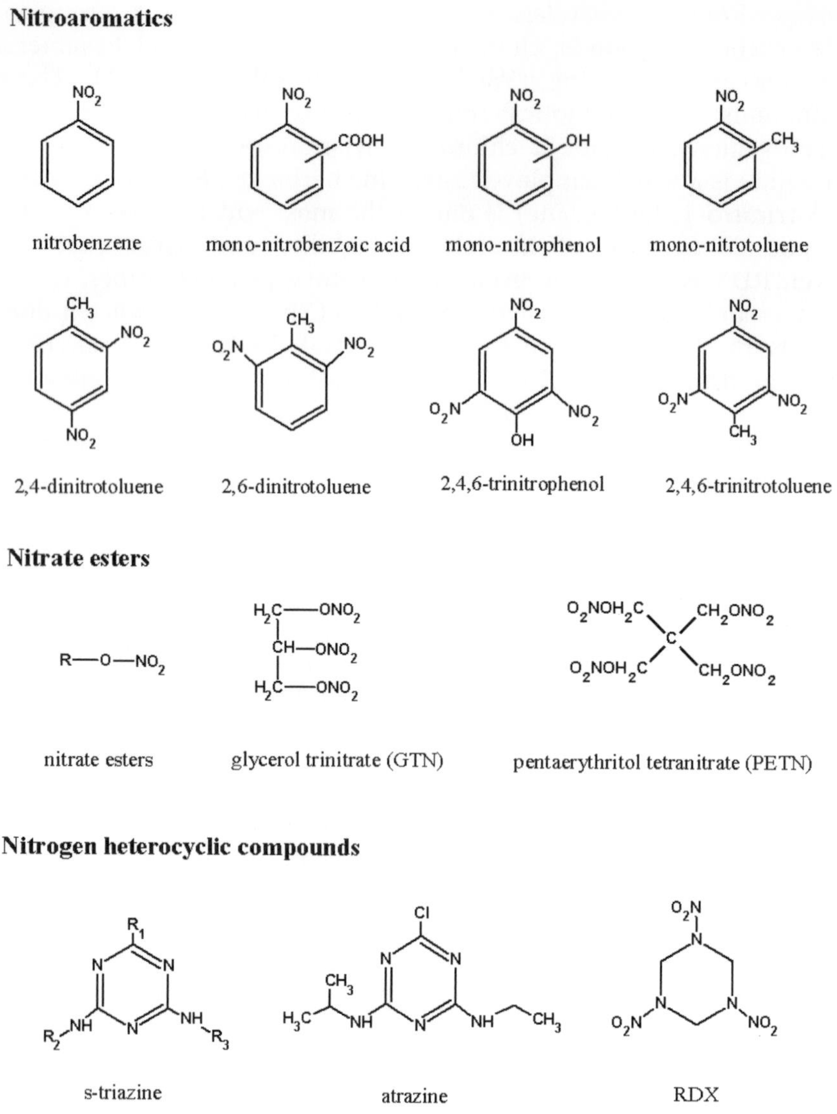

Nitroaromatics

nitrobenzene | mono-nitrobenzoic acid | mono-nitrophenol | mono-nitrotoluene

2,4-dinitrotoluene | 2,6-dinitrotoluene | 2,4,6-trinitrophenol | 2,4,6-trinitrotoluene

Nitrate esters

nitrate esters | glycerol trinitrate (GTN) | pentaerythritol tetranitrate (PETN)

Nitrogen heterocyclic compounds

s-triazine | atrazine | RDX

Fig. 1. Structure of different nitrogen-containing xenobiotics

Bacteria appear to use four main strategies to address the nitro-group under aerobic conditions (Nishino et al. 2000a; Ye et al. 2004):

1. Dioxygenation of the nitroaromatic ring with release of the nitro-group as nitrite and production of dihydroxy intermediates;
2. Monoxygenation to epoxides;

3. Formation of a hydride-Meisenheimer complex; and
4. Release of two molecules of ammonia, the first after partial reduction of the nitro-group and formation of hydroxylaminobenzene derivatives, and the second following rearrangement of the hydroxylaminobenzene to the corresponding catechol.

Nitrobenzene. Aerobic degradation of nitrobenzene (NB) is mediated by either the well-characterized dioxygenase pathway or the ubiquitous partial reductive pathway (Nishino et al. 2000a; Fig. 2). In the partial reductive pathway, as illustrated in *Pseudomonas pseudoalcaligenes* JS45 (Nishino and Spain 1993), NB is reduced through nitrosobenzene, with no aniline production, to hydroxylaminobenzene, which is mainly rearranged to aminophenol by a mutase. Nadeau et al. (2003) described three types of observed rearrangement mechanisms of hydroxylamino aromatic compounds. A dioxygenase-mediated *meta*-ring cleavage then occurs to produce 2-aminomuconic semialdehyde with subsequent release of ammonia. The initial reductive enzyme, an NB-inducible and oxygen-insensitive nitroreductase consists of a flavoprotein with a tightly bound FMN cofactor (Somerville et al. 1995). An intramolecular transfer reaction rearranges hydroxylaminobenzene to 2-aminophenol mediated by hydroxylaminobenzene mutase and the 2-aminophenol is converted to muconic semialdehyde in a dioxygenase-mediated step followed by deamination to 4-oxalocrotonic acid (He and Spain 1997; He et al. 2000). Zhao et al. (2001) observed a similar partial reductive metabolism in 3-nitrophenol-grown *Pseudomonas putida* 2NP8 strain where degradation of nitrobenzene occurred through hydroxylaminobenzene followed by the release of ammonia.

Dioxygenase-mediated reactions appear to be widely exploited by microbes for degradation of nitroaromatics and the biocatalytic mechanism has been extensively investigated to determine variations in enzymatic properties among different species and strains. Dioxygenases have broad substrate specificity and they attack the 1,2 positions instead of the 2,3 positions to yield a nitrohydrodiol which is spontaneously decomposed to catechol with the liberation of nitrite. Nishino and Spain (1995) observed that 154 strains out of 155 contained the above-mentioned partial reductive pathway, while one *Comamonas* sp. JS765 appeared to be able to mineralize NB via a dioxgenation pathway. The genes of some of the enzymes have been characterized. The dioxygenase enzyme system of *Comamonas* sp. strain JS765 has four components including reductase$_{NBZ}$, fererredoxin$_{NBZ}$, oxygenase$_{NBZ\alpha}$ and oxygenase$_{NBZ\beta}$, with the gene designations *nbzAa*, *nbzAb*, *nbzAc and nbzAd.*

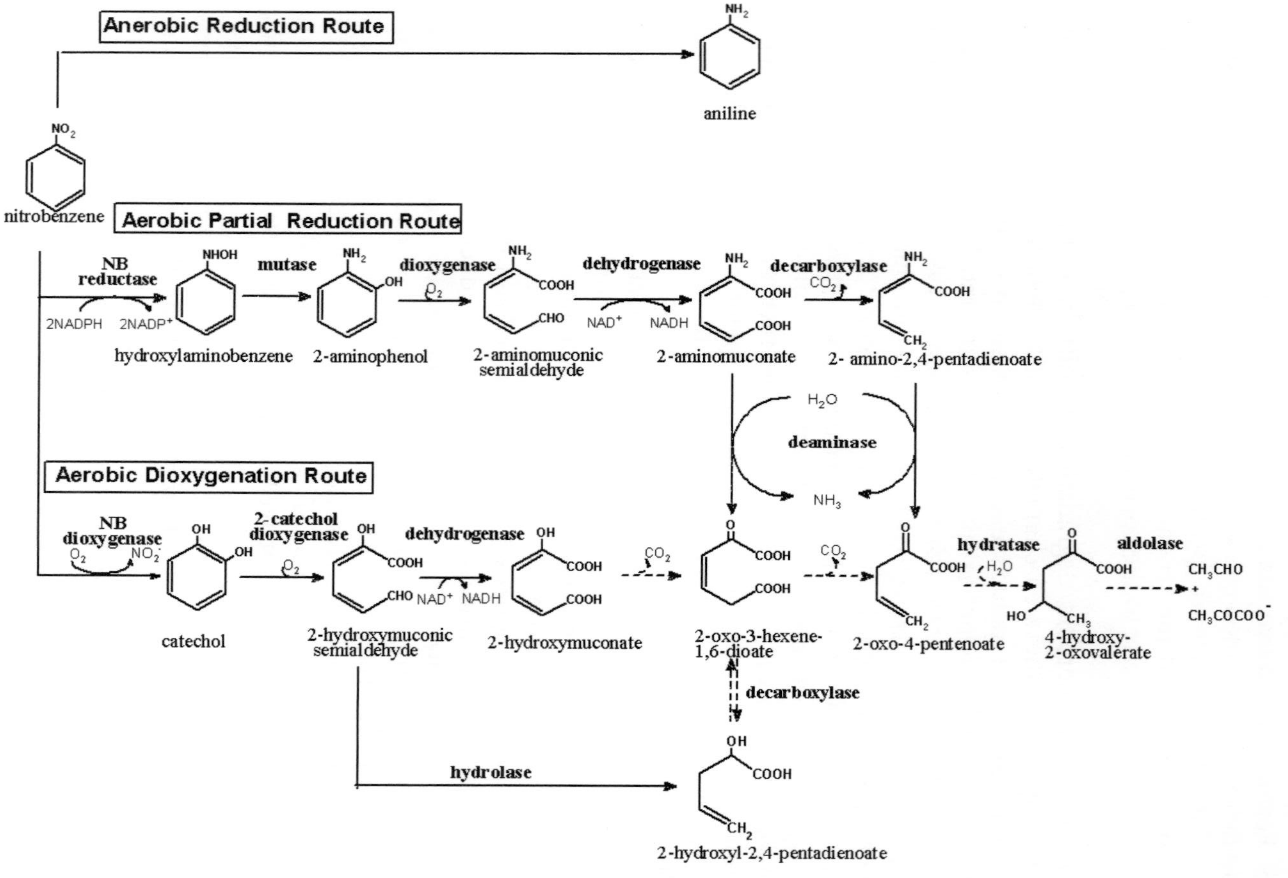

Fig. 2. Nitrobenzene metabolism. (Ye et al. 2004, with kind permission of Kluwer Academic Publishers)

Nitobenzene dioxygenase is inducible and exhibits high homology with the 2-nitrotoluene dioxygenase from *Pseudomonas* sp. strain JS42 (Lessner et al. 2002). While *Pseudomonas pseudoalcaligenes* JS45 (partial reductive pathway), *Comamonas* sp. JS765 (initial dioxygenase attack) and *Pseudomonas* sp. AP-3 (alternative metabolism of 2-aminophenol) exhibit different characteristics with respect to upstream metabolism, all three strains utilized the same *meta*-ring cleavage in the downstream pathways, suggesting the biodegradation activities of the three strains share a common evolutionary origin (He and Spain 1999). Genetic evidence of high similarity between the 2-aminophenol-1,6-diooxygenase from *P. pseudoalcaligenes* JS45 and the ring fission dioxygenase from *Pseudomonas* sp. AP-3 appears to support this latter conclusion.

For bioremediation application, nitrobenzene-degrading bacteria can carry out non-specific nitrobenzene reduction to the more soluble but less toxic aniline, which can be relatively easily mineralized through sequential anaerobic-aerobic treatment (Dickel et al. 1993). Use of a reductive consortium in combination with oxidative *Comamonas acidovorans* strain, supplied the reductive and oxidative activities to mineralize nitrobenzene (Peres et al. 1998). In a combined electron beam–biological treatment, nitrobenzene was effectively transformed to nitrophenols, which were further degraded using a nitrophenol-degrading mixed culture (Zhao and Ward 2001). In a sequential zero valent iron and aerobic biotreatment, nitrobenzene was transformed in the presence of granular iron to aniline, which was readily biodegraded.

Nitrobenzoates. Nitrobenzoates are used as substrate for biocatalytic transformations and are intermediates in the degradation of nitrotoluene. Like most nitroaromatic compounds, nitrobenzoates can be transformed to aminoaromatics under anaerobic conditions (Razo-Flores et al. 1999). Reductive biodegradation of 2-nitrobenzoate under aerobic conditions, with the release of the nitro-group as ammonia, has been described. This oxygen insensitive reduction route in *Arthrobacter protophormiae* RKJ100 converts 2-nitrobenzoate to 2-hydroxylaminaobenzoate and ammonia under aerobic conditions (Chauhan et al. 2000). *Pseudomonas fluorescens* KU-7 had a similar reductive pathway (Hasegawa et al. 2000). 2-Nitrobenzoate is converted to 2-hydroxylaminobenzoate by an NADPH-dependent reductase, which is then rearranged and converted to 3-hydroxyanthranilate followed by ring cleavage with no anthranilate detected.

The mechanism for degradation of 4-nitrobenzoate was similar to that of 2-nitrobenzoate, where the initial nitroreductase-mediated attack transforms 4-nitrobenzoate to 4-hydroxylaminobenzoate followed by the release of ammonia and formation of protocatechuate. This metabolic pattern has been found in *Comamonas acidovorans* NBA-10 (Groenewegen et al. 1992) and in some *Pseudomonas* species (Haigler and Spain 1993; Yabannavar and Zylstra 1995; Hughes and Williams 2001). *Burkholderia cepacia* PB4 and *Ralstonia paucula* SB4, two 4-aminobenzoate-grown strains, partially degraded 4-nitrobenzoate to protocatechuate and two dead-end products, the inducer 4-aminobenzoate was not found as an intermediate in the pathway. However, the strains containing an enzyme system capable of degrading nitroaromatic and aminoaromatic mixtures simultaneously may have potential practical applications in bioremediation of nitroaromatics-contaminated sites (Peres et al. 2001).

Biodegradation of 3-nitrobenzoate has been reported in different bacteria. While 3-nitrobenzoate is initially degraded by dioxygenase attack in *Pseudomonas* strain JS51 and *Comamonas* strain JS46 (Nadeau and Spain 1995), it was found to be mediated by two monooxygenation steps producing 3-hydrobenzoate leading finally to protocatechuate in *Nocardia* (Cartwright and Cain 1958). In both cases, formation of nitrite demonstrated oxidative biodegradation.

Nitrotoluenes. Nitrotoluenes are sometimes subject to non-specific nitro-reduction to form amino derivatives which could be the predominant degradative reaction prior to aromatic ring cleavage. 2-Nitrotoluene is degraded by *Pseudomonas* strain JS42 by dioxygenation via 3-methylcatechol with the release of nitrite leading to ring cleavage (Parales et al. 1998). The enzyme, nitrotoluene-2,3-dioxygenase, was characterized as having broad specificity towards aromatic compounds and is capable of catalyzing other dioxygenation and monooxygenation reactions. The system has been cloned and sequenced and appears to be a three-component dioxygenase similar to 2,4-DNT-dioxygenase with a different substrate profile. *Pseudomonas* strains TW3 and 4NT degrade 4-nitrotoluene via 4-nitrobenzoate followed by the partial reductive pathway. The initial oxidative reaction of the methyl group, which transforms nitotoluenes into more oxidized nitroaromatics, is believed to be mediated by the TOL plasmid. In one of the latter two strains, this oxidation reaction was NAD^+-dependent, whereas in the other it was not. Toluene and 4-nitrotoluene appear to induce the same genes in strain TW3 whose products were responsible for converting the hydrocarbon side chains to carboxylic acids (James and Williams 1998).

A *Psuedomonas* sp. mineralized 2,4-DNT with a nitrite removal pathway involving dioxygenase and monooxygenase enzymes (Suen and Spain 1993). *Burkholderia cepacia* strain JS850 and

Hydrogenophaga paleronii strain JS863 mineralized 2,4-DNT in the same way, but degraded 2,6-DNT in a different way (Nishino et al. 2000b). When a combination of 2,4- and 2,6-DNT were used as sole source of carbon and nitrogen, dioxygenation of the 2,6-DNT to 3-methyl-4-nitrocatechol (MNC) was the initial reaction, accompanied by the release of nitrite. MNC was then transformed by *meta*-ring cleavage. Although 2,4-DNT-degrading strains could also convert 2,6-DNT to MNC, further catabolism was halted at that point. The pathway for 2,4-DNT degradation was different from that for 2,6-DNT degradation. The gene encoding this dioxygenase showed a nucleotide sequence similar to the a-subunit among nitroarene dioxygenases.

The genes for the initial dioxygenases involved in 2,4-DNT and 2,6-DNT degradation were found to be closely related, but the enzymes are produced at low constitutive levels (Spain 1995; Nishino et al. 2000a). After initial dioxygenation, the two pathways appear to diverge.

Lendenmann and Spain (1998) demonstrated simultaneous degradation of 2,4-DNT and 2,6-DNT in an aerobic biofilm with mineralization rates of 98 and 94%, for 2,4- and 2,6-DNT, respectively. The release of nitrogen as nitrate indicated oxidative bacterial activity.

Although bacteria capable to degrading nitrotoluenes are widely distributed at contaminated sites, contaminant removal rates are often quite low. This raises the question as to how bioremediation might be improved and as to whether biostimulation approaches might be effective (Nishino et al. 1999). A mixed culture, maintained on crude oil and not previously exposed to TNT exhibited high TNT-transforming activity in both growing and resting cells (Popesku et al. 2003). Molasses augmentation to TNT-contaminated soil and water in constructed wetlands was superior to wood chips in promoting accelerated removal of TNT (Gerth et al. 2003). In most current reports, the reductive mechanism predominates in TNT degradation. The three nitro-groups with a nucleophilic aromatic ring structure make TNT vulnerable to reductive attack, but resistant to oxygenase attack from aerobic organisms (Lenke et al. 2000). Reduction of a nitro-substituent of TNT can lead to formation of aminodinitrotoluene (ADNT) isomers. However, Johnson et al. (2001) have shown that these ADNT isomers may be transformed by nitroarene dioxygenases, opening the possibility for extensive microbial TNT degradation. The genetics of nitroarene gene expression have been investigated in *Comomonas* and *Acidovorax* strains (Lessner et al. 2003). TNT could be reduced by carbon monoxide dehydrogenase from *Clostridium thermoaceticum* (Huang et al. 2000), and by the manganese-dependent peroxidase (MnP) from the white-rot fungus *Phlebia radiata* (Van Aken et al. 1999). *Enterobacter cloacae* PB2, which contains pentaerythritol tetranitrate (PETN) reductase could grow slowly on 2,4,6-TNT under aerobic conditions as the sole nitrogen source without production of dinitrotoluene as an intermediate (French et al. 1998). This strain catalyzed conversion of the TNT via a hydride-Meisenheimer complex with the nitro group released as nitrite.

Nitrophenols. In anoxic environments, mononitrophenols are subjected to reductive degradation to the corresponding aminophenol, which can be further converted to non-aromatic products and can possibly be mineralized completely (Uberoi and Bhattacharya 1997; Razo-Flores et al. 1999).

In comparison to other nitroaromatics, mononitrophenol degradation pathways are relatively diverse among microorganisms, with the 2-nitrophenol degradation pathway being the most straightforward. In *Pseudomonas putida* B2, 2-nitrophenol is attacked by a monooxygenase, forming nitrite and catechol, with further degradation of the latter by 1,2-dioxygenase-mediated *ortho*-ring cleavage (Zeyer et al. 1985). *P. putida* B2 degradation of 3-nitrophenol involves a different reductive pathway possibly via 3-hydroxylaminophenol, and with eventual conversion to 1,2,4-benzenetriol followed by ring cleavage (Meulenberg et al. 1996). 4-Nitrophenol can be degraded in three monooxidative pathways through different intermediates as characterized in a variety of species including *Moraxella* (Spain and Gibson 1991), *Bacillus* (Kadiyala et al. 1998), *Arthrobacter* (Bhushan et al. 2000) and *Pseudomonas* (Chauhan et al. 2000; Qureshi and Purohit 2002). In *Moraxella* sp. the predominant pathway for 4-NP degradation is via hydroquinone, possibly through 4-nitrocatechol. This pathway appears to be more prevalent in gram-negative bacteria. Other degradative pathways involve hydroxylation at either the 2- or 3-ring position by monooxygenation to form 4-nitroresorcinol or 4-nitrocatechol. *Bacillus sphaericus* JS905 may have a similar metabolic pathway (Kadiyala et al. 1998) with 4-nitrocatechol as intermediate. A two-component, NADH-dependent monooxygenase, comprising a flavoprotein reductase and oxygenase, appeared to be involved.

Hydroxylation of 2-nitrophenol and 4-nitrophenol at the nitro-group with release of nitrite leads to formation of an *ortho-* or *para*-dihydroxybenzene. In *Arthrobacter protophormiae* strain RKJ100 4-NP is converted through hydroquinone and 4-nitrocatechol via the 1,2,4-benzenetriol and 2-hydroxy-1,4-benzoquinone pathway (Chauhan and Jain 2000). The aromatic flavoprotein monooxygenase pentachlorophenol-4-monooxygenase of *Sphingomonas* spp UG30 mediates the degradation of 4-nitrophenol via 4-nitrocatechol to 1,2,4-benzenetriol (Leung et al. 1999).

Zhao and Ward (1999) described a mixed culture, enriched on 3-mononitrophenol, that degraded both nitrophenol and nitrobenzene. The mixed culture contained two isolates identified as *Comamonas testosteroni* and *Acidovorax delafieldii.* 3-Nitrophenol-induced cells of *Pseudomonas putida* 2NP8 degraded a wide range of nitroaromatic substrates, with rapid transformation of most mono- and di-nitroaromatics and with stoichiometric release of ammonia.

Under aerobic conditions, the initial oxygenation of 2,4,6-trinitrophenol usually occurs with the mono- and to some extent di-nitrophenol. As the number of nitro-groups increases, the reductive reaction becomes the dominant initial mechanism (Rieger and Knackmuss 1995). Mineralization of TNP by *Nocardioides* sp. strain CB22-2 (Behrend and Heesche-Wagner 1999) and *Rhodococcus erythropolis* (Rieger et al. 1999) involves a hydride-Meisenheimer complex of TNP and 2,4-dinitrophenol and to some extent 2,6-dinitrophenol. The Meisenheimer complex is recognized as a key intermediate of denitration in microbial degradation of TNP. The genes encoding the NADPH-dependent F_{420}(subscript) reductase and the hydride transferase II responsible for converting the nitrophenols to the Meisenheimer complexes were amplified from 2,4-DNP-degrading *Rhodococcus* species and were shown through heterologous expression to be involved in 2,4-DNP degradation (Heiss et al. 2003).

4 Nitrate Esters

Both regio- and non-regioselective removal of nitrogen from nitrate esters of glycerol has been observed. The nitrogen moieties have been found to be released as nitrite, nitrate or nitric oxide, depending on the microbial strain involved. As would be expected, cellulolytic organisms participate in the biodegradation of nitrocelluloses. Nitrogen removal from pentaerythritol tetranitrate is mediated by reductive cleavage of the nitrate esters with formation of nitrate or nitrite.

Glycerol Trinitrate. Glycerol trinitrate (GTN) is generally inhibitory to bacterial growth and information on its biodegradation is scarce in the literature (White and Snape 1993). Nitrogen removal from GTN to glycerol dinitrate (GDN) and from GDN to glycerol mononitrate (GMN) is regioselective. Although random attack of the three ester bonds of GTN would be expected to yield a ratio for 1,3-GDN:1,2-GDN of 1:2, a high ratio (2:1) was found in the *Geotrichum candidum*-mediated transformation (Ducrocq et al. 1989) and even higher in *Pseudomonas* sp. R1-NG1(White et al. 1996a). GTN conversion to GDN appears to be

C-2 regioselective since 1,3-GDN was the major isomeric product in both cases. In contrast in the formation of GMN, there appears to be a lack of significant regioselectivity in the *G. candidum* transformation (Ducrocq et al. 1989). In the further conversion from GDN to GMN by *Pseudomonas* R1-NG1, there appeared to be a regiospecific preference for C-1 but not for C-2 (White et al. 1996a).

Although sequential denitrations of GTN through GDN, GMN and glycerol have been demonstrated, the mechanism for enzymatic removal of nitrogen from nitrate esters has yet to be definitively established (Fig. 3). The nitro-group from GTN is transformed into nitrite (NO_2^-) and not nitrate (NO_3^-) by *Pseudomonas* R1-NG1 (White et al. 1996a). *Phanerochaete chrysosporium* converts GTN and its derivatives GDN and GMN concurrently with the formation of nitric oxide (NO) and nitrite (NO_2^-) (Servent et al. 1991), while *Agrobacterium radiobacter* removal of GTN and PETN appeared to be mediated by an NADH-

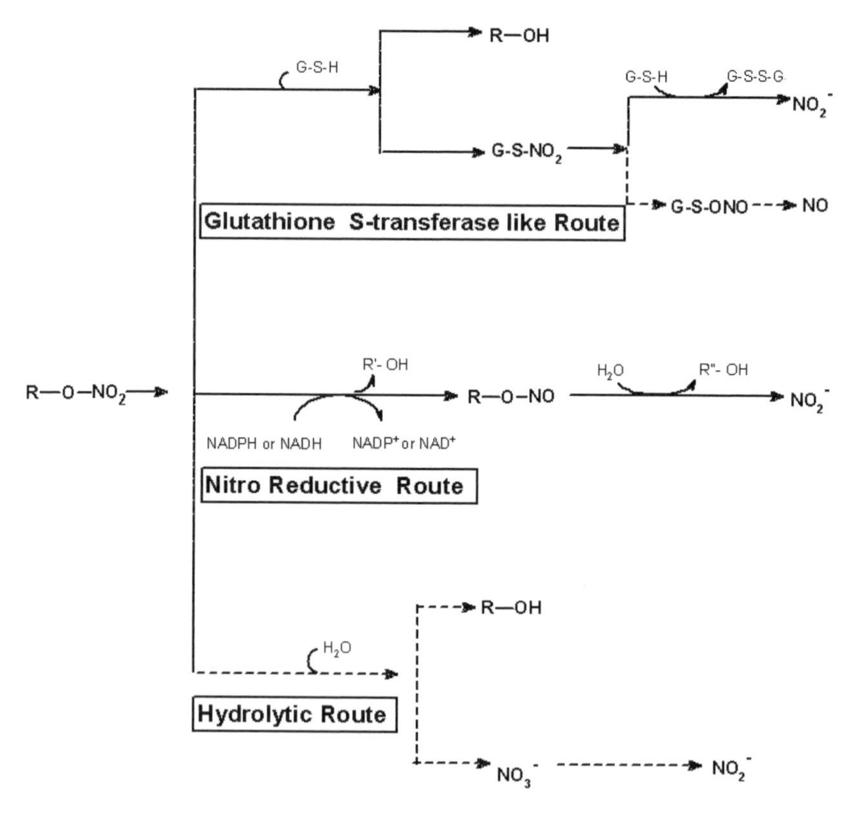

Fig. 3. Enzymatic removal of nitrate esters. (Ye et al. 2004, with kind permission of Kluwer Academic Publishers)

dependent nitroreductase, with release of nitrogen as nitrite (White et al. 1996b). In cell extracts of *Bacillus thuringiensis/cereus* and *Enterobacter agglomerans*, GTN conversion involved production of nitrate and possibly followed by reduction of nitrate to nitrite (Meng et al. 1995). No cofactor was required in this transformation. Sequential denitrations have also been observed in aerobic and anaerobic mixed culture systems (Bhaumik et al. 1997; Accashian et al. 1998).

The GTN-contaminant soil isolates, *P. putida*, *Arthrobacter* sp., *Rhodococcus* sp and *Klebsiella* sp. were able to degrade GTN sequentially, with the latter *Klebsiella* sp. alone exhibiting regioselectivity for the C-2 position of GTN. The *Rhodococcus* sp. converted GMN to glycerol and thus is the first reported strain with the capacity to mineralize GTN (Marshall and White 2001).

Nitrocellulose. Nitrocellulose is another nitrate ester which is generally non-toxic to humans, but its explosive property and air-polluting effect makes it hazardous (White and Snape 1993). Nitrocellulose is generally very resistant to microbial degradation. Sharma et al. (1995) has demonstrated aerobic degradation by a combination of cellulolytic and denitrifying fungi and other researchers have shown that degradation can occur under anaerobic conditions (Duran et al. 1995, Freeman et al. 1996).

Pentaerythritol Tetranitrate. Pentaerythritol tetranitrate (PETN), another nitrated compound incorporated into explosives, is generally recalcitrant to microbial attack. Most of the recent research has focused on PETN reductase. *Enterobacter cloacae* PB2, isolated from explosive-contaminated soil, aerobically transforms PETN by removal of two of the four nitrogen atoms converting them to nitrites (Binks et al. 1996). The PETN reductase from this strain was an NADPH-dependent flavoprotein containing non-covalently bound aerobically FMN.

Several similar non-covalently linked FMN-dependent reductases, capable of catalyzing the NADPH-dependent reductive cleavage of nitrate esters to form alcohol and nitrate, have been cloned and sequenced (Williams and Bruce 2000). *XenA* and *xen*B genes encoding nitroester reductases from *Pseudomonas putida* IIB and *Pseudomonas fluorescens* I-C were characterized (Blehert et al. 1997, 1999). These enzymes exhibit biochemical properties and sequences similar to old yellow enzyme (OYE), an FMN-containing flavoprotein from several yeasts (Williams and Bruce 2002). This family of xenobiotic reductases contains seven homologous bacterial flavoproteins, which can reduce a variety of nitroaromatic compounds including TNT and picric acid (French et al. 1998; Blehert et al. 1999). Khan et al. (2002) have recently established a PETN reductase model with mechanistic similarities with OYE and demonstrated a relationship and clear difference in reactivity between OYE and PETN reductase using kinetic and structural data.

The further exploration and characterization of these interesting nitrate ester-transforming enzymes, perhaps combined with molecular protein engineering strategies, will surely lead to the emergence of new biocatalysts for bioremediation of N-containing xenobiotics.

5 Compounds Containing Nitrogen-Heterocyclic Ring Structures

Atrazine can be biodegraded by a series of biochemical processes, including N-dealkylation, dechlorination and ring cleavage. Little is known regarding biodegradation of RDX (hexahydro-1, 3,5-trinitro-1,3,5-triazine). Its three weak —N—N— bonds make it vulnerable to enzymatic attack at the nitro-groups, potentially leading to destabilization of the inner C—N bonds, causing rapid ring cleavage (Hawari et al. 2000a, b; Bhushan et al. 2002a). However, the nitro groups are often not actively metabolized and hence RDX can persist over long periods of time in the environment (Gong et al. 2001).

Atrazine. Several bacterial strains have been found which can utilize *s*-triazine as nitrogen source, in most cases through oxidative N-dealkylation of the side chain rather than through ring cleavage. Although side chains appear accessible to microbial attack, ring substituents can impede the transformation and less heavily substituted and nonchlorinated *s*-triazines were found to be more biodegradable (Cook et al. 1985). The mineralization of *s*-triazines is generally very slow. Dead-end metabolites with the *s*-triazine ring structure may still remain after transformation by *Rhodococcus* (Behki et al. 1993; Nagy et al. 1995), *Phanerochaete chrysosporium* (Mougin et al. 1994), and *Nocardioides* (Topp et al. 2000). Very little is known regarding the mechanism of *s*-triazine ring degradation. Cyanuric acid, having the simplest triazine structure, was formed during transformation of atrazine by *Pseudomonas* (Karns 1999) and *Arthrobacter aurescens* TC1 (Strong et al. 2002).

Klebsiella pneumoniae and *Pseudomonas* sp. strains A and D used atrazine as nitrogen source through successive deamination to cyanuric acid followed by apparent hydrolytic ring cleavage (Cook et al. 1985; Karns and Eaton 1997; Karns 1999). The *s*-triazine degradation plasmid pPDL12 from *Klebsiella pneumoniae* 99 could be an agent for the dissemination of the *s*-triazine degradation gene(s) between bacteria (Karns and Eaton 1997). Karns (1999) described the gene sequence (*trz*D) encoding the *s*-triazine ring cleavage enzyme, cyanuric acid amidohydrolase from *Pseudomonas* sp. strain NRRLB-12227.

An atrazine-degrading *Pseudomonas* sp. ADP, isolated by mixed culture enrichment, degraded the atrazine ring at a high rate with release of $^{14}CO_2$ and has become a reference strain for atrazine degradation (Mandelbaum et al. 1995). The plasmid pADP-1 from this strain containing genes *atzABC*, responsible for *s*-triazine degradation prior to the formation of cyanuric acid, has been completely sequenced to elucidate the catabolic pathway involved in atrazine degradation (Wackett et al. 2002; Ralebitso et al. 2002). Genes *atzABC*, have been cloned and characterized from different genera (Shao and Behki 1995; De Souza et al. 1998; Sadowsky and Wackett 2000; Martinez et al. 2001; Clausen et al. 2002; Rousseaux et al. 2002; Shapir et al. 2002). The genes *atzA* from *Pseudomonas* sp. ADP and *triA* from NRRLB-12227 were found to be structurally similar (98% similarity), but functionally different (Seffernick et al. 2001), where *atzA* is responsible for dehalogenation and *triA* for deamination (Fig. 4). The deaminohydrolase in ADP strain also differed from *s*-triazine hydrolyase in *Nocardioides* sp. (Topp et al. 2000). Recently, Piutti et al. (2003) have genetically characterized the atrazine degradation pathway in *Nocardioides* sp. SP12. In general, it appears that the atrazine catabolic pathways may be a result of recruitment of different genes in microbial communities rather than a genetic element encoding the entire pathway (Rousseaux et al. 2002).

RDX. Among different anaerobic conditions examined, a sulfate-reducing consortium containing *Desulfovibrio* utilized RDX as sole nitrogen source and eliminated it within 12 days with concurrent release of NH_3 (Boopathy et al. 1998). The biotransformation of RDX was inhibited under nitrate-reducing conditions (Freeman and Sutherland 1998). RDX was rapidly reduced to nitroso metabolites, only after the nitrate was depleted by excess electron donors as in the case of ethanol. A Fe^0 filing system-amended municipal anaerobic sludge generated H_2 as electron donor to enhance the reduction of RDX (Oh et al. 2001). Methanogenic cultures augmented with hydrogen or electron donors that result in hydrogen production enhanced anaerobic degradation of RDX and other nitrogen-containing explosives (Adrian et al. 2003). Pudge et al. (2003) investigated the anaerobic transformation of RDX by *Enterobacter cloacae* ATCC 43560 in a two-phase partitioning bioreactor.

The formation of methylenedinitramine as a key RDX anaerobic ring cleavage metabolite was first demonstrated by Halasz et al. (2002). A

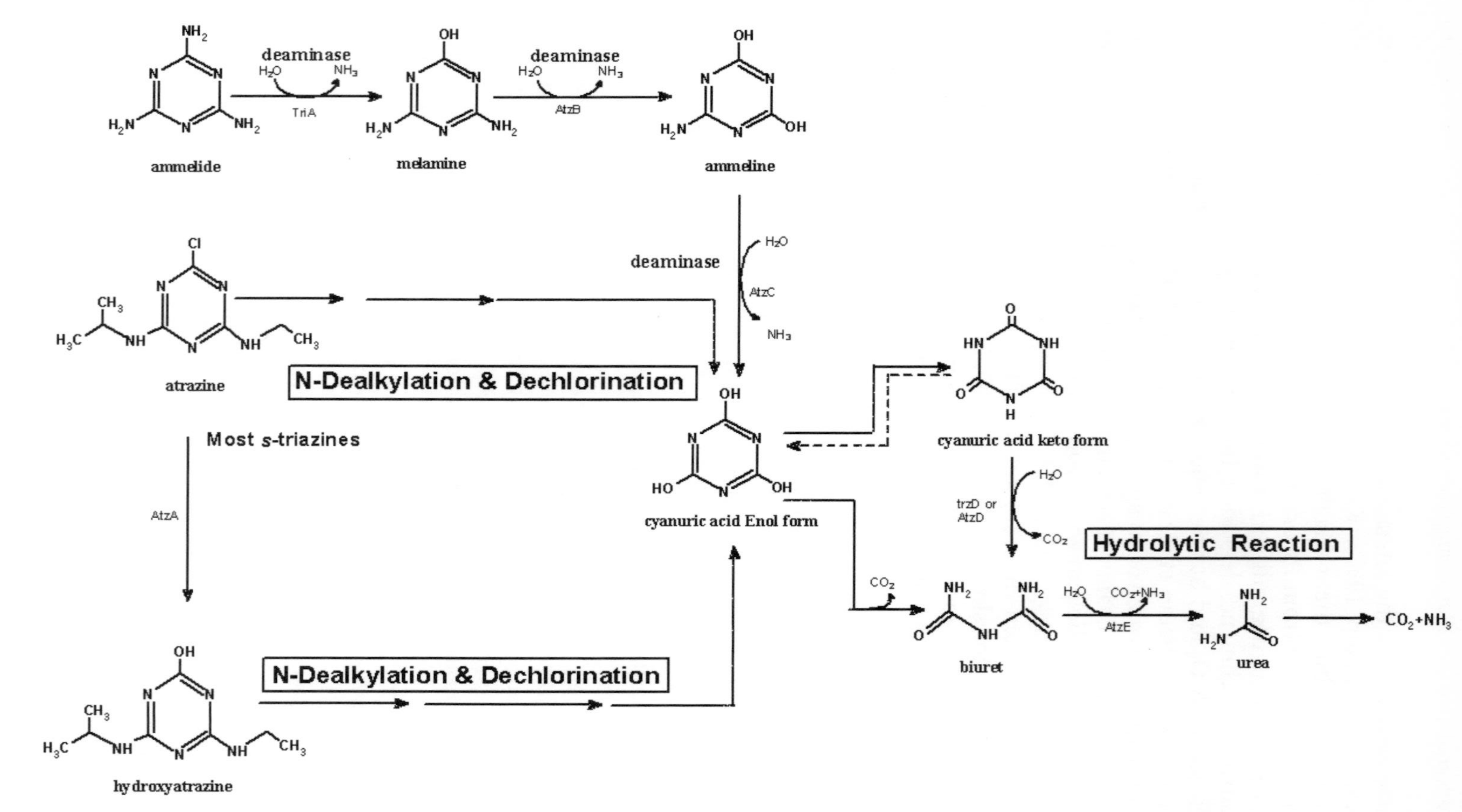

Fig. 4. *s*-Triazine metabolism. (Ye et al. 2004, with kind permission of Kluwer Academic Publishers)

novel route for hydroxylation of the RDX C—N bond to produce methylenedinitramine and bis(hydroxymethyl) as two ring cleavage intermediates was suggested. A facultative anaerobic bacterium *Klebsiella pneumoniae* strain SCZ-1 completely degraded RDX and transformed RDX and hexahydro-1-nitroso-3,5-dinitro-1,3,5-triazine (MNX) at very similar rates, indicating a possible common mechanism of initial attack (Zhao et al. 2002).

Type I nitroreductase, from bacteria *Morganella morganii* B2 and *Enterobacter cloacae* 96-3, an oxygen-insensitive enzyme, appears to facilitate reduction of the nitro-group of RDX or TNT under both aerobic and anaerobic conditions (Kitts et al. 2000). Under anaerobic conditions, strain B2 also carries out an RDX-cyclic nitramine reductive reaction. Unlike nitrate/nitrite reductases, nitroreductases do not reduce inorganic nitrate or nitrite and, therefore, do not provide a source of reduced nitrogen for cell growth. In a two-electron reduction, an NADPH-dependent nitrate reductase from A*spergillus niger* reduced the RDX nitro-group and formed MNX and methylenedinitramine stoichiometrically (Bhushan et al. 2002b).

The anaerobe *Clostridium kluyveri* uses a flavoenzyme, diaphorase with NADH as electron donor in transformation of RDX (Bhushan et al. 2002a). The initial enzymatic attack was oxygen sensitive and formed RDX•– whose spontaneous denitration generates nitrite and the free radical RDX•. FMN appears to play a key role in this net two redox equivalent transferring system. Methylenedinitramine, NO_2^-, HCHO, NH_4^+, and N_2O were observed as metabolites, but nitroso intermediates were not detected.

Under aerobic conditions, *Stenotrophomonas maltophilia* PB1 can utilize RDX as sole nitrogen source for growth forming two intermediates (MW of 167 and 171), the latter apparently related to the proven anaerobic intermediate methylenedinitramine (MW 171) (Binks et al. 1995; Hawari 2000). *Rhodococcus* sp. DN22 sequentially removed two equivalents of nitrogen as nitrite from RDX to form a cyclohexenyl- and then a cyclohexadienyl intermediate (Coleman et al. 1998; Fournier et al. 2002). The formation of CO_2 and HCHO, possibly through a hydrolytic ring cleavage, followed the denitrations. Not all the RDX was mineralized and a dead-end product $C_2H_5N_3O_3$ was detected. The *Rhodococcus* sp. DN22 strain has a plasmid encoding cytochrome P-450 enzyme, which could potentially be transferred between bacteria (Coleman et al. 2002). A cytochrome P-450 2B4 from rabbit liver catalyzed transformation of RDX in a manner similar to cytochrome P-450 from *Rhodococcus* sp. DN22. Both enzymes generated the same metabolite 4-nitro-2,4-diazabutanal (Bhushan et al. 2003).

Rhodococcus rhodochrous 11Y, isolated from an explosive-contaminated site, can completely mineralize RDX via denitration and ring cleavage under aerobic conditions (Seth-Smith et al. 2002). The gene which conferred the latter RDX degradation ability was constitutively expressed as cytochrome P-450-like gene *xplA*. The role of the flavodoxin domain in XplA as a redox agent, possible use of XplB as a reductase and evidence that the cytochrome P-450 inhibitor metyrapone reduced the RDX removal rate provide evidence in support of the function of cytochrome P-450 in the aerobic degradation of RDX.

Phanerochaete chrysosporium used RDX as nitrogen source with about 60% mineralization in 60 days (Sheremata and Hawari 2000; Sheremata et al. 2001). The only intermediate detected was MNX with one nitrogen in the final product N_2O originating from the RDX ring and the other from a ring substituent, suggesting ring cleavage.

Although biodegradation of RDX has been studied under both aerobic and anaerobic conditions, the enzymatic pathway(s) still remain obscure. Correlations between biodegradation of RDX and other *s*-triazines such as atrazine are not well understood. For example, in *Rhodococcus* sp. DN22 strain, atrazine was only a moderate inducer of the P-450 enzyme, which induces RDX biodegradation (Coleman et al. 2002). *S. maltophila* PB1, which used RDX as nitrogen source, was unable to use cyanuric acid as nitrogen source, but it can grow on other *s*-triazines as the limiting nitrogen source (Binks et al. 1995). The RDX ring may be cleaved without complete denitration, whereas in atrazine dealkylation to cyanuric acid precedes ring cleavage.

6 Conclusions

Exploration of new metabolic pathways has considerably improved our understanding of the genes and enzymes responsible for key biotransformation reactions of N-containing xenobiotics. Development of strategies to allow more recalcitrant compounds to serve as growth substrates for microorganisms should be the current focus of research. Another interesting direction would be the possibility of producing value-added products from nitroaromatic wastes by developing effective biocatalysts that utilize relatively inexpensive nitroaromatic waste feedstocks, which can potentially reduce waste at source and reduce complications in bioremediation efforts. Increasing commercial interest in bioremediation and advances in our understanding of the molecular basis of biodegradation, are expected to produce more rational approaches in developing new technologies.

References

Accashian JV, Vinopal RT, Kim B-J, Smets BF (1998) Aerobic growth on nitroglycerin as the sole carbon nitrogen, and energy source by a mixed bacterial culture. Appl Environ Microbiol 64:3300–3304

Adrian NR, Arnett CM, Hickey RF (2003) Stimulating the anaerobic biodegradation of explosives by the addition of hydrogen or electron donors that produce hydrogen. Water Res 37:3499–3507

Behki R, Topp E, Dick W, Germon P (1993) Metabolism of the herbicide atrazine by a *Rhodococcus* strain. Appl Environ Microbiol 59:1955–1959

Behrend C, Heesche-Wagner K (1999) Formation of hydride-Meisenheimer complexes of picric acid (2,4,6-trinitrophenol) and 2,4-dinitrophenol during mineralization of picric acid by *Nocardioides* sp. Strain CB 22-2. Appl Environ Microbiol 65:1372–1377

Bell LS, Devlin JF, Gillham RW, Binning PJ (2003) A sequential zero valent iron and aerobic biodegradation treatment system for nitrobenzene. J Contam Hydrol 66: 201–217

Bhaumik S, Christodoulatos C, Korfiatis GP, Brodman BW (1997) Aerobic and anaerobic biodegradation of nitroglycerin in batch and packed bed bioreactors. Water Sci Technol 36:139–146

Bhushan B, Chauhan A, Samanta SK, Jain RK (2000) Kinetics of biodegradation of *p*-nitrophenol by different bacteria. Biochem Biophys Res Commun 274:626–630

Bhushan B, Halasz A, Spain J, Thiboutot S, Ampleman G, Hawari J (2002a) Biotransformation of hexahydro-1,3,5-trinitro-1,3,5-triazine catalysed by a NAD(P)H: Nitrate oxidoreductase from *Aspergillus niger*. Environ Sci Technol 36:3104–3108

Bhushan B, Halasz A, Spain J, Hawari J (2002b) Diaphorase catalysed biotransformation of RDX via N-denitration mechanism. Biochem Biophys Res Commun 296: 779–784

Bhushan B, Trott S, Spain JC, Halasz A, Paquet L, Hawari J (2003) Biotransformation of hexahydro-1,3,5-trinitro-1,3,5-triazine (RDX) by a rabbit liver cytochrome P450: insight into the mechanism of RDX biodegradation by *Rhodococcus* sp. strain DN22. Appl Environ Microbiol 69:1347–1351

Binks PR, Nicklin S, Bruce NC (1995) Degradation of hexahydro-1,3,5-trinitro-1,3,5-triazine (RDX) by *Stenotrophomonas maltophilia* PB1. Appl Environ Microbiol 61: 1318–1322

Binks PR, French CE, Nicklin S, Bruce NC (1996) Degradation of pentaerythritol tetranitrate by *Enterobacter cloacae* PB2. Appl Environ Microbiol 62:1214–1219

Blehert DS, Knoke KL, Fox BG Chambliss GH (1997) Regioselectivity of nitroglycerin denitration by flavoprotein nitroester reductase purified from two *Pseudomonas* species. J Bacteriol 179:6254–6263

Blehert DS, Fox BG, Chambliss GH (1999) Cloning and sequence analysis of two *Pseudomonas flavoprotein* xenobiotic reductases. J Bacteriol 181:6254–6263

Boopathy R, Gurgas M, Ullian J, Manning JF (1998) Metabolism of explosive compounds by sulfate-reducing bacteria. Curr Microbiol 37:127–131

Cartwright NJ, Cain RB (1958) Bacterial degradation of the nitrobenzoic acids. Biochem J 71:248–261

Chauhan A, Jain RK (2000) Degradation of *o*-nitrobenzoate via anthranilic acid (o-aminobenzoate) by *Arthrobacter protophormiae*: a plasmid-encoded new pathway. Biochem Biophys Res Commun 267:236–244

Chauhan A, Chakraborti AK, Jain RK (2000) Plasmid-encoded degradation of *p*-nitrophenol and 4-nitrocatechol by *Arthrobacter protophormiae*. Biochem Biophys Res Commun 270:733–740

Clausen GB, Larsen L, Johnsen K, De Lipthay JR, Aamand J (2002) Quantification of the atrazine-degrading *Pseudomonas* sp. strain ADP in aquifer sediment by quantitative competitive polymerase chain reaction. FEMS Microbiol Ecol 41:221–229

Coleman VN, Nelson DR, Duxbury T (1998) Aerobic biodegradation of hexahydro-1,3,5-trinitro-1,3,5-triazine (RDX) as a nitrogen source by a *Rhodococcus* sp. strain DN22. Soil Biol Biochem 30:1159–1167

Coleman VN, Spain JC, Duxbury T (2002) Evidence that RDX biodegradation by *Rhodococcus* sp. strain DN22 is plasmid-borne and involves a cytochrome P-450. J Appl Microbiol 93:463–472

Cook AM, Beilstein P, Grossenbacher H, Hütter R (1985) Ring cleavage and degradative pathway of cyanuric acid in bacteria. Biochem J 231:25–30

De Souza ML, Wackett LP, Sadowaky MJ (1998) The *atz*ABC genes encoding atrazine catabolism are located on a self-transmissible plasmid in *Pseudomonas* sp. strain ADP. Appl Environ Microbiol 64:2323–2326

Dickel O, Huang W, Knackmuss HJ (1993) Biodegradation of nitrobenzene by a sequential anaerobic -aerobic process. Biodegradation 4:187–194

Ducrocq C, Servy C, Lenfant M (1989) Bioconversion of glyceryl trinitrate into mononitrates by *Geotrichum candidum*. FEMS Microbiol Lett 65:219–222

Duran M, Kim BJ, Speece RE (1995) Anaerobic biotransformation of nitrocellulose. Waste Manage 14:481–487

EPA (2003) Numeric criteria. http://oaspub.epa.gov/wqsdatabase/wqsi_epa_criteria.rep_parameter

Fournier D, Halasz A, Spain J, Fiurasek P, Hawari J (2002) Determination of key metabolites during biodegradation of hexahydro-1,3,5-trinitro-1,3,5-triazine with *Rhodococcus* sp. strain DN22. Appl Environ Microbiol 68:66–172

Freeman DL, Caenepeel BM, Kim RJ (1996) Biotransformation of nitrocellulose under methanogenic conditions. Water Sci Technol 34:327–334

Freeman DL, Sutherland KW (1998) Biodegradation of hexahydro-1,3,5-trinito-1,3,5-triazine (RDX) under nitrate-reducing condition. Water Sci Technol 38: 33–40

French CE, Nicklin S, Bruce NC (1998) Aerobic degradation of 2,4,6-trinitrotoluene by *Enterobacter cloacae* PB2 and by pentaerythritol tetranitrate reductase. Appl Environ Microbiol 64:2864–2868

Gerth A, Hebner A, Thomas H (2003) Natural remediation of TNT-contaminated water and soil. Acta Biotechnol 23: 143–150

Gong P, Hawari J, Thiboutot S, Ampleman G, Sunahara GI (2001) Ecotoxicological effects of hexahydro-1,3,5-trinitro-1,3,5-triazine on soil microbial activities. Environ Toxicol Chem 20:947–951

Groenewegen PEJ, Breeuwer P, Van Helvoort JMLM, Langenhoff AAM, De Vries FP, De Bont JAM (1992) Novel degradative pathway of 4-nitrobenzoate in *Comamonas acidovorans* NBA-10. J Gen Microbiol 138:1599–1605

Haigler BE, Spain JC (1991) Biotransformation of nitrobenzene by bacteria containing toluene degradation pathway. Appl Environ Microbiol 57:3156–3162

Haigler BE, Spain JC (1993) Biodegradation of 4-nitrotoluene by *Pseudomonas* sp. strain 4NT. Appl Environ Microbiol 59:2239–2243

Haigler BE, Johnson GR, Suen W-C, Spain JC (1999) Biochemical and genetic evidence for *meta*-ring cleavage of 2,4,5-trihydroxytoluene in *Burkholderia* sp. strain DNT. J Bacteriol 181:965–972

Halasz A, Spain J, Paquet L, Beaulieu C, Hawari J (2002) Insights into the formation and degradation mechanisms of methylenedinitramine during the incubation of RDX with anaerobic sludge. Environ Sci Technol 36:633–638

Hasegawa Y, Muraki T, Tokuyama T, Iwaki H, Tatsuno M, Lau PCK (2000) A novel degradative pathway of 2-nitrobenzoate via 3-hydroxyanthranilate in *Pseudomonas fluorescens* strain KU-7. FEMS Microbiol Lett 190:185–190

Hawari J (2000) Biodegradation of RDX and HMX: from basic research to field application. In: Spain JC, Highes JB, Knackmuss H-J (eds) Biodegradation of nitroaromatic compounds and explosives. Lewis, Boca Raton. pp 277–300

Hawari J, Beaudet S, Halasz A, Thiboutot S, Ampleman G (2000a) Microbial degradation of explosives: biotransformation versus mineralization. Appl Microbiol Biotechnol 54:605–618

Hawari J, Halasz A, Sheremata T, Beaudet S, Groom C, Paquet L, Rhofir C, Ampleman G, Thiboutot S (2000b) Characterization of metabolites during biodegradation of hexahydro-1,3,5-trinitro-1,3,5-triazine (RDX) with municipal anaerobic sludge. Appl Environ Microbiol 66:2652–2657

He Z, Spain JC (1997) Studies of the catabolic pathway of degradation of nitrobenzene by *Pseudomonas pseudoalcaligenes* JS45: removal of the amino group from 2-aminomuconic semialdehyde. Appl Environ Microbiol 63:4839–4843

He Z, Spain JC (1999) Compparison of the downstream pathways for degradation of nitrobenzene by *Pseudomonas pseudoalcaligenes* JS45 (2-aminophenol pathway) and by *Comamonas* sp. JS765 (catechol pathway). Arch Microbiol 171: 309–316

He Z, Nadeau LJ, Spain JC (2000) Characterization of hydroxylaminobenzene mutase from pNBZ139 cloned from *Pseudomonas pseudoalcaligenes* JS45 A highly associated SDS-stable enzyme catalyzing an intramolecular transfer of hydroxy groups. Eur J Biochem 267:1110–1116

Heiss G, Trachtmann N, Abe Y, Takeo M, Knackmuss H (2003) Homologous npdG1 genes in 2,4-dinitrophenol- and 4-nitrophenol-degrading *Rhodococcus* spp. Appl Environ Microbiol 69: 2748–2754

Huang S, Lindahl PA, Wang C, Bennett GN, Rudolph FB, Hugnes JB (2000) 2,4,6-Trinitrotoluene reduction by carbon monoxide dehydrogenase from *Clostridium thermoaceticum.* Appl Environ Microbiol 66:1474–1478

Hughes MC, Williams PA (2001) Cloning and characterization of the *pnb* genes, encoding enzymes for 4-nitrobenzoate catabolism in *Pseudomonas putida* TW3. J Bacteriol 183:1225–1232

James KD, Williams PA (1998) *ntn* Genes determining the early steps in the divergent catabolism of 4-nitrotoluene and toluene in *Pseudomonas* sp. strain TW3. J Bacteriol 180:2043–2049

Johnson GR, Spain JC (2003) Evolution of catabolic pathways for synthetic compounds: bacterial pathways for degradation of 2,4-dinitrotoluene and nitrobenzene. Appl Microbiol Biotechnol 62: 110–123

Johnson GR, Smets BF, Spain JC (2001) Oxidative transformation of aminodinitrotoluene isomers by multicomponent dioxygenases. Appl Environ Microbiol 67: 5460–5466

Kadiyala V, Smets BF, Chandran K, Spain JC (1998) High affinity p-nitrophenol oxidation by *Bacillus sphaericus* JS905. FEMS Microbiol Lett 166:115–120

Karns JS (1999) Gene sequence and properties of an *s*-triazine ring-cleavage enzyme from *Pseudomonas* sp. strain NRRLB-12227. Appl Environ Microbiol 65:3512–3517

Karns JS, Eaton RW (1997) Genes encoding *s*-triazine degradation are plasmid-born in *Klebsislla pneumoniae* strain 99. J Agric Food Chem 45:1017–1022

Khan H, Harris RJ, Barna T, Craig DH, Bruce NC, Munro AW, Moody PCE, Scrutton NS (2002) Kinetic and structural basis of reactivity of pentaerythriol tetranitrate reductase with NADPH, 2-cyclohexenone, nitroesters and nitroaromatic explosives. J Biol Chem 227:21906–21912

Kitts CL, Green CE, Otley RA, Alvarez MA, Unkefer PJ (2000) Type I nitroreductases in soil enterobacteria reduce TNT (2,4,6-trinitrotoluene) and RDX (hexahydro-1,3,5-trinitro-1,3,5-triazine). Can J Microbiol 46:278–282

Lendenmann U, Spain JC (1998) Simultaneous biodegradation of 2,4-dinitrotoluene and 2,6-dinitrotoluene in an aerobic fluidized-bed biofilm reactor. Environ Sci Technol 32:82–87

Lenke H, Achtnich C, Knackmuss H-J (2000) Perspectives of bioelimination of polynitroaromatic compounds. In: Spain JC, Highes JB, Knackmuss H-J (eds) Biodegradation of nitroaromatic compounds and explosives, Lewis, Boca Raton, pp 91–126

Lessner DJ, Johnson GR, Parales RE, Spain JC, Gibson DT (2002) Molecular characterization and substrate specificity of nitrobenzene dioxygenase from *Comamonas* sp. JS765. Appl Environ Microbiol 68:634–641

Lessner DJ, Parales RE, Narayan S, Gibson DT (2003) Expression of the nitroarene dioxygenase genes in *Comomonas* sp. strain JS765 and *Acidovorax* sp. strain JS42 is induced by multiple aromatic compounds. J Bacteriol 185:3895–3904

Leung KT, Campbell S, Gan Y, White DC, Lee H, Trevors JT (1999) The role of the *Sphingomonas* sp. UG30 pentachlorophenol-4-monooxygenase in *p*-nitrophenol degradation. FEMS Microbiol Lett 173:247–253

Mandelbaum RT, Allan D, Wackett LP (1995) Isolation and characterization of a *Pseudomonas* sp. that mineralizes the s-triazine herbicide atrazine. Appl Environ Microbiol 61:1451–1457

Marshall SJ, White GF (2001) Complete denitration of nitroglycerin by bacteria isolated from a washwater soakway. Appl Environ Microbiol 67:2622–2626

Martinez B, Tomkins J, Wackett L, Wing R, Sadowsky MJ (2001) Complete nucleotide sequence and organization of the atrazine catabolic plasmid pADP-1 from *Pseudomonas* sp. strain ADP. J Bacteriol 183:5684–5697

Meng M, Sun W-Q, Geelhaar LA, Kumar G, Patel AR, Payne GF, Speedie MK, Stacy JR (1995) Denitration of glycerol trinitrate by resting cells and cell extracts of *Bacillus thuringiensis/cereus* and *Enterobacter agglomerans*. Appl Environ Microbiol 61: 2548–2553

Meulenberg R, Pepi M, De Bont JAM (1996) Degradation of 3-nitrophenol by *Pseudomonas putida* B2 occurs via 1,2,4-benzenetriol. Biodegradation 7:303–311

Mougin C, Laugero C, Asther M, Dubroca J, Frasse P, Asther M (1994) Biotransformation of the herbicide atrazine by the white rot fungus *Phanerochaete chrysosporium*. Appl Environ Microbiol 60:705–708

Nadeau LJ, Spain JC (1995) Bacterial degradation of *m*-nitrobenzoic acid. Appl Environ Microbiol 61:840–843

Nadeau LJ, He Z, Spain JC (2003) Bacterial conversion of hydroxylamino aromatic compounds by both lyase and mutase enzymes involves intramolecular transfer of hydroxyl groups. Appl Environ Microbiol 69: 2786–2793

Nagy I, Compernoll Ghys K, Vanderleyden J, De Mot R (1995) A single cytochrome P-450 system is involved in degradation of the herbicide EPTC (s-ethyl dipropylthiocarbamate) and atrazine by *Rhodococcus* sp. strain NI86/21. Appl Environ Microbiol 61:2056–2060

Nishino SF, Spain JC (1993) Degradation of nitrobenzene by a *Pseudomonas pseudocalcaligens*. Appl Environ Microbiol 59:2520–2525

Nishino SF, Spain JC (1995) Oxidative pathway for the biodegradation of nitrobenzene by *Comamonas* sp. strain JS765. Appl Environ Microbiol 61:2308–2313

Nishino SF, Spain JC, Lenke H, Knackmuss H-J (1999) Mineralization of 2,4-and 2,6-dinitrotoluene in soil slurries. Environ Sci Technol 33:1060–1064

Nishino SF, Spain JC, He Z (2000a) Strategies for aerobic degradation of nitroaromatic compounds by bacteria s process discovery to field application. In: Spain JC, Highes

JB, Knackmuss H-J (eds) Biodegradation of nitroaromatic compounds and explosives. Lewis, Boca Raton, pp 9–61

Nishino SF, Paoli GC, Spain JC (2000b) Aerobic degradation of dinitrotoluenes and pathway for bacterial degradation of 2,6-dinitrotoluene. Appl Environ Microbiol 66:2139–2147

Oh B-T, Just CL, Alvarez PJJ (2001) Hexahydro-1,3,5-trinitro-1,3,5-triazine mineralization by zerovalent iron and mixed anaerobic culture. Environ Sci Technol 35: 4341–4346

Parales JV, Parales RE, Resnick SM, Gibson DT (1998) Enzyme specificity of 2-nitrotoluene 2,3-dioxygenase from *Pseudomonas* sp. strain JS42 is determined by the C-terminal region of the alpha subunit of the oxygenase component. J Bacteriol 180:1194–1199

Pennington JC, Brannon JM (2002) Environmental fate of explosives. Thermochim Acta 384:163–172

Peres CM, Haveau H, Agathos SN (1998) Biodegradation of nitrobenzene by its simultaneous reduction into aniline and mineralization of the aniline formed. Appl Microbiol Biotechnol 49:343–349

Peres CM, Russ R, Lenke H, Agathos SN (2001) Biodegradation of 4-nitrobenzoate, 4-aminobenzoate and their mixtures: new strains, unusual metabolites and insight into pathway regulation. FEMS Microbiol Ecol 37:151–159

Piutti S, Semon E, Landry D, Hartmann A, Dousset S, Lichtfouse E, Topp E, Soulas G, Martin-Laurent F (2003) Isolation and characterisation of *Nocardioides* sp. SP12, an atrazine-degrading bacterial strain possessing the gene trzN from bulk- and maize rhizosphere soil. FEMS Microbiol Lett 221:111–117

Price CB, Brannon JM, Yost S (1998) Technical report IRRP-98-2, US Army Engineer Waterways Experiment Station, Vicksburg, MS

Popesku JT, Singh A, Zhao J-S, Hawari J, Ward OP (2003) High TNT-transforming activity by a mixed culture acclimated and maintained on crude-oil-containing media. Can J Microbiol 49:362–366

Pudge IB, Daugulis AJ, Dubois C (2003) The use of *Enterobacter cloacae* ATCC 43560 in the development of a two-phase partitioning bioreactor for the destruction of hexahydro-1,3,5-trinitro-1,3,5-s-triazine (RDX). J Biotechnol 100:65–75

Qureshi AA, Purohit HJ (2002) Isolation of bacterial consortia for degradation of p-nitrophenol from agricultural soil. Ann Appl Biol 140:159–162

Ralebitso TK, Senior E, Van Verseveld HW (2002) Microbial aspects of atrazine degradation in natural environments. Biodegradation 13:11–19

Razo-Flores E, Lettinga G, Field JA (1999) Biotransformation and biodegradation of selected nitroaromatics under anaerobic conditions. Biotechnol Prog 15:358–365

Rieger PG, Knackmuss H-J (1995) Basic knowledge and perspectives on biodegradation of 2,4,6-trinitrotoluene and related nitroaromatic compounds in contaminated soil. In: Spain JC (ed) Biodegradation of nitroaromatic compounds. Plenum Press, New York, pp 1–18

Rieger PG, Sinnwell V, Preub A, Francke W, Knackmuss H-J (1999) Hydride-Meisenheimer complex formation and protonation as key reactions of 2,4,6-trinitrophenol biodegradation by *Rhodococcus erythropolis.* J Bacteriol 181:1189–1195

Rousseaux S, Soulas G, Hartmann A (2002) Plasmid localization of atrazine-degrading genes in newly described *Chelatobacter* and *Arthrobacter* strains. FEMS Microbiol Ecol 41:69–75

Russ R, Walters DM, Knackmuss H-J (2000) Identification of genes involved in picric acid and 2,4-dinitrophenol degradation by mRNA differential display. In: Spain JC, Highes JB, Knackmuss H-J (eds) Biodegradation of nitroaromatic compounds and explosives, Lewis, Boca Raton, pp 127–143

Sadowsky MJ, Wackett LP (2000) Genetics of atrazine and *s*-triazine degradation by *Pseudomonas* sp. strain ADP and other bacteria. In: Hall JC, Hoagland RE, Zablotowicz RM (eds) Pesticide biotransformation in plants and microorganisms: similarities and divergences. ACS Symposium Series, American Chemical Society, Washington, DC, pp 268–281

Seffernick JL, De Souza M, Sadowsky MJ, Wackett L (2001) Melamine deaminase and atrazine chlorohydrolase: 98 percent identical but functionally different. J Bacteriol 183:2405–2410

Servent D, Ducrocq C, Henry Y, Guissani A, Lenfant M (1991) Nitroglycerin metabolism by *Phanerochaete chrysosporium:* evidence for nitric oxide and nitrite formation. Biochim Biophys Acta 1074:320–325

Seth-Smith HMB, Rosser S, Basran A, Travis ER, Dabbs ER, Nicklin S, Bruce NC (2002) Cloning sequencing, and characterization of the hexahydro-1,3,5-trinitro-1,3,5-triazine degradation gene cluster from *Rhodococcus rhodochrous.* Appl Environ Microbiol 68:4764–4771

Shao ZQ, Behki R (1995) Cloning of the genes for degradation of the herbicides EPTC (S-ethyl dipropylthiocarbamate) and atrazine from *Rhodococcus* sp. strain TE1. Appl Environ Microbiol 61:2061–2065

Shapir N, Osborne JP, Johnsin G, Sadowsky MJ, Wackett LP (2002) Purification substrate range, and metal center of *Atz*C: N-isopropylammelide aminohydrolase involved in bacterial atrazine metabolism. J Bacteriol 184:5376–5384

Sharma A, Sundaram ST, Zhang Y-Z, Brodman BW (1995) Nitrocellulose degradation by a coculture of *Sclertium rolfsii* and *Fusarium solani.* J Ind Microbiol 15:1–4

Sheremata TW, Hawari J (2000) Mineralization of RDX by the white rot fungus *Phanerochaete chrysosporium* to carbon dioxide and nitrous oxide. Environ Sci Technol 34:3384–3388

Sheremata TW, Halasz A, Paquet L, Thiboutot S, Ampleman G, Hawari J (2001) The fate of the cyclic nitramine explosive RDX in natural soil. Environ Sci Technol 35: 1037–1040

Somerville CC, Nishino SF, Spain JC (1995) Purification and characterization of nitrobenzene nitroreductase from *Pseudomonas pseudoalcaligenes* JS45. J Bacteriol 177:3837–3482

Spain JC (1995) Bacterial degradation of nitroaromatic compound under aerobic conditions. In: Spain JC (ed) Biodegradation of nitroaromatic compounds, Plenum Press, New York, pp 19–35

Spain JC, Gibson DT (1991) Pathway for biodegradation of *p*-nitrophenol in a *Moraxella* sp. Appl Environ Microbiol 57:812–819

Strong LC, Rosendahl C, Johnson G, Sadowsky MJ, Wackett LP (2002) *Arthrobacter aurescens* TC1 metabolizes diverse *s*-triazine ring compound. Appl Environ Microbiol 68:5973–5980

Suen WC, Spain JC (1993) Cloning and characterization of *Pseudomonas* sp. strain dinitrotoluene genes for 2,4-dinitrotoluene degradation. J Bacteriol 175:1831–1837

Topp E, Mulbry WM, Zhu H, Nour SM, Cuppels D (2000) Characterization of *s*-triazine herbicide metabolism by a *Nocardioides* sp. isolated from agricultural soils. Appl Environ Microbiol 66:3134–3141

Uberoi V, Bhattacharya SK (1997) Toxicity and degradability of nitrophenol in anaerobic system. Water Environ Res 69:146–156

Van Aken B, Hofrichter M, Scheibner K, Hatakka AI, Naveau H, Agathos SN (1999) Transformation and mineralization of 2,4,6-trinitrotoluene (TNT) by manganese peroxidase from the white-rot *Basidiomycete phlebiaradiata*. Biodegradation 10: 83–91

Wackett LP, Hershberger CD (2001) Evolution of catabolic enzymes and pathways. In: Wackett LP, Hershberger CD (eds) Biocatalysis and biodegradation: microbial transformation of organic compounds, ASM Press, Washington, DC, pp 115–134

Wackett LP, Sadowsky MJ, Martinez NS (2002) Biodegradation of atrazine and related *s*-triazine compounds: from enzymes to field studies. Appl Microbiol Biotechnol 58:39–45

Wang J, Zhou J-T, Zhang J-S, Zhang A-L, Hong LU (2001) Aerobic degradation of nitrobenzene by *Pseudomonas* sp. JX165 and its intact cells. China Environ Sci 21: 144–147

White GF, Snape JR (1993) Microbial cleavage of nitrate esters: defusing the environment. J Gen Microbiol 139:1947–1957

White GF, Snape JR, Nicklin S (1996a) Bacterial biodegradation of glycerol trinitrate. Int Biodeter Biodegradation 19:77–82

White GF, Snape JR, Nicklin S (1996b) Biodegradation of glycerol, trinitrate and pentaerythritol tetranitrate by *Agrobacterium radiobacter.* Appl Environ Microbiol 62: 637–642

Williams RE, Bruce NC (2000) Strategies for aerobic degradation of nitroaromatic compounds by bacteria s process discovery to field application. In: Spain JC, Highes JB, Knackmuss H-J (eds) Biodegradation of nitroaromatic compounds and explosives, Lewis, Boca Raton, pp 161–184

Williams RE, Bruce NC (2002) New uses for an old enzyme-the old yellow enzyme family of flavoenzymes. Microbiology 148:1607–1614

Yabannavar AV, Zylstra G (1995) Cloning and characterization of the genes for p-nitrobenzoate degradation from *Pseudomonas pickettii* YH105. Appl Environ Microbiol 61:4284–4290

Ye J, Singh A, Ward OP (2004) Biodegradation of nitroaromatics and other nitrogen-containing xenobiotics. World J Microbiol Biotechnol (in press)

Yinon J (1990) Toxicity and metabolism of explosive. CRC Press, Ann Arbor

Zeyer J, Wasserfallen A, Timmis KN (1985) Microbial mineralization of ring-substituted anilines through an *ortho*-cleavage pathway. Appl Environ Microbiol 50:447–453

Zhao J-S, Ward OP (1999) Microbial degradation of nitrobenzene and mono-nitrophenol by bacteria enriched from municipal activated sludge. Can J Microbiol 45:427–432

Zhao J-S, Ward OP (2001) Substrate selectivity of a 3-nitrophenol-induced metabolic system in *Pseudomonas putida* 2NP8 transforming nitroaromatic compounds into ammonia under aerobic conditions. Appl Environ Microbiol 67:1388–1391

Zhao J-S, Ward OP, Lubicki P, Cross JD, Huck P (2001) Process for degradation of nitrobenzene: combining electron beam irradiation with biotransformation. Biotechnol Bioeng 73:306–312

Zhao J-S, Halasz A, Paquet L, Beaulieu C, Hawari J (2002) Biodegradation of hexahydro-1,3,5-trinitro-1,3,5-triazine and its mononitroso derivative hexahydro-1-nitroso-3,5-dinitro-1,3,5-triazine by *Klebsiella pneumoniae* strain SCZ-1 isolated from an anaerobic sludge. Appl Environ Microbiol 68:5336–5341

Zylstra GJ, Bang S-W, Newman LM, Perry LL (2000) Microbial degradation of mononitrophenols and mononitrobenzoates. In: Spain JC, Highes JB, Knackmuss H-J (eds) Biodegradation of nitroaromatic compounds and explosives. Lewis, Boca Raton, pp 145–160

8 Aromatic Hydrocarbon Dioxygenases*

Rebecca E. Parales[1] and Sol M. Resnick[2]

1 Role and Significance of Aromatic Hydrocarbon Dioxygenases in Biodegradation

Since the initial discovery of toluene dioxygenase (TDO) by David Gibson and coworkers (Yeh et al. 1977; Subramanian et al. 1979), aromatic hydrocarbon dioxygenases have been reported to catalyze the initial reaction in the bacterial biodegradation of a diverse array of aromatic and polyaromatic hydrocarbons, aromatic acids, chlorinated aromatic, and heterocyclic aromatic compounds. To date, more than 100 aromatic compound dioxygenases have been described in the literature based on biological activity or nucleotide sequence identity. These enzymes are cofactor-requiring multicomponent heteromultimeric proteins (EC 1.14.12) that catalyze the initial activation through reductive dihydroxylation of their substrates, and are distinct from aromatic ring-cleavage (or ring-fission) dioxygenases (EC 1.13.11), which act on the catechol intermediates in many of the same catabolic pathways. This chapter will focus primarily on the aromatic ring-hydroxylating dioxygenases (Rieske non-heme iron dioxygenases; EC 1.14.12), which initiate an attack on aromatic hydrocarbons, heterocycles and related compounds carrying various substituents (Cl, NO_2). The aspects described will relate to dioxygenase discovery, classification, enzymology, structure, electron transport, mechanism and applications.

Much of our understanding of the structure and function of aromatic hydrocarbon dioxygenases comes from studies of naphthalene dioxy-

[1] Section of Microbiology, 226 Briggs Hall, University of California, Davis, California 95616, USA, e-mail: reparales@ucdavis.edu, Tel: +1-530-7545233, Fax: +1-530-7529014
[2] Biotechnology R&D, The Dow Chemical Company, San Diego, California 92121, USA

* We dedicate this chapter to Professor David T. Gibson, whose pioneering efforts have advanced our understanding of aromatic hydrocarbon dioxygenases.

Soil Biology, Volume 2
Biodegradation and Bioremediation
(ed. by. A. Singh and O. P. Ward)

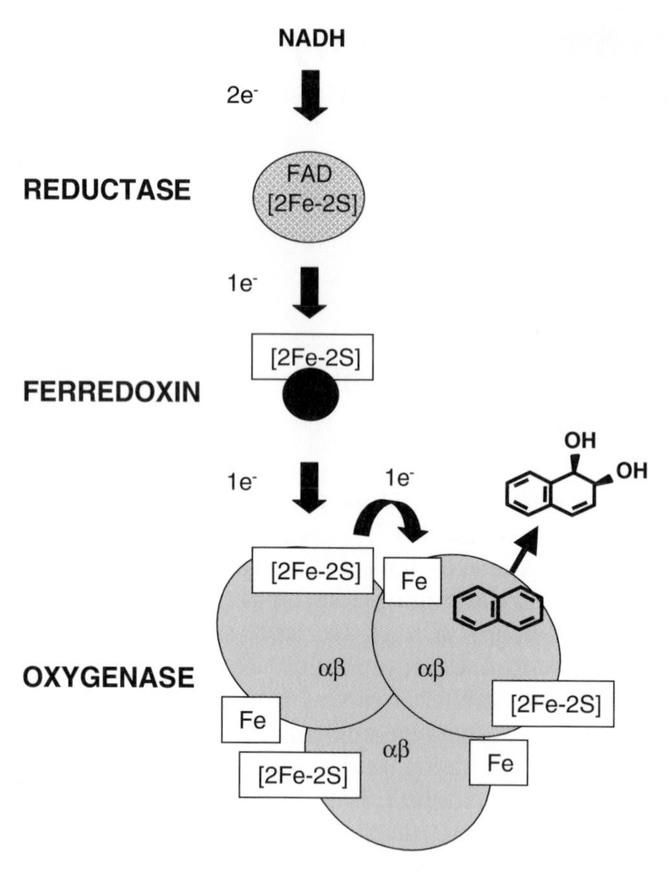

Fig. 1. Reaction catalyzed by the three-component naphthalene dioxygenase (NDO) system. The iron-sulfur flavoprotein reductase transfers electrons from NADH to the Rieske [2Fe-2S] ferredoxin. An electron is then transferred from the ferredoxin to one of the Rieske centers in the oxygenase. The reduced oxygenase catalyzes the addition of both atoms of O_2 to the substrate naphthalene to form enantiomerically pure (+)-naphthalene *cis*-(1*R*,2*S*)-dihydrodiol

genase (NDO; Fig. 1) and TDO. These and related enzymes are important for the catabolism of a wide range of environmental pollutants and are distributed among a variety of Gram negative and Gram positive bacteria capable of degrading key classes of aromatic hydrocarbon pollutants.

2 Isolation of Aromatic Hydrocarbon Dioxygenases from the Environment

The majority of aromatic ring-hydroxylating dioxygenases have been identified from isolated microorganisms capable of growth on specific aromatic hydrocarbons. Methods for their isolation typically include selective enrichment and subsequent plating of subcultures on minimal medium containing the aromatic substrate. Substrates can be supplied in the vapor phase (Gibson 1976), incorporated into the agar media or sprayed as an insoluble layer onto the plate surface (Kiyohara et al. 1982), depending on the solubility and vapor pressure of the compound. The source of inocula selected for enrichment often includes contaminated environments with a history of previous exposure to environmental pollutants such as creosote, gasoline, or refined petroleum products. A number of colorimetric indicators have been established for detecting the activities of dioxygenases or of downstream enzymes required for aromatic hydrocarbon degradation. These include the well-established conversion of indole to indigo (Ensley et al. 1983), a reaction catalyzed by many aromatic hydrocarbon dioxygenases, and the conversion of indole carboxylic acids to indigo by certain aromatic acid dioxygenases (Eaton and Chapman 1995). The detection of ring-fission products formed from catechols derived through the oxidation of parent hydrocarbon substrates (e.g., dibenzothiophene, dibenzofuran, biphenyl) has been used to identify coupled activities of ring-hydroxylating dioxygenases, *cis*-dihydrodiol dehydrogenases, and *meta* ring-cleavage dioxygenases (Kodama et al. 1973).

Approaches involving gene-based discovery of aromatic hydrocarbon dioxygenases can be employed for screening libraries prepared directly from the environment, or from enrichments or pure cultures. Such methods were successfully applied in the discovery of hydrolases (Gray et al. 2003). Several challenges exist for the application of these techniques to multicomponent aromatic hydrocarbon dioxygenases. Activity-based screening requires that each protein is present and sufficiently expressed for detection of activity. Homology-based hybridization can limit the identification of genetic diversity in favor of homologue re-discovery. Hybridization of multiple components can be complicated by the "mosaic" genetic organization of the encoding genes (Kim and Zylstra 1999) and the application of substrate-independent discovery techniques ultimately requires a demonstration of new activity. An alternative approach was used to identify novel PAH degradation genes by screening the genomic DNA of isolates for lack of hybridization to standard dioxygenase gene probes (Zylstra et al. 1997).

3
Classification and Relationships of Aromatic Ring-Hydroxylating Dioxygenases

Rieske non-heme iron oxygenases have been classified in a number of ways. An early classification system was based on the components of the electron transfer chains present in the ten characterized Rieske non-heme iron oxygenase systems (Batie et al. 1991). Two-component (reductase, oxygenase) and three-component (reductase, ferredoxin, oxygenase) enzyme systems were represented, and the classes were further subdivided based on the number of proteins comprising the oxygenase, the type of flavin moiety (FAD or FMN) present in the reductase, the presence or absence of an iron-sulfur center in the reductase, whether a ferredoxin was involved, and the type of iron-sulfur center (plant vs. Rieske) present in the ferredoxin. This classification system worked well when only a small number of enzymes were known. However, as the number and diversity of these enzyme systems have grown, new enzymes with redox partners that do not fit into the original classification have emerged. Werlen et al. (1996) proposed a classification system based on sequence alignments of the available oxygenase α subunits, designating four families (naphthalene, toluene/benzene, biphenyl, and benzoate/toluate). Since the oxygenase is the catalytic component and the α subunit plays a major role in determining substrate specificity, these classifications are based on the catalytic activity of the enzymes. It was clear from this report that related dioxygenases typically coded for enzymes with similar substrates. However, what this study missed was the relationship of the α_n dioxygenases (those in the Batie system class IA) to those containing β subunits. Nakatsu et al. (1995) clarified this point and demonstrated that these α_n oxygenases form a separate lineage.

Two new classification systems based on the phylogeny of the oxygenase α subunits have been recently proposed (Gibson and Parales 2000; Nam et al. 2001). Four distinct families (Fig. 2) or groups emerged from the phylogenetic analyses and, in general, enzymes with similar native substrates cluster together. One exception is group I, or the phthalate family, which consists of all of the Rieske non-heme iron oxygenases that contain only α subunits. Substrates for this diverse group of enzymes include several aromatic acids such as phthalate, *p*-toluene sulfonate, and phenoxybenzoate, but also include carbazole and 2-oxo-1,2-dihydroquinoline. Group II, or the benzoate family represents a cluster of enzymes with activities toward various aromatic acids. Naphthalene, phenanthrene, and nitroarene dioxygenases occupy group

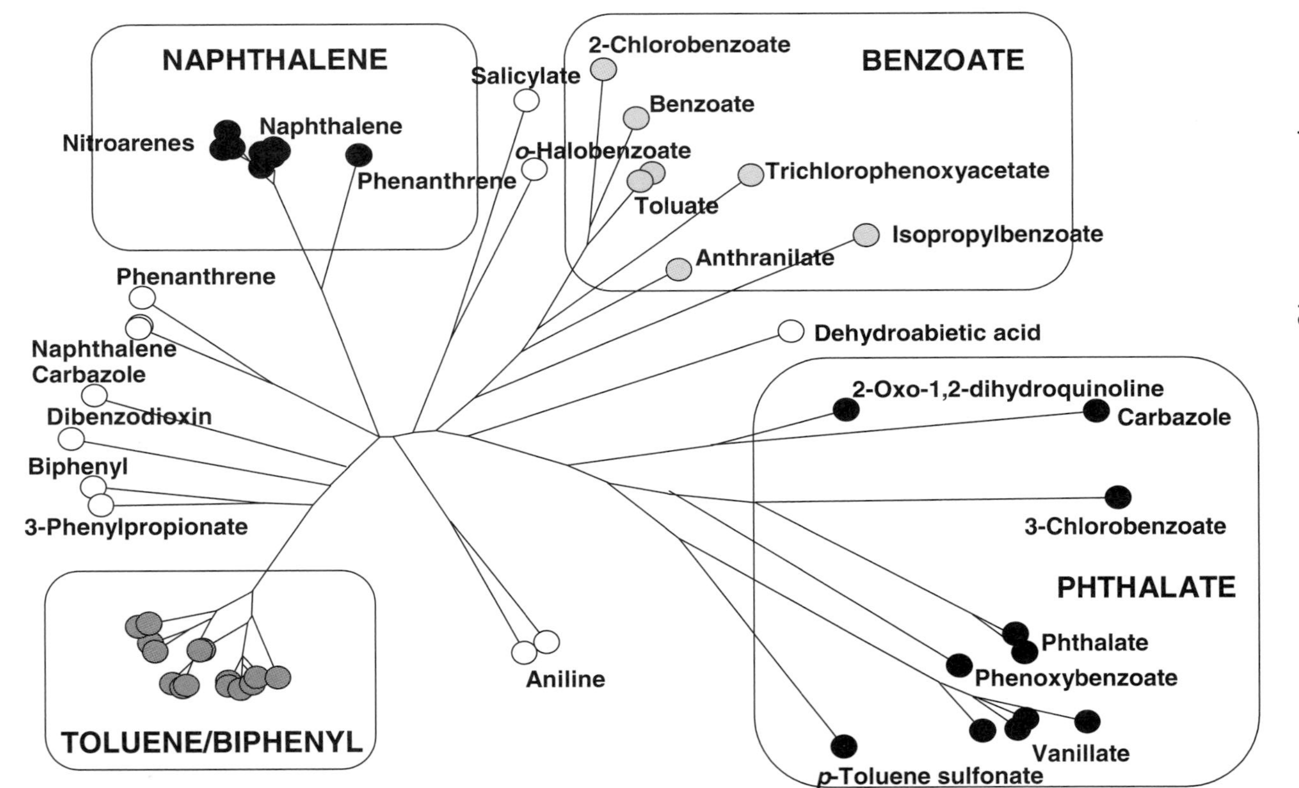

Fig. 2. Phylogenetic tree of α subunits of aromatic ring-hydroxylating dioxygenases. Many of the proteins fall into four main families, the benzoate, phthalate, naphthalene, and toluene/biphenyl families. Reproduced from Gibson and Parales (2000) with permission from the publisher

III or the naphthalene family, and the monocyclic aromatic hydrocarbons, biphenyl, and chlorinated aromatic hydrocarbon dioxygenases are tightly clustered in group IV or the toluene/biphenyl family. There are now a number of enzymes that are quite distantly related to those found in the core of these four groups or families (Fig. 2). It will be interesting to see whether newly identified enzymes will result in the development of more distinct lineages or if the divisions will be further blurred.

4 Enzymology of Aromatic Hydrocarbon Dioxygenases

4.1 General Reaction Catalyzed

Aromatic ring-hydroxylating dioxygenases utilize molecular oxygen as a required substrate, adding both atoms of O_2 to the aromatic ring of the compound. In general, this reaction is the most difficult in the degradation of aromatic hydrocarbons, and the addition of hydroxyl groups to the highly stable aromatic ring structure poises the molecule for further oxidation and eventual ring cleavage. To date, several dioxygenases have been purified and studied in detail. Some of the most in-depth studies have been carried out with the NDO system (Fig. 1). In this chapter, we will focus primarily on NDO as a model system and provide additional results obtained with other related enzyme systems. All three NDO protein components have been purified (Ensley and Gibson 1983; Haigler and Gibson 1990a, b). The reductase is a monomer of approximately 35 kDa. It contains one molecule of FAD and a plant-type iron-sulfur center, and can accept electrons from either NADH or NADPH (Haigler and Gibson 1990b; Simon et al. 1993). The ferredoxin is a monomer of approximately 11.4 kDa that contains a Rieske iron-sulfur center (Haigler and Gibson 1990a; Simon et al. 1993). The oxygenase consists of α and β subunits in an $\alpha_3\beta_3$ configuration (Kauppi et al. 1998). Each α subunit contains a Rieske [2Fe-2S] center and mononuclear Fe^{2+} at the active site. Individual subunits of the oxygenase were purified and reconstituted (Suen and Gibson 1993, 1994). Both subunits were essential for activity, a result similar to those seen with TDO (Jiang et al. 1999) and biphenyl dioxygenase (Hurtubise et al. 1996). The reaction requires the transfer of electrons from NAD(P)H to the reductase, then to the ferredoxin and finally to the Rieske center of the oxygenase. Two electrons are necessary to complete the reaction cycle.

4.2 The Structure of NDO and Electron Transport Proteins

The structures of ferredoxin reductases from three different dioxygenase systems revealed striking differences. The benzoate dioxygenase reductase from *Acinetobacter* sp. strain ADP1 (Karlsson et al. 2002) and the phthalate dioxygenase reductase (Correll et al. 1992) are members of the $NADP^+$-ferredoxin reductase superfamily. Each is composed of three domains, for binding NAD, [2Fe-2S], and flavin, although the arrangement of the domains differs in the two proteins. Modeling studies have suggested that the NDO reductase structure will be similar to that of benzoate dioxygenase reductase (Karlsson et al. 2002). In contrast, the biphenyl dioxygenase reductase from *Pseudomonas* sp. strain KKS102, which exhibited low sequence identities with other aromatic ring-hydroxylating dioxygenase reductase components (Kikuchi et al. 1994), was found to be evolutionarily related to glutathione reductases (Senda et al. 2000). Although these proteins belong to two distinct reductase families, the NAD- and FAD-binding domains are quite similar (Karlsson et al. 2002).

The structure of the biphenyl dioxygenase ferredoxin component from *Burkholderia* sp. strain LB400 (Colbert et al. 2000) revealed that the Rieske center is located near the surface of the protein. The positioning of the Rieske center may have implications for docking with the reductase and oxygenase components to allow electron transfer. Cross-linking studies with the three TDO components revealed the involvement of electrostatic interactions between proteins and argued against the formation of a ternary complex (Lee 1998). These results imply that in the three-component enzyme systems, the reductase and oxygenase alternately bind the ferredoxin at the same location near the exposed Rieske center to carry out the necessary electron transfers.

The crystal structure of the oxygenase component of NDO has been determined in the presence and absence of the substrates (Kauppi et al. 1998; Carredano et al. 2000; Karlsson et al. 2003). The trimer of $\alpha\beta$ heterodimers contains three distinct and apparently independent active sites at the junctions of adjacent α subunits. Each α subunit contains a Rieske [2Fe-2S] center coordinated by two histidine and two cysteine residues (Cys81, His83, Cys101, His104), and mononuclear iron coordinated by His208, His213, Asp362 and a water molecule (Fig. 3; Kauppi et al. 1998). This 2-his-1-carboxylate motif for the coordination of the mononuclear iron has been recently identified in a variety of non-heme iron-containing enzymes that catalyze a wide range of reaction types (Lange and Que 1998; Que 2000). The substitution of Asp362 by alanine resulted in a completely inactive form of NDO,

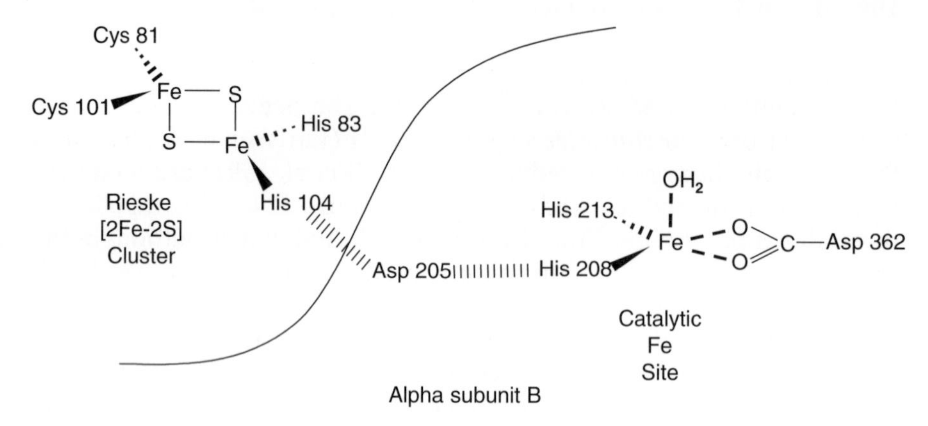

Fig. 3. Structure of the NDO active site located at the junction between two α subunits. Shown are the Rieske [2Fe-2S] center and mononuclear iron at the active site. See text for details

demonstrating that this iron-coordinating residue is essential (Parales et al. 2000a). Site-directed mutagenesis studies with TDO demonstrated that the conserved residues Glu214, Asp219, His222 and His228 are essential for enzyme activity, and were suggested to be involved in coordinating the mononuclear iron (Jiang et al. 1996). The substitution of these histidines in benzene dioxygenase also resulted in inactive enzymes (Butler and Mason 1997). His222 and His228 correspond to the iron ligands His208 and His213 in NDO. Glu214 corresponds to Glu200 in NDO, and this residue appears to provide an important contact between adjacent α subunits by forming a salt link with Arg84 (Kauppi 1997). Asp205 in NDO, which corresponds to Asp219 in TDO, is hydrogen bonded to His104 of the Rieske center and His208 of the catalytic iron center. Asp205 appears to provide an important link in the electron transfer pathway between the Rieske center and mononuclear iron in adjacent α subunits in NDO (Fig. 3; Kauppi et al. 1998). This hypothesis is supported by site-directed mutagenesis studies in which Asp205 was replaced by Ala, Glu, Gln or Asn residues (Parales et al. 1999). The modified proteins were almost completely inactive although they appeared to be identical in structure to the wild-type enzyme. In addition, benzene, an efficient uncoupling agent for wild-type NDO (Lee 1999), did not stimulate oxygen uptake by the purified Gln205-containing enzyme, suggesting that electrons are not transferred to the iron at the active site in the mutant protein (Parales et al. 1999).

4.3
Roles of the α and β Subunits of the Oxygenase

Among all of the Rieske non-heme iron oxygenases identified to date, two types of oxygenase structures are known: those consisting of only α subunits, and those with both α and β subunits. Most of the enzymes that catalyze the oxidation of aromatic hydrocarbons, chlorinated aromatics, nitroaromatics, and heterocyclic aromatic compounds contain both α and β subunits. Exceptions include a carbazole dioxygenase (Nam et al. 2002) and a 3-chlorobenzoate dioxygenase (Nakatsu et al. 1995). Based on studies in which the individual α and β subunits from different dioxygenases were substituted, it appears that the α subunits control substrate specificity in NDO and in the closely related enzymes 2-nitrotoluene dioxygenase and 2,4-dinitotoluene dioxygenase (Parales et al. 1998a, b) as well as in various biphenyl dioxygenases (Mondello et al. 1997; Barriault et al. 2001; Zielinski et al. 2002). Similar results were reported with biphenyl dioxygenase hybrids, TDO-tetrachlorobenzene dioxygenase hybrids, and benzene-biphenyl dioxygenase hybrids (Tan and Cheong 1994; Beil et al. 1998; Zielinski et al. 2002). Together with the crystal structure data for NDO, which indicated that no β subunit residues are located near the active site (Kauppi et al. 1998), these studies are consistent with a structural role for the β subunit. On the other hand, other studies have suggested that the β subunits play a role determining substrate specificity in toluene, biphenyl and toluate dioxygenases (Harayama et al. 1986; Hirose et al. 1994; Hurtubise et al. 1998; Chebrou et al. 1999). Thus it appears that the β subunit plays primarily a structural role in most dioxygenases, but is capable of modulating substrate specificity in others.

4.4
Enzyme Mechanism

As isolated, the native form of NDO contains an oxidized Rieske center and mononuclear ferrous iron. Spectroscopic studies with phthalate dioxygenase, an enzyme that catalyzes the *cis*-dihydroxylation of phthalate, demonstrated that the iron is in a six-coordinate octahedral configuration in the absence of substrate, but this configuration changes to five-coordinate when the substrate is bound (Tsang et al. 1989; Gassner et al. 1993; Pavel et al. 1994). This result is in contrast to crystal structure data for NDO, which indicates that the coordination of iron does not change significantly upon substrate binding (Carredano et al. 2000; Karlsson et al. 2003). Single turnover studies with NDO indicated that 0.85 units of product were formed per electron added, and one Rieske center and mononuclear iron were oxidized during the reaction (Wolfe et al. 2001). Recently, evidence for a peroxide-mediated turnover with the native NDO has been reported. During the reaction, the correct product was produced and mononuclear iron was oxidized. The observation that the reaction stopped after one turnover suggested that the release of product required the input of an additional electron (Wolfe and Lipscomb 2003). The oxidation state of the iron during catalysis is currently unknown. Early studies suggested an attack by a ferric-peroxo intermediate in the monooxygenation and *cis*-dihydroxylation reactions catalyzed by 4-methoxybenzoate monooxygenase (Bernhardt and Kuthan 1981). More recent work with NDO supports this possibility (Carredano et al. 2000; Karlsson et al. 2003; Lee 1999). The participation of a high valent iron-oxo species, such as those in the cytochrome $P450_{cam}$ (Schlichting et al. 2000) and methane monooxygenase (Lipscomb 1994) reaction cycles has also been proposed (Que and Ho 1996; Wolfe et al. 2001; Wolfe and Lipscomb 2003). Although some of the reaction types catalyzed by ring-hydroxylating dioxygenases are also catalyzed by other types of enzymes, the *cis*-dihydroxylation reaction is unique to bacterial aromatic ring-hydroxylating dioxygenases. For example, cytochromes P450 also catalyze monooxygenation reactions, but are incapable of catalyzing *cis*-dihydroxylations. This unique reaction has implications for the mechanism of the enzyme that may be explained by the recent observation of an oxygen bound side-on to iron at the active site of the NDO (Fig. 4; Karlsson et al. 2003). The substrates naphthalene and indole bind in an elongated cleft in the active site of NDO, approximately 4 Å from the mononuclear iron (Carredano et al. 2000; Karlsson et al. 2003). In the absence of bound substrate, dioxygen binds side-on approximately 2.2 Å from the iron and slightly closer to the iron when indole is present (Karlsson et al. 2003). These crystal structures suggest a reaction mechanism in which both atoms of molecular oxygen react in a concerted fashion with the carbon atoms of the substrate double bond. This mechanism would explain the specificity of this class of enzymes in catalyzing *cis*-dioxygenation reactions. In addition, the position of the product of naphthalene oxidation, naphthalene *cis*-1,2-dihydrodiol bound at the active site is consistent with a mechanism where both atoms of oxygen react in concert with the double-bonded carbon atoms of the substrate (Karlsson et al. 2003).

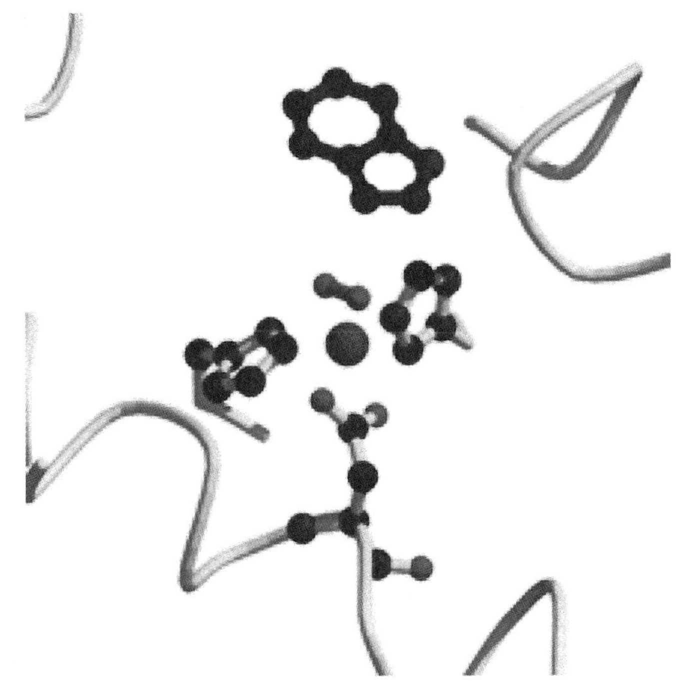

Fig. 4. The substrates O_2 and indole are shown bound at the active site of NDO. The five-member ring of indole is positioned for attack by dioxygen bound side-on to mononuclear iron. Amino acid side chains shown are the iron-coordinating ligands His208, His213, and Asp362. Figure kindly provided by S. Ramaswamy, University of Iowa

5 Substrate Range, Applications and Enzyme Engineering

5.1 Range of Reactions Catalyzed

As a group, aromatic ring hydroxylating dioxygenases are capable of initiating oxidative attack on a broad range of substrates (Resnick et al. 1996; Boyd et al. 2002). Individual enzymes themselves display remarkable diversity in both the numbers of substrates oxidized and the types of reactions catalyzed. Much of our understanding of the mechanism of substrate oxidation has been based on studies with strains expressing TDO and NDO, and in some cases with purified enzymes. NDO is known to catalyze more than 75 different oxidations by reactions including *cis*-dihydroxylation (dioxygenation), benzylic monohydroxylation (monooxygenation, Wackett et al. 1988; Resnick et al. 1994), sul-

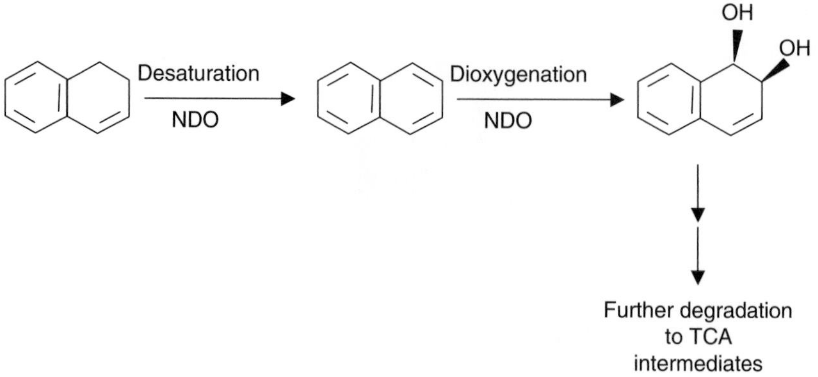

Fig. 5. Initial steps in a potential pathway for the degradation of 1,2-dihydronaphthalene via NDO-catalyzed desaturation and dioxygenation reactions

fur oxidation (sulfoxidation, Lee et al. 1995), *O*- and *N*-dealkylation (Resnick and Gibson 1993) and desaturation (Gibson et al. 1995; Torok et al. 1995). The substrates and products – including the absolute stereochemistry and enantiopurity of chiral products, when known – have been reviewed (Resnick et al. 1996) and can be also accessed online via the University of Minnesota Biocatalysis/Biodegradation Database (UM-BDD; http://umbbd.ahc.umn.edu). With respect to biodegradation, activities other than *cis*-dihydroxylation serve to extend the substrate range by yielding oxidation products that can be further metabolized. For example, NDO catalyzes desaturation of 1,2-dihydronaphthalene (Torok et al. 1995) to yield naphthalene, which can be mineralized by the enzymes of a catabolic pathway (Fig. 5). Similar reactions are plausible for the desaturation of indan to indene.

Perhaps the most impressive versatility of a dioxygenase is exemplified by TDO: this enzyme has overlapping but often distinct specificity from NDO and is known to catalyze over 100 oxidation reactions (http://umbbd.ahc.umn.edu) on monocyclic aromatics, fused and linked aromatics, and aromatic heterocyclic compounds (Boyd and Sheldrake 1998; Hudlicky et al. 1999). As with NDO, multiple reaction types are catalyzed by TDO, including dioxygenation, monooxygenation (Wackett et al. 1988), sulfoxidation (Allen et al. 1995; Lee et al. 1995) and desaturation (Torok et al. 1995). In addition to oxidizing an extensive range of aromatic substrates, TDO from strain *Pseudomonas putida* F1 oxidizes a variety of halogenated and nonhalogenated aliphatic olefins. Dioxygenation was observed for substrates that lacked a halogen substituent on sp2 carbon atoms while monooxygenation of halogenated substrates containing one or more unsubstituted allylic methyl groups yielded allylic alcohols (Lange and Wackett 1997).

Dioxygenase-catalyzed dechlorination has also been demonstrated for chlorinated benzenes and biphenyls. In this case, dioxygenation at a chlorine-substituted carbon resulted in a concomitant elimination of chloride (Haddock et al. 1995; Werlen et al. 1996). Similar reactions were shown with nitroaromatic substrates and resulted in nitrite elimination (An et al. 1994; Spanggord et al. 1991). In both cases, the displacement reactions result in the formation of dihydroxylated (catecholic) products that are activated for further metabolism.

5.2 Enzyme Engineering and Applications

The concept of engineering bacterial enzymes and pathways to more efficiently degrade environmental pollutants is not new (Timmis et al. 1989, 1994; Minshull 1995; Pieper et al. 1996; Pieper and Reineke 2000). A number of research laboratories have generated engineered forms of different dioxygenases using a variety of methods, based on both rational design and random mutagenesis strategies. These studies have led to the development of enzymes with novel or enhanced activities, and have provided insights into the control of substrate specificity.

With only sequence comparisons available as a guide, Mondello and coworkers generated hybrid and site-directed mutant forms of biphenyl dioxygenase to identify specific regions and amino acid residues in the α subunits that control substrate specificity (Erickson and Mondello 1993; Mondello et al. 1997). These and other studies (Kimura et al. 1997) resulted in engineered enzymes with enhanced polychlorinated biphenyl (PCB) degradation abilities and identified the residue at position 377 in the *Burkholderia* sp. The LB400 enzyme is critical for the oxidation of 2,5,2′,5′-tetrachlorobiphenyl. A TDO-biphenyl dioxygenase hybrid enzyme was shown to have an enhanced ability to degrade trichloroethylene (TCE; Furukawa et al. 1994). Kinetic analysis of the purified hybrid protein demonstrated that it had a lower K_m for TCE and a higher catalytic efficiency than the wild-type TDO (Maeda et al. 2001). Another series of hybrid constructions and site-directed mutagenesis experiments led to the identification of a single amino acid substitution (M220A) in the α subunit of TDO that conferred the ability to dechlorinate 1,2,4,5-tetrachlorobenzene (Beil et al. 1998). The crystal structure of NDO revealed the presence of 17 amino acid residues lining the substrate-binding pocket (Carredano et al. 2000). This information was used to design a series of variants of the enzyme with one or more amino acid substitutions near the active site iron. The results of these studies (Parales et al. 2000a, b; Yu et al. 2001) identified phenylalanine 352 as a critical amino acid for determining the regioselectivity and enantioselectivity of the enzyme.

Using DNA shuffling (Zhao and Arnold 1997), variant biphenyl dioxygenases with enhanced abilities to degrade single-ring aromatic hydrocarbons, PCBs, and heterocyclic aromatic compounds have been generated (Kumamaru et al. 1998; Furukawa 2000; Suenaga et al. 2001a, b; Barriault et al. 2002). A similar study used directed-evolution techniques to specifically engineer variants of TDO that were capable of converting 4-picoline to 3-hydroxy-4-picoline (Sakamoto et al. 2001). The same type of directed evolution strategy was utilized in order to generate a form of TDO that produced the chiral starting substrate for the production of indinavir (Crixivan), a HIV-1 protease inhibitor used in the treatment of AIDS (Zhang et al. 2000). The desired reaction was the conversion of indene to (–)-*cis*-(1*S*,2*R*)-indandiol in high enaniomeric excess and with minimal side product formation. After three rounds of mutagenesis, variants that produced significantly more *cis*-indandiol relative to the undesired by-product indenol were obtained. However, the stereoselectivity was changed to favor the production of the undesired (+)-*cis*-indandiol enantiomer (Zhang et al. 2000).

Strain construction efforts also provide examples where expression of multiple activities in a single microorganism results in new degradative phenotypes. A pathway for 2-chlorotoluene degradation was assembled by expressing the *todC1C2BA* genes encoding TDO from *P. putida* F1 with the upper TOL pathway from pWWO of *Pseudomonas putida* strain mt-2 encoding benzylalcohol dehydrogenase (XylB) and benzaldehyde dehydrogenase (XylC) (Haro and de Lorenzo 2001). The expression of seven genes encoding the cytochrome P450cam monooxygenase and TDO resulted in a *Pseudomonas* strain capable of metabolizing polyhalogenated compounds through sequential reductive and oxidative reactions. In this strain, cytochrome P450cam monooxygenase catalyzed the reduction of polyhalogenated ethanes, under low oxygen tension, to products that were oxidized by TDO (Wackett et al. 1994). Cloning of the *tod* genes encoding TDO into the chromosome of *Deinococcus radiodurans* resulted in a strain capable of growth and pollutant degradation in highly irradiating environments (Lange et al. 1998). Expression of the Hg (II) resistance gene (*merA*) with the *tod* gene cluster in *Deinococcus radiodurans* demonstrated the integration of multiple remediation functions into single engineered strains for amelioration of mixed radioactive waste containing organic pollutants and ionic mercury (Brim et al. 2000).

A cooxidation strategy for bioremediation of trichloroethylene (TCE) has been used in the field (Hopkins and McCarty 1995). TCE is a widely used solvent and is a United States Environmental Protection Agency (EPA) priority pollutant. TDO and toluene/phenol monooxygenases can cooxidize TCE (Nelson et al. 1987; Wackett and Gibson 1988), and studies with purified TDO indicated that the major oxidation products formed from [^{14}C]trichloroethylene were formic acid and glyoxylic acid (Li and Wackett 1992). In the field study, phenol or toluene and either oxygen or hydrogen peroxide were added as cosubstrates to stimulate in situ TCE degradation.

6 Conclusions

It is clear that aromatic hydrocarbon dioxygenases play an important role in the biodegradation of a variety of environmental pollutants, both in the natural environment (natural attenuation) and in engineered bioremediation systems. Recent structural and mechanistic information coupled with enzyme improvement and strain construction approaches have allowed the development of engineered microorganisms with new and enhanced degradation capabilities.

Acknowledgments. Research in the author's laboratory is supported by the Strategic Environmental Research and Development Program (REP). We thank S. Ramaswamy and Juan Parales for providing figures and David Gibson for helpful comments on the manuscript.

References

Allen CCR, Boyd DR, Dalton H, Sharma ND, Haughey SA, McMordie RAS, McMurray BT, Sheldrake GN, Sproule K (1995) Sulfoxides of high enantiopurity from bacterial dioxygenase-catalyzed oxidation. J Chem Soc Chem Commun:119–120

An D, Gibson DT, Spain JC (1994) Oxidative release of nitrite from 2-nitrotoluene by a three-component enzyme system from *Pseudomonas* sp. strain JS42. J Bacteriol 176: 7462–7467

Barriault D, Plante M-M, Sylvestre M (2002) Family shuffling of a targeted *bphA* region to engineer biphenyl dioxygenase. J Bacteriol 184:3794–3800

Barriault D, Simard C, Chatel H, Sylvestre M (2001) Characterization of hybrid biphenyl dioxygenases obtained by recombining *Burkholderia* sp. strain LB400 *bphA* with the homologous gene of *Comamonas testosteroni* B-356. Can J Microbiol 47:1025–1032

Batie CJ, Ballou DP, Corell CC (1991) Phthalate dioxygenase reductase and related flavin-iron-sulfur containing electron transferases. In: Müller F (ed) Chemistry and biochemistry of flavoenzymes. CRC Press, Boca Raton, pp 543–556

Beil S, Mason JR, Timmis KN, Pieper DH (1998) Identification of chlorobenzene dioxygenase sequence elements involved in dechlorination of 1,2,4,5-tetrachlorobenzene. J Bacteriol 180:5520–5528

Bernhardt F-H, Kuthan H (1981) Dioxygen activation by putidamonooxin: the oxygen species formed and released under uncoupling conditions. Eur J Biochem 120: 547–555

Boyd DR, Sharma ND, Modyanova LV, Carroll JG, Malone JF, Allen CCR, Hamilton JTG, Gibson DT, Parales RE, Dalton H (2002) Dioxygenase-catalyzed *cis*-dihydroxylation of pyridine-ring systems. Can J Chem 80:589–600

Boyd DR, Sheldrake GN (1998) The dioxygenase-catalysed formation of vicinal *cis*-diols. Nat Prod Rep 15:309–324

Brim H, McFarlan SC, Fredrickson JK, Minton KW, Zhai M, Wackett LP, Daly MJ (2000) Engineering *Deinococcus radiodurans* for metal remediation in radioactive mixed waste environments. Nat Biotechnol 18:85–90

Butler CS, Mason JR (1997) Structure-function analysis of the bacterial aromatic ring-hydroxylating dioxygenases. Adv Microbial Physiol 38:47–84

Carredano E, Karlsson A, Kauppi B, Choudhury D, Parales RE, Parales JV, Lee K, Gibson DT, Eklund H, Ramaswamy S (2000) Substrate binding site of naphthalene 1,2-dioxygenase: functional implications of indole binding. J Mol Biol 296:701–712

Chebrou H, Hurtubise Y, Barriault D, Sylvestre M (1999) Catalytic activity toward chlorobiphenyls of purified recombinant His-tagged oxygenase component of *Rhodococcus globerulous* strain P6 biphenyl dioxygenase and of chimeras derived from it and their expression in *Escherichia coli* and in *Pseudomonas putida.* J Bacteriol 181:4805–4811

Colbert CL, Couture MM-J, Eltis LD, Bolin JT (2000) A cluster exposed: structure of the Rieske ferredoxin from biphenyl dioxygenase and the redox properties of Rieske Fe-S proteins. Structure 8:1267–1278

Correll CC, Batie CJ, Ballou DP, Ludwig ML (1992) Phthalate dioxygenase reductase: a modular structure for electron transfer from pyridine nucleotides to [2Fe-2S]. Science 258:1604–1610

Eaton RW, Chapman PJ (1995) Formation of indigo and related compounds from indolecarboxylic acids by aromatic acid-degrading bacteria: chromogenic reactions for cloning genes encoding dioxygenases that act on aromatic acids. J Bacteriol 177: 6983–6988

Ensley BD, Gibson DT (1983) Naphthalene dioxygenase: purification and properties of a terminal oxygenase component. J Bacteriol 155:505–511

Ensley BD, Ratzkin BJ, Osslund TD, Simon MJ, Wackett LP, Gibson DT (1983) Expression of naphthalene oxidation genes in *Escherichia coli* results in the biosynthesis of indigo. Science 222:167–169

Erickson BD, Mondello FJ (1993) Enhanced biodegradation of polychlorinated biphenyls after site-directed mutagenesis of a biphenyl dioxygenase gene. Appl Environ Microbiol 59:3858–3862

Furukawa K (2000) Engineering dioxygenases for efficient degradation of environmental pollutants. Curr Opin Biotechnol 11:244–249

Furukawa K, Hirose J, Hayashida S, Nakamura K (1994) Efficient degradation of trichloroethylene by a hybrid aromatic ring dioxygenase. J Bacteriol 176:2121–2123

Gassner GT, Ballou DP, Landrum GA, Whittaker JW (1993) Magnetic circular dichroism studies on the mononuclear ferrous active site of phthalate dioxygenase from *Pseudomonas cepacia* show a change of ligation state on substrate binding. Biochemistry 32:4820–4825

Gibson DT (1976) Initial reactions to the bacterial degradation of aromatic hydrocarbons. Zentralbl Bakt Hyg I Abt Orig B 162:157–168

Gibson DT, Parales RE (2000) Aromatic hydrocarbon dioxygenases in environmental biotechnology. Curr Opin Biotechnol 11:236–243

Gibson DT, Resnick SM, Lee K, Brand JM, Torok DS, Wackett LP, Schocken MJ, Haigler BE (1995) Desaturation, dioxygenation and monooxygenation reactions catalyzed by naphthalene dioxygenase from *Pseudomonas* sp. strain 9816-4. J Bacteriol 177: 2615–2621

Gray KA, Richardson TH, Robertson DE, Swanson PE, Subramanian MV (2003) Soil-based gene discovery: a new technology to accelerate and broaden biocatalytic applications. Adv Appl Microbiol 52:1–27
Haddock JD, Horton JR, Gibson DT (1995) Dihydroxylation and dechlorination of chlorinated biphenyls by purified biphenyl 2,3-dioxygenase from *Pseudomonas* sp. strain LB400. J Bacteriol 177:20–26
Haigler BE, Gibson DT (1990a) Purification and properties of ferredoxin$_{NAP}$, a component of naphthalene dioxygenase from *Pseudomonas* sp. strain NCIB 9816. J Bacteriol 172:465–468
Haigler BE, Gibson DT (1990b) Purification and properties of NADH-ferredoxin$_{NAP}$ reductase, a component of naphthalene dioxygenase from *Pseudomonas* sp. strain NCIB 9816. J Bacteriol 172:457–464
Harayama S, Rekik M, Timmis KN (1986) Genetic analysis of a relaxed substrate specificity aromatic ring dioxygenase, toluate 1,2-dioxygenase, encoded by TOL plasmid pWWO of *Pseudomonas putida*. Mol Gen Genet 202: 226–234
Haro MA, de Lorenzo V (2001) Metabolic engineering of bacteria for environmental applications: construction of *Pseudomonas* strains for biodegradation of 2-chlorotoluene. J Biotechnol 85:103–113
Hirose J, Suyama A, Hayashida S, Furukawa K (1994) Construction of hybrid biphenyl (*bph*) and toluene (*tod*) genes for functional analysis of aromatic ring dioxygenases. Gene 138:27–33
Hopkins GD, McCarty PL (1995) Field evaluation of in-situ aerobic cometabolism of trichloroethene and three dichloroethene isomers using phenol and toluene as the primary substrates. Environ Sci Technol 29:1628–1637
Hudlicky T, Gonzalez D, Gibson DT (1999) Enzymatic dihydroxylation of aromatics in enantioselective synthesis: Expanding asymmetric methodology. Aldrichim Acta 32:35–62
Hurtubise Y, Barriault D, Sylvestre M (1996) Characterization of active recombinant His-tagged oxygenase component of *Comamonas testosteroni* B-356 biphenyl dioxygenase. J Biol Chem 271:8152–8156
Hurtubise Y, Barriault D, Sylvestre M (1998) Involvement of the terminal oxygenase β subunit in the biphenyl dioxygenase reactivity pattern toward chlorobiphenyls. J Bacteriol 180:5828–5835
Jiang H, Parales RE, Lynch NA, Gibson DT (1996) Site-directed mutagenesis of conserved amino acids in the alpha subunit of toluene dioxygenase: potential mononuclear non-heme iron coordination sites. J Bacteriol 178:3133–3139
Jiang H, Parales RE, Gibson DT (1999) The α subunit of toluene dioxygenase from *Pseudomonas putida* F1 can accept electrons from reduced ferredoxin$_{TOL}$ but is catalytically inactive in the absence of the β subunit. Appl Environ Microbiol 65:315–318
Karlsson A, Beharry ZM, Eby M, Coulter ED, Neidle EL, Kurtz DMJ, Eklund H, Ramaswamy S (2002) X-ray crystal structure of benzoate 1,2-dioxygenase reductase from *Acinetobacter* sp. strain ADP1. J Mol Biol 318:261–272
Karlsson A, Parales JV, Parales RE, Gibson DT, Eklund H, Ramaswamy S (2003) Crystal structure of naphthalene dioxygenase: Side-on binding of dioxygen to iron. Science 299:1039–1043
Kauppi B (1997) PhD, Thesis. Swedish University of Agricultural Sciences, Uppsala
Kauppi B, Lee K, Carredano E, Parales RE, Gibson DT, Eklund H, Ramaswamy S (1998) Structure of an aromatic ring-hydroxylating dioxygenase-naphthalene 1,2-dioxygenase. Structure 6:571–586
Kikuchi Y, Nagata Y, Hinata M, Kimbara K, Fukuda M, Yano K, Takagi M (1994) Identification of the *bphA4* gene encoding ferredoxin reductase involved in biphenyl

and polychlorinated biphenyl degradation in *Pseudomonas* sp. strain KKS102. J Bacteriol 176:1689–1694

Kim E, Zylstra GJ (1999) Functional analysis of genes involved in biphenyl, naphthalene, phenanthrene, and *m*-xylene degradation by *Sphingomonas yanoikuyae* B1. J Ind Microbiol Biotechnol 23:294–302

Kimura N, Nishi A, Goto M, Furukawa K (1997) Functional analyses of a variety of chimeric dioxygenases constructed from two biphenyl dioxygenases that are similar structurally but different functionally. J Bacteriol 179:3936–3943

Kiyohara H, Nagao K, Yana K (1982) Rapid screen for bacteria degrading water-insoluble, solid hydrocarbons on agar plates. Appl Environ Microbiol 43:454–457

Kodama K, Umehara K, Shimizu K, Nakatani S, Minoda Y, Yamada K (1973) Identification of microbial products from dibenzothiophene and its proposed oxidation pathway. Agric Biol Chem 37:45–50

Kumamaru T, Suenaga H, Mitsuoka M, Watanabe T, Furukawa K (1998) Enhanced degradation of polychlorinated biphenyls by directed evolution of biphenyl dioxygenase. Nat Biotechnol 16:663–666

Lange CC, Wackett LP (1997) Oxidation of aliphatic olefins by toluene dioxygenase:enzyme rates and product identification. J Bacteriol 179:3858–3865

Lange CC, Wackett LP, Minton KW, Daly MJ (1998) Engineering a recombinant *Deinococcus radiodurans* for organopollutant degradation in radioactive mixed waste environments. Nat Biotechnol 16:929–933

Lange SJ, Que LJ (1998) Oxygen activating nonheme iron enzymes. Curr Opin Chem Biol 2:159–172

Lee K (1998) Involvement of electrostatic interactions between the components of toluene dioxygenase from *Pseudomonas putida* F1. J Microbiol Biotechnol 8: 416–421

Lee K (1999) Benzene-induced uncoupling of naphthalene dioxygenase activity and enzyme inactivation by production of hydrogen peroxide. J Bacteriol 181:2719–2725

Lee K, Brand JM, Gibson DT (1995) Stereospecific sulfoxidation by toluene and naphthalene dioxygenases. Biochem Biophys Res Commun 212:9–15

Li S, Wackett LP (1992) Trichloroethylene oxidation by toluene dioxygenase. Biochem Biophys Res Commun 185:443–451

Lipscomb JD (1994) Biochemistry of the soluble methane monooxygenase. Annu Rev Microbiol 48:371–399

Maeda T, Takahashi Y, Suenaga H, Suyama A, Goto M, Furukawa K (2001) Functional analyses of Bph-Tod hybrid dioxygenase, which exhibits high degradation activity toward trichloroethylene. J Biol Chem 276:29833–29838

Minshull J (1995) Cleaning up our own back yard: developing new catabolic pathways to degrade pollutants. Curr Biol 2:775–780

Mondello FJ, Turcich MP, Lobos JH, Erickson BD (1997) Identification and modification of biphenyl dioxygenase sequences that determine the specificity of polychlorinated biphenyl degradation. Appl Environ Microbiol 63:3096–3103

Nakatsu CH, Straus NA, Wyndham RC (1995) The nucleotide sequence of the Tn*5271* 3-chlorobenzoate 3,4-dioxygenase genes (*cbaAB*) unites the class IA oxygenases in a single lineage. Microbiology 141:485–495

Nam J-W, Nojiri H, Yoshida T, Habe H, Yamane H, Omori T (2001) New classification for oxygenase components involved in ring-hydroxylating oxygenations. Biosci Biotechnol Biochem 65:254–263

Nam J-W, Nojiri H, Noguchi H, Uchimura H, Yoshida T, Habe H, Yamane H, Omori T (2002) Purification and characterization of carbazole 1,9a-dioxygenase, a three component dioxygenase system of *Pseudomonas resinovorans* strain CA10. Appl Environ Microbiol 68:5882–5890

Nelson MJK, Montgomery SO, Mahaffey WR, Pritchard PH (1987) Biodegradation of trichloroethylene and involvement of an aromatic biodegradative pathway. Appl Environ Microbiol 53:949–954

Parales JV, Parales RE, Resnick SM, Gibson DT (1998a) Enzyme specificity of 2-nitrotoluene 2,3-dioxygenase from *Pseudomonas* sp. strain JS42 is determined by the C-terminal region of the α subunit of the oxygenase component. J Bacteriol 180:1194–1199

Parales RE, Emig MD, Lynch NA, Gibson DT (1998b) Substrate specificities of hybrid naphthalene and 2,4-dinitrotoluene dioxygenase enzyme systems. J Bacteriol 180:2337–2344

Parales RE, Parales JV, Gibson DT (1999) Aspartate 205 in the catalytic domain of naphthalene dioxygenase is essential for activity. J Bacteriol 181:1831–1837

Parales RE, Lee K, Resnick SM, Jiang H, Lessner DJ, Gibson DT (2000a) Substrate specificity of naphthalene dioxygenase: effect of specific amino acids at the active site of the enzyme. J Bacteriol 182:1641–1649

Parales RE, Resnick SM, Yu CL, Boyd DR, Sharma ND, Gibson DT (2000b) Regioselectivity and enantioselectivity of naphthalene dioxygenase during arene *cis*-dihydroxylation: control by phenylalanine 352 in the α subunit. J Bacteriol 182:5495–5504

Pavel EG, Martins LJ, Ellis WRJ, Solomon EI (1994) Magnetic circular dichroism studies of exogenous ligand and substrate binding to the non-heme ferrous active site in phthalate dioxygenase. Chem Biol 1:173–183

Pieper DH, Reineke W (2000) Engineering bacteria for bioremediation. Curr Opin Biotechnol 11:262–270

Pieper DH, Timmis KN, Ramos JL (1996) Designing bacteria for the degradation of nitro- and chloroaromatic pollutants. Naturwissenschaften 83:201–213

Que LJ (2000) One motif- many different reactions. Nat Struct Biol 7:182–184

Que LJ, Ho RYN (1996) Dioxygen activation by enzymes with mononuclear non-heme iron active sites. Chem Rev 96:2607–2624

Resnick SM, Gibson DT (1993) Biotransformation of anisole and phenetole by aerobic hydrocarbon-oxidizing bacteria. Biodegradation 4:195–203

Resnick SM, Torok DS, Lee K, Brand JM, Gibson DT (1994) Regiospecific and stereoselective hydroxylation of 1-indanone and 2-indanone by naphthalene dioxygenase and toluene dioxygenase. Appl Environ Microbiol 60:3323–3328

Resnick SM, Lee K, Gibson DT (1996) Diverse reactions catalyzed by naphthalene dioxygenase from *Pseudomonas* sp. strain NCIB 9816. J Ind Microbiol 17:438–457

Sakamoto T, Joern JM, Arisawa A, Arnold FH (2001) Laboratory evolution of toluene dioxygenase to accept 4-picoline as a substrate. Appl Environ Microbiol 67:3882–3887

Schlichting I, Berendzen J, Chu K, Stock AM, Maves SA, Benson DE, Sweet RM, Ringe D, Petsko GA, Sligar SG (2000) The catalytic pathway of cytochrome P450cam at atomic resolution. Science 287:1651

Senda T, Yamada T, Sakurai N, Kubota M, Nishizaki T, Masai E, Fukuda M, Mitsui Y (2000) Crystal structure of NADH-dependent ferredoxin reductase component in biphenyl dioxygenase. J Mol Biol 304:397–410

Simon MJ, Osslund TD, Saunders R, Ensley BD, Suggs S, Harcourt A, Suen W-C, Cruden DL, Gibson DT, Zylstra GJ (1993) Sequences of genes encoding naphthalene dioxygenase in *Pseudomonas putida* strains G7 and NCIB 9816-4. Gene 127: 31–37

Spanggord RJ, Spain JC, Nishino SF, Mortelmans KE (1991) Biodegradation of 2,4-dinitrotoluene by a *Pseudomonas* sp. Appl Environ Microbiol 57:3200–3205

Subramanian V, Liu T-N, Yeh W-K, Gibson DT (1979) Toluene dioxygenase: purification of an iron-sulfur protein by affinity chromatography. Biochem Biophys Res Commun 91:1131–1139

Suen W-C, Gibson DT (1993) Isolation and preliminary characterization of the subunits of the terminal component of naphthalene dioxygenase from *Pseudomonas putida* NCIB 9816-4. J Bacteriol 175:5877–5881

Suen W-C, Gibson DT (1994) Recombinant *Escherichia coli* strains synthesize active forms of naphthalene dioxygenase and its individual α and β subunits. Gene 143: 67–71

Suenaga H, Goto M, Furukawa K (2001a) Emergence of multifunctional oxygenase activities by random priming recombination. J Biol Chem 276:22500–22506

Suenaga H, Mitsuoka M, Ura Y, Watanabe T, Furukawa K (2001b) Directed evolution of biphenyl dioxygenase: emergence of enhanced degradation capacity for benzene, toluene, and alkylbenzenes. J Bacteriol 183:5441–5444

Tan H-M, Cheong C-M (1994) Substitution of the ISPα subunit of biphenyl dioxygenase from *Pseudomonas* results in a modification of the enzyme activity. Biochem Biophys Res Commun 204:912–917

Timmis KN, Rojo F, Ramos JL (1989) Design of new pathways for the catabolism of environmental pollutants. Adv Appl Biotechnol 4:61–82

Timmis KN, Steffan RJ, Unterman R (1994) Designing microorganisms for the treatment of toxic wastes. Annu Rev Microbiol 48:525–557

Torok DS, Resnick SM, Brand JM, Cruden DL, Gibson DT (1995) Desaturation and oxygenation of 1,2-dihydronaphthalene by toluene and naphthalene dioxygenase. J Bacteriol 177:5799–5805

Tsang H-T, Batie CJ, Ballou PD, Penner-Halm JE (1989) X-ray absorption spectroscopy of the [2Fe-2S] Reiske cluster in *Pseudomonas cepacia* phthalate dioxygenase: determination of core dimentions and iron ligand. Biochemistry 28:7233–7240

Wackett LP, Gibson DT (1988) Degradation of trichloroethylene by toluene dioxygenase in whole cell studies with *Pseudomonas putida* F1. Appl Environ Microbiol 54:1703–1708

Wackett LP, Kwart LD, Gibson DT (1988) Benzylic monooxygenation catalyzed by toluene dioxygenase from *Pseudomonas putida*. Biochemistry 27:1360–1367

Wackett LP, Sadowsky MJ, Newman LM, Hur H-G, Li S (1994) Metabolism of polyhalogenated compounds by a genetically engineered bacterium. Nature 368:627–629

Werlen C, Kohler H-P, van der Meer JR (1996) The broad substrate chlorobenzene dioxygenase and *cis*-chlorobenzene dihydrodiol dehydrogenase of *Pseudomonas* sp. strain P51 are linked evolutionarily to the enzymes for benzene and toluene degradation. J Biol Chem 271:4009–4016

Wolfe MD, Lipscomb JD (2003) Hydrogen peroxide-coupled *cis*-diol formation catalyzed by naphthalene 1,2-dioxygenase. J Biol Chem 278:829–835

Wolfe MD, Parales JV, Gibson DT, Lipscomb JD (2001) Singe turnover chemistry and regulation of O_2 activation by the oxygenase component of naphthalene dioxygenase. J Biol Chem 276:1945–1953

Yeh W-K, Gibson DT, Liu T-N (1977) Toluene dioxygenase: a multicomponent enzyme system. Biochem Biophys Res Commun 78:401–410

Yu C-L, Parales RE, Gibson DT (2001) Multiple mutations at the active site of naphthalene dioxygenase affect regioselectivity and enatioselectivity. J Ind Microbiol Biotechnol 27:94–103

Zhang N, Stewart BG, Moore JC, Greasham RL, Robinson DK, Buckland BC, Lee C (2000) Directed evolution of toluene dioxygenase from *Pseudomonas putida* for

improved selectivity toward *cis*-indandiol during indene bioconversion. Metab Eng 2:339–348
Zhao H, Arnold FH (1997) Optimization of DNA shuffling for high fidelity recombination. Nucleic Acids Res 6:1307–1308
Zielinski M, Backhaus S, Hofer B (2002) The principle determinants for the structure of the substrate-binding pocket are located within a central core of a biphenyl dioxygenase α subunit. Microbiology 148:2439–2448
Zylstra GJ, Kim E, Goyal AK (1997) Comparative molecular analysis of genes for polycyclic aromatic hydrocarbon degradation. Genet Eng 19:257–269

9 Bacterial Reductive Dehalogenases

Marc B. Habash,[1] Jack T. Trevors,[2] and Hung Lee[2]

1 Introduction

Toxic halogenated organic compounds can originate from both natural (Gribble 1994, 1996) and industrial sources (Fetzner 1998; Mohn and Tiedje 1992). Many industrially produced polyhalogenated organic compounds such as pentachlorophenol (PCP), polychlorinated biphenyls (PCBs), perchloroethylene (PCE), trichloroetheylene (TCE) and dichloromethane (DCM) are significant environmental and human health problems due to their toxicity and persistence in the environment. However, despite their toxicity, some bacteria have evolved mechanisms to metabolize these compounds. A key group of bacterial enzymes that initiate the degradation of halo-organic compounds is the dehalogenases. Many types of dehalogenases have been described. They include monooxygenases, dioxygenases, dehydrohalogenases, reductive dehalogenases, hydrolytic dehalogenases, thiolytic dehalogenases, and methyl transferases (Fetzner 1998). This chapter examines bacterial reductive dehalogenases that catalyze the substitution of a halogen substituent, typically a chlorine, bromine or iodine atom, with a hydrogen atom.

Reductive dehalogenation requires a reduced electron donor, a role fulfilled by a reduced organic substrate or H_2. If the electron donor provides sufficient reducing power, the removal of a chlorine atom can occur as a one-step reaction where the electron donor provides both the two electrons and single proton necessary for the reaction (Mohn and Tiedje 1992). Alternatively, if the electron donor does not have sufficient reducing power, the halogen atom is removed in a two-step reaction where the electron donor provides two electrons and the surrounding

[1]The Centre for Infection and Biomaterials Research, Hospital for Sick Children, 555 University Ave., Toronto, Ontario, M561X8, Canada
[2]Department of Environmental Biology, University of Guelph, Guelph, Ontario, N1G 2W1, Canada, e-mail: hlee@uoguelph.ca, Tel: +1-519-8244120 Ext. 3828, Fax: +1-519-8370442

Soil Biology, Volume 2
Biodegradation and Bioremediation
(ed. by. A. Singh and O. P. Ward)

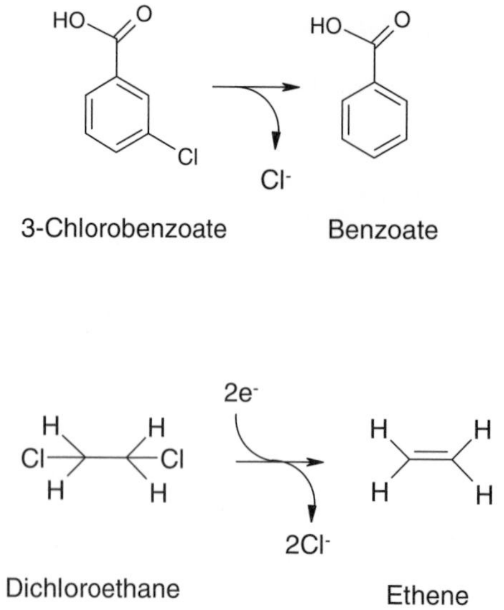

Fig. 1. General reductive dehalogenation mechanisms. (1) Substitution of a chlorine by hydrogen atom during the reductive chlorination of 3-chlorobenzoate to benzoate; and (2) removal of two chlorine atoms on adjacent carbons, resulting in the formation of a double bond in the reductive chlorination of 2-chloroethane to ethene

solvent, such as water provides the proton (Mohn and Tiedje 1992). The two-step reaction has been shown during reductive dechlorination of PCB (2,3,4,5,6 pentachlorobiphenyl) by vitamin B_{12} (Assaf-Anid et al. 1992) and in solvent kinetic studies with the reductive dehalogenase from *Desulfitobacterium chlororespirans* (Krasotkina et al. 2001). Overall, dehalogenation reactions can occur via the removal of either one or two halogen atoms (Fig. 1). In the former, a single halogen atom can be replaced by hydrogen (Mohn and Tiedje 1992), while in the latter, the removal of two adjacent halogen atoms results in the formation of a double bond between the carbons initially bonding the halogen groups (Mohn and Tiedje 1992).

Reductive dehalogenases are incorporated as part of three different microbial pathways: carbon metabolism, energy conservation and co-metabolism. The focus of this chapter will be on carbon metabolism and energy conservation pathways where reductive dehalogenases degrade halogenated compounds specifically for the benefit of a microorganism. With respect to carbon metabolism, we will examine two of the best-studied bacterial reductive dehalogenases that are members of the glutathione transferase (GST) superfamily of enzymes (see Table 1). With respect to energy conservation, we will explore some of the best-studied reductive dehalogenases found in anaerobic microorganisms including the 3-chlorobenzoate, tetra- and tri-chloroethene and chlorophenol reductive dehalogenases (see Table 2).

Table 1. Some properties of selected reductive dehalogenases from aerobic bacteria involved in carbon metabolism of halogenated compounds

Bacteria	Enzyme	MW/Number of subunits	Cellular localization	Optimum pH and temperature	Cosubstrate	Substrates	Type of reaction	References
Sphingomonas paucimobilis UT26	2,5-DCHQ RD	40 kDa	Cytoplasm	nr	GSH	2,5 DCHQ, CHQ	Substitution via glutathione transfer	Miyauchi et al. (1998)
Sphingobium chlorophenolicum	TCHQ RD	58 kDa Dimer (2 × 28 kDa monomers)	Cytoplasm	7–8, 40 °C	GSH	TCHQ, TriCHQ	Substitution via glutathione transfer	Anandarajah et al. (2000), Kiefer and Copley (2002), Xun (1992), Xun et al. (1992)
Sphingomonas sp. UG30	TCHQ RD	58 kDa Dimer (2 × 28 kDa monomers)	Cytoplasm	8.7, 50 °C	GSH	TCHQ, TriCHQ	Substitution via glutathione transfer	Habash et al. (2002)
Methylobacterium dichloromethanicum DM 4	DCM RD	37.4 kDa Monomer	Cytoplasm	8.0–8.5	GSH	DCM	Substitution via glutathione transfer	La Roche and Leisinger (1990)
Methylophilus sp. DM 11	DCM RD	31.2 kDa Monomer	Cytoplasm	8.0–10.0	GSH	DCM	Substitution via glutathione transfer	Bader and Leisinger (1994), Vuilleumier et al. (2001)
Hyphomicrobium sp. DM 2	DCM RD	nr	Cytoplasm	8.0–8.5	GSH	DCM	Substitution via glutathione transfer	Kohler-Staub and Leisinger (1985), Vuilleumier et al. (2001)

Table 1. *Continued*

Bacteria	Enzyme	MW/Number of subunits	Cellular localization	Optimum pH and temperature	Cosubstrate	Substrates	Type of reaction	References
Methylorhabdus multivorans DM 13	DCM RD	nr	Cytoplasm	8.0–9.0	GSH	DCM	Substitution via glutathione transfer	Vuilleumier et al. (2001)
Hyphomicrobium sp. GJ21	DCM RD	nr	Cytoplasm	7.5–9.0	GSH	DCM	Substitution via glutathione transfer	Vuilleumier et al. (2001)
Corynebacterium sepedonicum KZ-4	2,4-DCB-CoA RD	nr	Cytoplasm	nr	NADPH	2,4 DCB-CoA	Hydrolytic chloride elimination[a]	Romanov and Hausinger (1996)
Coryneform bacterium strain NTB-1	2,4-DCB-CoA RD	nr	Cytoplasm	nr	NADPH	2,4 DCB-CoA	Hydrolytic chloride elimination[a]	Romanov and Hausinger (1996)
Rhodopseudomonas palustris	3-CB-CoA RD	nr	Cytoplasm	nr	nr	3-CB-CoA	nr	Egland et al. (2001)

Abbreviations: 3-CB-CoA: 3-chlorobenzoyl CoA; 2,4 DCB CoA: 2,4 dichlorobenzoyl CoA; DCHQ: dichlorohydroquinone; DCM: dichloromethane; GSH: glutathione; nr: not reported; RD: reductive dehalogenase; TCHQ: tetrachlorohydroquinone; TriCHQ: trichlorohydroquinone.

[a] Speculated reaction, precise reaction is unclear.

Table 2. Some properties of purified reductive dehalogenases from anaerobic bacteria involved in dehalorespiration

Organism	Enzyme	MW and number of subunits	Cellular localization	Optimum pH, temperature	Substrates	Chromophore	References
Desulfomonile tiedjei DCB-1	3-CB RD	Heterodimer 64 and 37 kDa	Membrane-bound	7.2–7.5, 38°C	3-CB, PCE[a], TCE[a], PCP[a] and several other CPs[a]	Heme	Mohn and Kennedy (1992), Ni et al. (1995), Townsend and Suflita (1996)
Desulfitobacterium chlororespirans	Cl OHCB RD	50 kDa	Membrane-bound	6.8, 59°C	o-CP, PCP, Cl OHCB, 2,3-, 2,6-DCP, 3,5 DCl OHB, OH PCB	Corrinoid and 2 FeS clusters	Krasotkina et al. (2001)
Desulfitobacterium dehalogenans	o-CP RD	48 kDa Monomer	Membrane-bound	8.2, 52°C	Cl OHPA, 2 CP, 2,3-, 2,4-, 2,6-DCP, PCP, 2B-4CP, OH-PCB	Corrinoid and 2 FeS clusters	van de Pas et al. (1999), Wiegel et al. (1999)
Desulfitobacterium sp. strain PCE1	Cl OHPA RD	48 kDa Monomer	Membrane-associated[b]	nr	o-CP	Corrinoid and 2 FeS clusters	van de Pas et al. (2001b)
Desulfitobacterium sp. strain PCE1	PCE RD	48 kDa Monomer	Membrane-associated[b]	nr	PCE, TCE	Corrinoid and 2 FeS clusters	van de Pas et al. (2001b)
Desulfitobacterium frappieri TCE1	PCE RD	59 kDa Monomer	Membrane-associated[b]	nr	PCE, TCE	Corrinoid and 2 FeS clusters	van de Pas et al. (2001b)
Desulfitobacterium sp. strain PCE-S	PCE RD	65 kDa Monomer	Membrane associated on cytoplasmic side	7.2, 50°C	PCE, TCE	Corrinoid and 2 FeS clusters	Miller et al. (1998)
Desulfitobacterium sp. strain Y51	PCE RD	58 kDa Monomer	Periplasm	7.5, 37°C	PCE, TCE, hexa-, penta-, and tetra-CEAs	Corrinoid and 2 FeS clusters	Suyama et al. (2002)

Table 2. *Continued*

Organism	Enzyme	MW and number of subunits	Cellular localization	Optimum pH, temperature	Substrates	Chromophore	References
Dehalospirillum multivorans	PCE RD	57 kDa Monomer	Cytoplasm	8, 42 °C	PCE, TCE, chlorinated propenes	Corrinoid and 2 FeS clusters	Neumann et al. (1996), Neumann et al. (2002)
Clostridium bifermentans	PCE RD	35 and 35.7 kDa Dimer	Peripheral membrane	7.5, 35 °C	DCE, TCE, *cis*-DCE, *trans*-1,2 ethylene, 1,1 DC ethylene,1,2 C propane	Corrinoid and 2 FeS clusters	Okeke et al. (2001)
Desulfitobacterium hafniense	PCE RD	46.5 kDa	Membrane associated	nr	PCE	Corrinoid and 2 FeS clusters[c]	Christiansen et al. (1998)
Dehalobacter restrictus	PCE RD	61 kDa	Membrane associated on cytoplasmic side	nr	PCE	Corrinoid and 2 FeS clusters	Schumacher et al. (1997)
Dehalococcoides ethenogenes 195	PCE RD	51 kDa Monomer	Membrane-bound	nr	PCE	Corrinoid and 2 FeS clusters	Magnuson et al. (1998)
Dehalococcoides ethenogenes 195	TCE RD	61 kDa Monomer	Membrane-bound	nr	TCE, *cis*-DCE, *trans*-DCE, VC	Corrinoid and 2 FeS clusters	Magnuson et al. (1998)

Abbreviations: 2B-4CP: 2-bromo-4-chlorophenol; 3-CB: 3-chlorobenzoate; CEA: chloroethane; ClOHPA: chlorohyroxyphenylacetate; CP: chlorophenol; DCE: dichloroethene; DClOHB: dichlorohydoxybenzoate; DCP: dichlorophenol; nr: not reported; OH-PCB: hydroxylated polychlorinated biphenyl; PCE: tetrachloroethene; PCP: pentachlorophenol; RD: reductive dehalogenase; TCE: trichloroethene; VC: vinyl chloride.

[a] Co-metabolic dehalogenation induced by 3-chlorobenzoate.

[b] Nature of association with membrane was not determined (van de Pas et al. 2001b).

[c] Likely 2 FeS clusters (Smidt et al. 2000b), however, quantification of Fe and S content of *D. hafniense* indicates potential for three FeS clusters (Christiansen et al. 1998).

This chapter focuses on the biochemical, genetic and functional characteristics of these dehalogenases and, where possible, their regulation and proposed mechanisms of action.

In contrast to carbon metabolism or energy conservation pathways, co-metabolic or fortuitous dehalogenation processes do not directly benefit degrading microorganisms since they are unable to derive any metabolic energy from these reactions. In fact, many co-metabolic reactions consume energy. The only benefit to the degrading microorganisms may arise indirectly from detoxification of the substrate(s) being co-metabolically dehalogenated. Co-metabolic dehalogenation reactions occur as a result of the broad substrate specificity, or "promiscuity", of some microbial dehalogenases. Co-metabolic processes have been found in many methanogenic, acetogenic, sulfate-reducing and iron-reducing microorganisms and involve mainly haloalkanes. These reactions may involve different cofactors such as corrinoid/FeS or factor F_{430}, a nickel porphinoid co-factor found in methanogens (Holliger and Schraa 1994). Co-metabolic reactions have been reviewed elsewhere (Holliger and Schraa 1994; El Fantroussi et al. 1998; Fetzner 1998; Beaudette et al. 2002) and the reader is referred to those excellent reviews for additional information.

2
Reductive Dehalogenases Involved in Carbon Metabolism

Carbon catabolism is the enzymatic breakdown of larger organic compounds into smaller compounds. This vital process is a major source of energy for many microorganisms. However, most of the enzymes involved in carbon metabolism are unable to degrade halogenated organic compounds due to interference by the halogen groups. The metabolic breakdown of many halogenated organic compounds can only occur once the halogen atoms are removed, a role fulfilled by dehalogenases. Acting either alone or as part of a multiple enzyme dehalogenation pathway, reductive dehalogenases remove the halogen groups from a wide assortment of halogenated organic compounds. The dehalogenated organic compound(s) may then be mineralized to carbon dioxide and water through a microorganism's carbon metabolism pathway. Reductive dehalogenases that aid carbon metabolism have been observed in facultatively anaerobic organisms (McGrath and Harfoot 1997; Egland et al. 2001) and in several aerobic bacteria (Saber and Crawford 1985; Häggblöm et al. 1988, 1989; Schenk et al. 1989; Uotila et al. 1991; Xun et al. 1992; Masai et al. 1993; Romanov and Hausinger 1996; Leung et al. 1997; Miyauchi et al. 1998; Fetzner 1998;

Crawford and Ederer 1999; Kumari et al. 2002).What makes these enzymes particularly interesting to study in aerobic bacteria is that they are present in microorganisms that require oxygen to grow and divide, and yet, the presence of that same oxygen adversely affects the ability of the reductive dehalogenases to function properly.

The metabolism of a bromoalkane and various chloroalkanes by the facultative anaerobes *Rhodospirillum rubrum*, *Rhodospirillum photometricum*, and *Rhodopseudomonas palustris* has been observed (Egland et al. 2001; McGrath and Harfoot 1997). The mechanism of 3-chlorobenzoate (3-CB) degradation has been described for *Rhodopseudomonas palustris*. Firstly, 3-CB is metabolized via conversion to 3-CB-CoA, which is dechlorinated by an unknown reductive dehalogenase to benzoyl-CoA, which can then be used as a carbon source (Egland et al. 2001). Other examples of aerobic bacteria that generate CoA thioesters in the reductive degradation of chlorinated benzoates include *Corynebacterium sepedonicum* KZ-4 and Coryneform bacterium strain NTB-1 (Romanov and Hausinger 1996).

An extensively studied group of reductive dehalogenases in the catabolic pathways of aerobic microorganisms are the glutathione transferases (GSTs). GSTs, found in both prokaryotes and eukaryotes, catalyze the conjugation of glutathione to an electrophilic xenobiotic and/or endogenous compound. Currently, this group of enzymes is divided into three superfamilies: the canonical GSTs, the membrane-bound microsomal GSTs, and the fosfomycin-resistance GSTs (Armstrong 1998). Among these, the most relevant to reductive dehalogenation are the canonical or soluble GSTs. This group is divided into several classes that include the alpha (α), beta (β), delta (δ), epsilon (ε), kappa (κ), mu (μ), pi (π), theta (υ), sigma (σ), zeta (ζ), and omega (ω) subdivisions. However, the classification scheme of the soluble GSTs may be subject to change, with new characteristics being discovered in existing members and additional members being identified. Several soluble bacterial GSTs catalyze the removal of chlorine atoms from organic substrates as part of catabolic pathways. The best-known examples of these pathways include the biodegradation of dichloromethane and the recalcitrant xenobiotics PCP and lindane. Two well-studied bacterial soluble GSTs are tetrachlorohydroquinone (TCHQ) reductive dehalogenase and dichloromethane (DCM) dehalogenase.

2.1 Tetrachlorohydroquinone Reductive Dehalogenase

2.1.1 General Characteristics

Tetrachlorohydroquinone reductive dehalogenase (PcpC) is one of the best-studied reductive dehalogenases. Present in several PCP-degrading aerobic bacteria (Saber and Crawford 1985; Häggblöm et al. 1988, 1989; Uotila et al. 1991; Xun et al. 1992; Leung et al. 1997; Crawford and Ederer 1999), PcpC has been best characterized from *Sphingobium chlorophenolicum* ATCC 39723 (formerly *Flavobacterium*, Xun et al. 1992 and *Sphingomonas chlorophenolica* ATCC 39723, Anandarajah et al. 2000; McCarthy et al. 1996) and *Sphingomonas* sp. UG30 (Leung et al. 1997; Habash et al. 2002). The enzyme is a 58-kDa homodimer which is able to catalyze two very different glutathione-dependent reactions (Fig. 2): an isomerization (Anandarajah et al. 2000) and a reductive dechlorination (Xun et al. 1992; McCarthy et al. 1996; Anandarajah et al. 2000; Habash et al. 2002). In vivo, the isomerization reaction converts

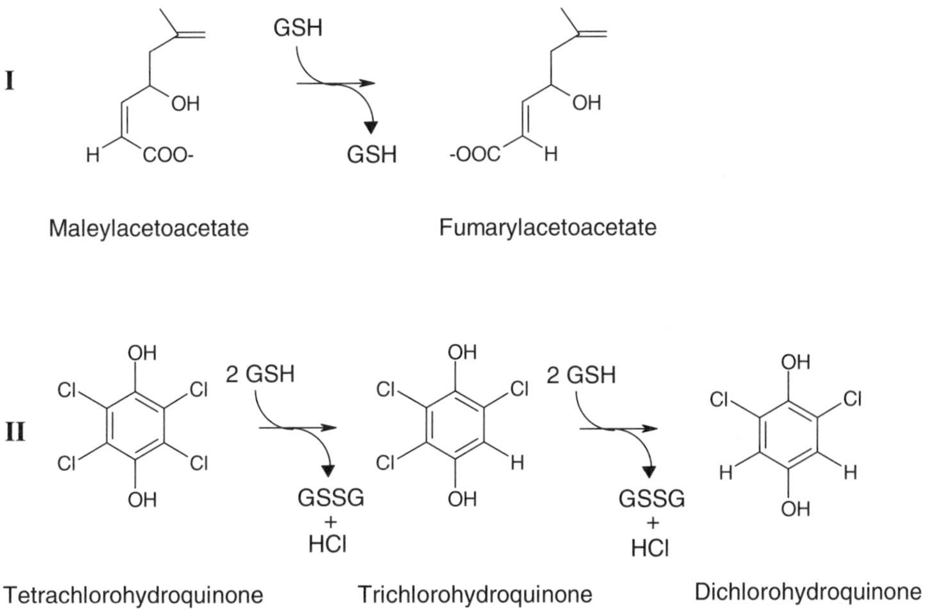

Fig. 2. The two glutathione-dependent reactions catalyzed by PcpC. (1) The isomerization of maleylacetoacetate to fumarylacetoacetate; and (2) the two reductive dechlorination steps forming dichlorohydroquinone from tetrachlorohydroquinone via trichlorohydroquinone. (Anandarajah et al. 2000)

maleylacetoacetone (MAA) to fumarylacetoacetone, an intermediate step in phenylalanine and tyrosine catabolism (Anandarajah et al. 2000). However, due to the instability of MAA, in vitro experiments examining PcpC isomerase activity utilized maleylacetone (Anandarajah et al. 2000), which is an acceptable MAA alternative (Seltzer 1973). The reductive dechlorination occurs in two successive steps. First, tetrachlorohydroquinone (TCHQ) is converted to trichlorohydroquinone (TriCHQ) by the replacement of a chlorine atom with hydrogen. Secondly, TriCHQ is converted to dichlorohydroquinone (DCHQ) with the replacement of a second chlorine atom with hydrogen (McCarthy et al. 1996; Xun et al. 1992). These two reactions form part of the PCP degradation pathway in these bacteria. Aromatic substrates attacked by PcpC must have two hydroxyl groups para to each other. Pentachlorophenol and tetrachlorophenol are competitive inhibitors of this reaction. However, the lack of para-substituted hydroxyl groups in these compounds renders them unable to be turned over by PcpC (Kiefer and Copley 2002).

2.1.2 Reaction Mechanism

A reaction mechanism for the *S. chlorophenolicum* PcpC (Fig. 3) has been proposed by McCarthy et al. (1996), Kiefer et al. (2002), and Kiefer and Copley (2002). The initial steps catalyzed by PcpC to conjugate reduced glutathione to TCHQ did not follow the typical nucleophilic substitution reaction seen in most GSTs (Kiefer et al. 2002). Rather, it has been shown that this reaction involves the ketonization of deprotonated TCHQ to form 2,3,5,6-tetrachloro-4-hydroxycyclohexa-2,4-dienone, followed by a 1,4-elimination reaction to form trichlorobenzoquinone (Kiefer and Copley 2002). Trichlorobenzoquinone then reacts with reduced glutathione to form a glutathione conjugate (Kiefer and Copley 2002).

In subsequent steps, glutathione is removed by formation of a mixed disulfide between glutathione and an N-terminal Cys from PcpC (McCarthy et al. 1996). The enzyme utilizes Cys13 in the glutathione binding site to act as a nucleophile to remove the conjugated glutathione during catalysis (McCarthy et al. 1996; Willett and Copley 1996). A second reduced glutathione is used to extract the enzyme-bound glutathione to form oxidized glutathione. This is only one of a few other known examples of GSTs that utilize two molecules of reduced glutathione (GSH) for the removal of a halogen group with the production of oxidized glutathione (Vuilleumier and Pagni 2002). The other examples, also bacterial enzymes, are the dichlorohydroquinone reductive

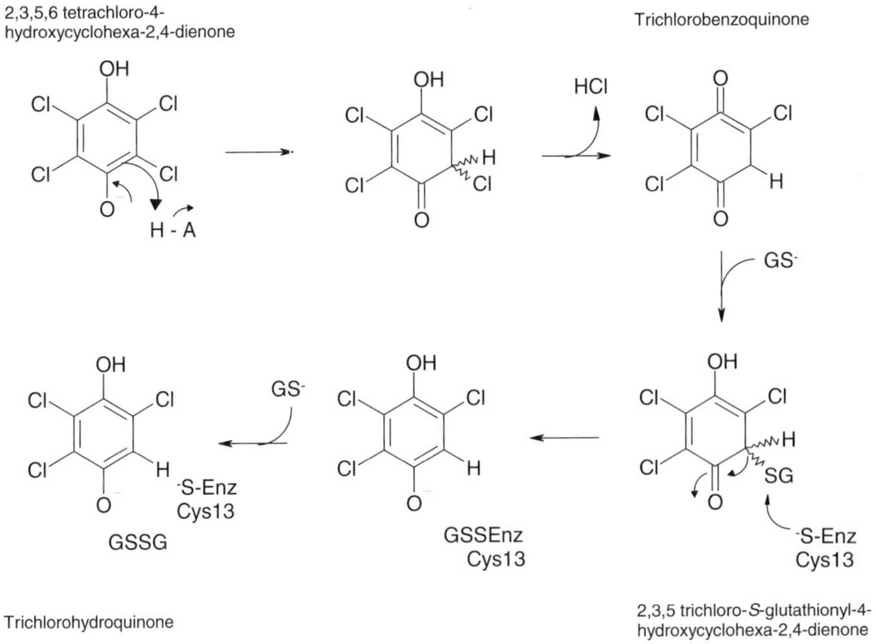

Fig. 3. Proposed mechanism for the removal of the first chlorine atom from tetrachlorohydroquinone (TCHQ) by PcpC as described in Kiefer et al. (2002) and McCarthy et al. (1996)

dehalogenase (LinD) from *Sphingomonas paucimobilis* strains UT26 (Miyauchi et al. 1998) and B90 (Kumari et al. 2002), and LigF from *Pseudomonas paucimobilis* SYK-6 (Masai et al. 1993).

2.1.3 Molecular Characterization

PcpC is a member of the pathway that initiates pentachlorophenol degradation in *S. chlorophenolicum* (Fig. 4). All the genes encoding enzymes in this pathway have been cloned, sequenced and the organization and regulation of the genes examined. The catabolic gene clusters were localized to two fragments. The first 24-kb fragment contains approximately 20 ORFs, 4 of which are involved in PCP degradation (Cai and Xun 2002). Two of the four genes have been identified. They are *pcpC*, encoding the GST responsible for reductive dechlorination of TCHQ and TriCHQ, and *pcpA*, encoding the ring cleaving 2,6-dichlorohydroquinone 1,2-dioxygenase. Two genes which have recently been identified are *pcpE*, encoding maleylacetate reductase, and *pcpM*, a

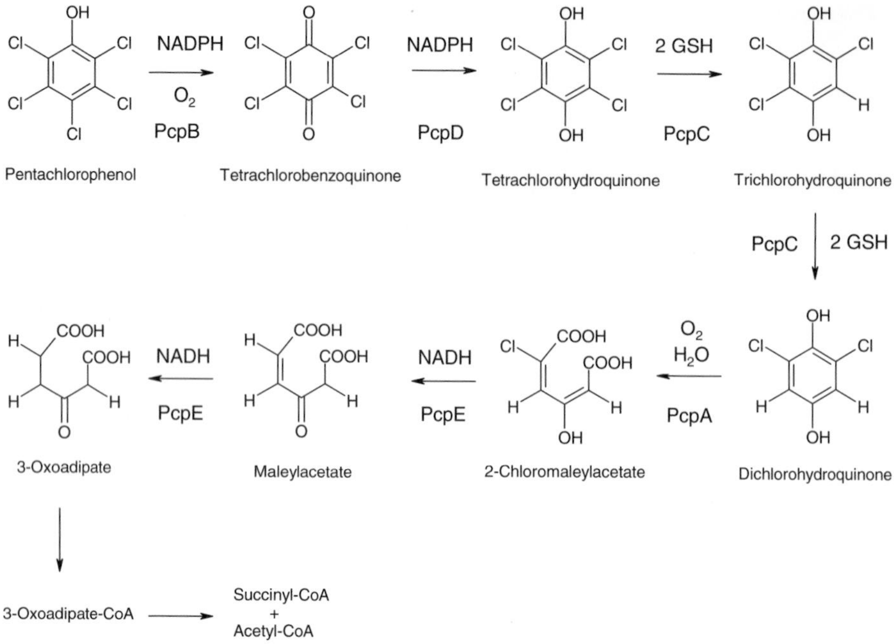

Fig. 4. The pentachlorophenol degradative pathway for *Sphingobium chlorophenolicum* ATCC 39723 and *Sphingomonas* sp. UG30. The pathway was adapted from http://umbbd.ahc.umn.edu/pcp/pcp_image_map.html. The PcpD reaction is described by Dai et al. (2003)

proposed regulatory gene whose role in PCP degradation remains unknown (Cai and Xun 2002). The second 8-kb fragment contains *pcpB*, encoding PCP-4-monooxygenase, *pcpD* encoding TCHQ reductase, and *pcpR*, a LysR type regulatory gene required for PCP degradation (Cai and Xun 2002).

The organization of the genes is unusual in that the five catabolic genes, *pcpA*, *pcpB*, *pcpC*, *pcpD* and *pcpE*, are localized in four discreet locations in the genome. Another interesting feature is the expression of *pcpA*, *pcpB*, *pcpD* and *pcpE* which are inducible by PCP while the expression of *pcpC* is constitutive. Two reasons have been proposed to explain the constitutive expression of the *pcpC* gene. Firstly, the PcpC may have other essential functions in the cell that have not been identified (Cai and Xun 2002). Another explanation for the constitutive nature of PcpC could be from its proposed evolution from a maleylacetoacetate isomerase (Copley 2000). Maleylacetoacetate isomerase is a central metabolic pathway enzyme involved in tryrosine metabolism, the constitutive expression of central metabolic pathway genes is not unusual. Alternatively, maleylacetoacetate isomerase expression was regulated by elements able to recognize the presence of tyrosine or one of its metabolites. However, during the evolution of PcpC, and the PCP degradative pathway, these maleylacetoacetate isomerase regulatory elements would have failed to recognize PCP to initiate production of PcpC. The spontaneous mutation of the maleylacetoacetate regulatory elements and the subsequent selection of constitutively expressed *pcpC* would be a possible solution (Cai and Xun 2002).

2.1.4
Structural Characterization: Comparison to Other GSTs

Structural data on the canonical GST classes have been recently reviewed (Sheehan et al. 2001). Despite a low level of sequence homology, GSTs from all classes have a basic scaffold. Each monomer of the enzyme consists of two domains. The N-terminal domain consists of the first 80 amino acids and takes the form of a highly conserved thioredoxin fold that is responsible for GSH binding (Sheehan et al. 2001). The second domain, made up of α-helices, is primarily responsible for binding the second substrate (Sheehan et al. 2001). The second domain can be more variable owing to the broad substrate range of the different classes of GST. The two domains are joined by a short amino acid linker sequence.

PcpC is presently classified as a member of the zeta class of the GST superfamily based on sequence analyses (Anandarajah et al. 2000; Habash et al. 2002) and substrate range (Anandarajah et al. 2000). To date, PcpC has not yet been crystallized for 3-D structure determination. Since the GSTs are built around a common structure, an examination of other crystallized members of the same class may provide important insights into the structural features of PcpC necessary for activity. One example of a crystallized zeta class GST is the human maleylacetoacetate isomerase, GSTZ1 (Polekhina et al. 2001). One of the amino acids proposed to be important for activity is an N-terminal Ser which stabilizes GSH in the glutathione-binding domain (Board et al. 1997, Polekhina et al. 2001). Although this residue normally points away from the active site, it has been suggested that on GSH binding, a conformational change may occur to allow it to orient in the proper direction, as has been shown in the π GST class (Oakley et al. 1997).

The biochemical (maleylacetone isomerization and dichloroacetic acid metabolism) and primary sequence similarities of PcpC to GSTZ1 indicate that PcpC may have evolved from an ancient pathway of tyrosine catabolism (Anandarajah et al. 2000; Polekhina et al. 2001). Relatively recent mutations would have allowed PcpC to accept the substrates TCHQ and TriCHQ, which are derived from the degradation of PCP, a xenobiotic compound that has only been in the environment for about 66 years. Since *S. chlorophenolicum* also possesses at least one other bona fide maleylacetoacetate isomerase, it was suggested that a gene duplication followed by mutagenesis of the duplicated gene may have resulted in the initial development of PcpC to be recruited for use in the PCP degradation pathway. However, the exact evolutionary origin of PcpC is not known.

The *S. chlorophenolicum* PcpC has been suggested to be an immature enzyme as it suffers from severe substrate inhibition (Copley 1998, 2000; Anandarajah et al. 2000). In vitro, this has hampered efforts to accurately determine the kinetic parameters of the wild-type enzyme (McCarthy et al. 1996; Anandarajah et al. 2000; Kiefer and Copley 2002; Kiefer et al. 2002), though the substrate concentration where this inhibition occurs has not been reported. In vivo, the impact of the substrate inhibition is difficult to evaluate. It has been estimated that in vivo concentrations of TCHQ in *Sphingobium chlorophenolicum* cells exposed to 0.94 mM PCP for 50 min reached 9 μM, while cytoplasmic concentrations were lower at 2 μM (McCarthy et al. 1997). If substrate inhibition does not occur at such low concentrations, then there is likely to be no selective pressure for the cells to evolve an enzyme with less substrate inhibition.

In comparison, the highly homologous (94% amino acid identity) PcpC from *Sphingomonas* sp. UG30 does not exhibit substrate inhibition at TCHQ concentration as high as 300 μM (Habash et al. 2002). The PcpC proteins from *S. chlorophenolicum* ATCC 39723 and UG30 differ by only 13 amino acids. It is not known which of these amino acid residues are responsible for such large differences in the susceptibility to substrate inhibition. A potential area to examine is a group of amino acids likely to be found within the TCHQ binding site and includes three amino acid changes (Thr163Val, Val169Ile and Lys173Ala) from the *S. chlorophenolicum* to the UG30 PcpC sequence. In the UG30 PcpC, this region is more hydrophobic based on a hydropathy plot (Fig. 5), and represents the only real distinct region of hydrophobicity difference between the two PcpC enzymes. The greater hydrophobicity of the UG30 PcpC may allow for better accommodation of the substrate in the binding site. Other significant differences between the two PcpC enzymes include a 10 °C higher temperature optimum and approximately 2 pH units higher pH optimum for the UG30 PcpC compared to the *S. chlorophenolicum* PcpC (Habash et al. 2002). The amino acid residues responsible for these large functional differences are not known.

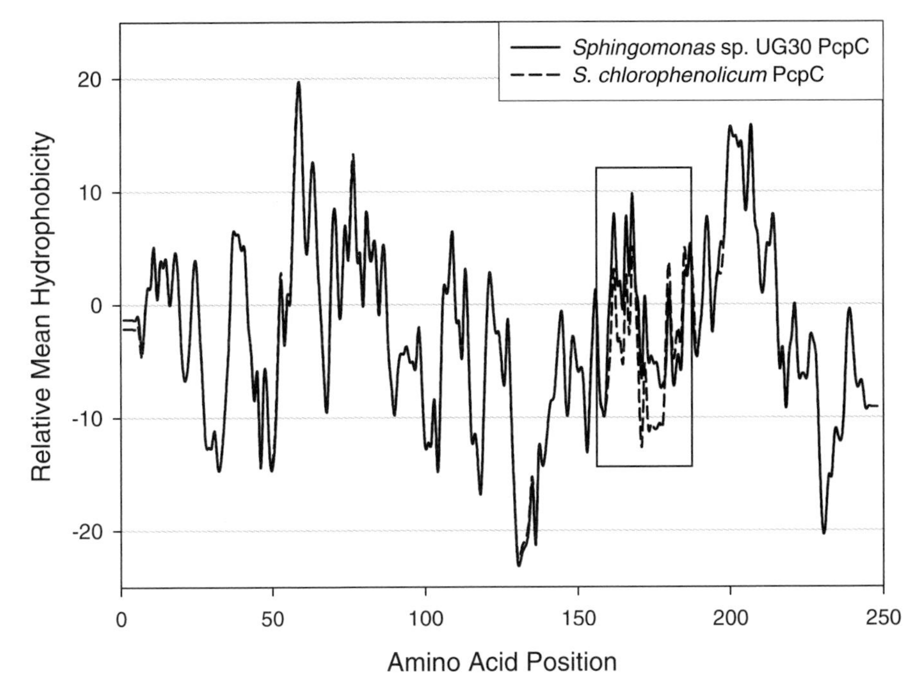

Fig. 5. Kyte and Doolittle hydropathy profiles of PcpC enzymes from *S. chlorophenolicum* ATCC 39723 and *Sphingomonas* sp. UG30 using a scan window of nine amino acids. The profiles for the two enzymes are identical except for the *boxed region* between amino acids 155–180 where the UG30 PcpC is more hydrophobic and may aid in better accommodation of the substrate in its active site

2.2 Dichloromethane (DCM) Dehalogenase

PcpC is an example of a reductive dehalogenase that works in concert with several other enzymes in a dehalogenation pathway involved in carbon metabolism. An example of a solitary reductive dehalogenase involved in carbon metabolism is the DCM dehalogenase. This enzyme dechlorinates DCM to formaldehyde, which is funneled into the carbon and energy metabolic pathways of methylotrophic bacteria (Leisinger et al. 1994; Leisinger and Braus-Stromeyer 1995). This enzyme has been reviewed previously (Leisinger et al. 1994; Leisinger and Braus-Stromeyer 1995; Vuilleumier 1997; Copley 1998; Vuilleumier and Pagni 2002). Here, we will provide an overview of some of the key features of this enzyme.

2.2.1
Enzyme Characteristics

Most studies on DCM dehalogenases have been done on enzymes from *Methylobacterium dichloromethanicum* DM 4 (previously *Methylobacterium* sp. strain DM 4, La Roche and Leisinger 1990) and *Methylophilus* sp. strain DM 11 (Scholtz et al. 1988; La Roche and Leisinger 1990, 1991; Bader and Leisinger 1994; Vuilleumier and Leisinger 1996; Marsh and Ferguson 1997; Gisi et al. 1998, 1999; Evans et al. 2000; Kayser et al. 2000, 2002; Kayser and Vuilleumier 2001; Wheeler et al. 2001). The primary sequences of these enzymes share 56% identity. They catalyze the dechlorination of DCM to formaldehyde in a two-step reaction (Fig. 6). In the first step, DCM dehalogenase catalyzes the conjugation of DCM with GSH, forming an unstable *S*-chloromethylglutathione with the loss of a chlorine atom. In the second step, the conjugated product spontaneously reacts with water to form *S*-hydroxymethylglutathione with the loss of a second chlorine atom (Wheeler et al. 2001). *S*-Hydroxymethylglutathione is further converted spontaneously to formaldehyde and GSH (Wheeler et al. 2001).

The two dehalogenases from strains DM 4 and DM 11 differ significantly in kinetic parameters and dehalogenase expression levels. For example, the DM 4 DCM dehalogenase exhibits approximately six-fold lower V_{max} and K_m values than the DM 11 dehalogenase. Thus, the DM 4 DCM dehalogenase works more slowly, but it has a greater affinity for DCM than the DM 11 dehalogenase. DCM dehalogenase can be expressed to a greater extent by strain DM 4 (15% of the total cell protein) relative to strain DM 11 (7% of the total cell protein) (Scholtz et al. 1988). These differences affect the conditions under which each microorganism functions best. *Methylobacterium dichloromethanicum* DM 4 grows and divides optimally under continuous culture and DCM-limited conditions (Gisi et al. 1998), while *Methylophilus* sp. strain DM 11 grows best in batch culture with an excess of

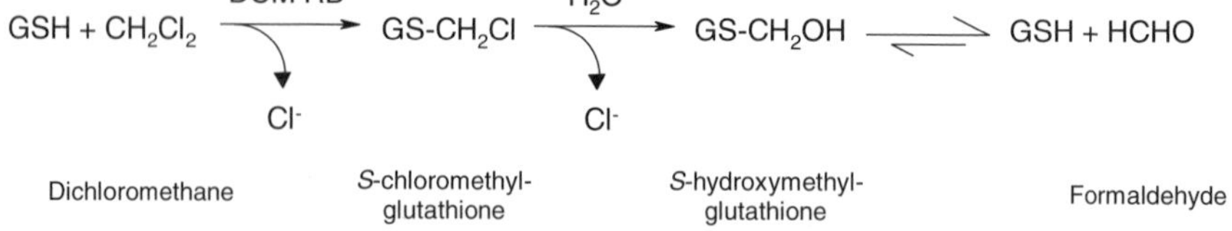

$$GSH + CH_2Cl_2 \xrightarrow[Cl^-]{DCM\ RD} GS\text{-}CH_2Cl \xrightarrow[Cl^-]{H_2O} GS\text{-}CH_2OH \rightleftharpoons GSH + HCHO$$

Dichloromethane — *S*-chloromethyl-glutathione — *S*-hydroxymethyl-glutathione — Formaldehyde

Fig. 6. Dechlorination of dichloromethane to formaldehyde by dichloromethane dehalogenase with glutathione (GSH) as a cofactor. Two reaction intermediates are formed: *S*-chloromethyl-glutathione and *S*-hydroxymethyl-glutathione. (Wheeler et al. 2001)

DCM (Gisi et al. 1998). These differences were postulated to be adaptive responses to conditions found in the natural environments from which each of the bacteria was isolated (Gisi et al. 1998). The above differences have been used to divide the DCM dehalogenases into two groups represented by the DM 4 enzyme (group A) and DM 11 enzyme (group B), respectively.

An interesting secondary reaction arising from the action of DCM dehalogenase is the formation of DNA adducts (Kayser and Vuilleumier 2001). One might reason that formaldehyde formed from DCM degradation may have mutagenic effects upon the degrading microorganism. However, formaldehyde is not mutagenic to *M. dichloromethanicum* DM 4 (Wheeler et al. 2001) and this strain is able to grow on DCM (Kayser et al. 2000). Subsequent studies found that it is the *S*-chloromethylglutathione that is problematic. In experiments using either ^{14}C-DCM or ^{35}S-GSH, the radiolabeled *S*-chloromethylglutathione produced by DCM dehalogenase formed adducts with DNA, resulting in DNA breakage (Kayser and Vuilleumier 2001). Subsequent experiments with whole cells of *Methylobacterium dichloromethanicum* DM 4 and its *polA* mutant (a DNA polymerase deficient mutant) verified that DCM degradation results in DNA damage (Kayser and Vuilleumier 2001). Although both strains produce similar levels of DCM dehalogenase, the *polA* mutant showed a greater frequency of DNA strand breaks (Kayser and Vuilleumier 2001). The results suggested that DNA polymerase I is needed to repair DNA breaks when *Methylobacterium dichloromethanicum* DM 4 is grown on DCM as a sole carbon and energy source (Kayser et al. 2000; Kayser and Vuilleumier 2001).

2.2.2 Molecular Characterization

DCM dehalogenase is an inducible enzyme encoded by the *dcmA* gene (La Roche and Leisinger 1991). In strain DM 4, *dcmA* is regulated by *dcmR*, a 30-kDa putative DCM-specific repressor that prevents the transcription of *dcmA* (La Roche and Leisinger 1991). In addition, in DM 4 and other DM 4-like methylotrophic bacteria the *dcmA* and *dcmR* genes have been localized to a 4.2-kb *Bam*H1 fragment (Schmid-Appert et al. 1997). The fragment is flanked by insertional sequence elements suggesting the DCM catabolic genes may potentially be transferred horizontally by transposition (Schmid-Appert et al. 1997).

An analysis of DCM-degrading bacteria retrieved from enrichment cultures and sludge from wastewater treatment facilities has shown similarity to DM 4 dehalogenase based on cross-reactivity with polyclonal antibodies raised against the DCM dehalogenase from DM 4 strain (Kohler-Staub et al. 1986) and *dcmA* hybridization tests (Scholtz et al. 1988). A more detailed examination of *dcmA* gene sequences has revealed slight variations among different DCM-degrading isolates, mainly as single nucleotide substitutions (Vuilleumier et al. 2001). From 15 DCM-degrading bacterial isolates, 8 different *dcmA* gene sequences were found, and this translates into 7 different DcmA amino acid sequences (Vuilleumier et al. 2001). A strain isolated from chlorinated aliphatic hydrocarbon-contaminated sludge in the Netherlands (GJ21) had the greatest number of substitutions, with 15 nucleotide substitutions corresponding to 9 amino acid changes (Vuilleumier et al. 2001). However, no significant differences in dehalogenase activity were noted for the seven different DCM dehalogenases when compared to the DCM dehalogenase from strain DM 4 (Vuilleumier et al. 2001). These results indicate that the sequence and enzymatic properties of DCM-degrading aerobic methylotrophic bacteria are well conserved (Vuilleumier et al. 2001).

2.2.3 Structural Characterization

In strain DM 4, *dcmA* encodes a 287 amino acid protein with a predicted MW of 37.4kDa (La Roche and Leisinger 1990). In contrast, the *dcmA* gene in strain DM 11 encodes a 267 amino acid protein with a predicted MW of 31.2kDa (Bader and Leisinger 1994). The DCM dehalogenases have been placed in the GST superfamily based on several lines of evidence. Site-directed mutagenesis studies have identified an essential Ser in the N-terminal cofactor-binding site, typical of the theta class GSTs (Vuilleumier and Leisinger 1996). A subsequent study identified another N-terminal amino acid (Ala27 for the DM 4 enzyme and Val18 for the DM 11 enzyme) important for glutathione binding (Vuilleumier et al. 1997). Mutations of Ala27 in DM 4 and Val18 in DM 11 are believed to cause conformational changes in the DCM dehalogenase, resulting in decreased affinity for glutathione (Vuilleumier et al. 1997). This requirement for glutathione supports the view that the dichloromethane dehalogenases are GSTs (Vuilleumier et al. 1997). Finally, a 3-D structural homology model of the DM 11 DCM dehalogenase predicted an affiliation with the glutathione binding site of the theta class GSTs (Marsh and Ferguson 1997).

3 Reductive Dehalogenases Involved in Energy Conservation

Some bacteria can couple the degradation of halogenated organic compounds to the production of cellular energy. This process, termed dehalorespiration, has only been found in anaerobic microorganisms and includes the well-studied dehalogenation reactions of tetrachloroethene (PCE), trichloroethene (TCE), chlorophenols (CP), and 3-chlorobenzoate (3-CB). The basis for energy conservation in bacteria is the development of an electrochemical gradient on either side of a membrane (a chemiosomotic gradient), produced by the translocation of protons and/or their utilization on one side of a membrane coupled with their production on the other side. This is associated with the passage of electrons through an electron transport chain present within the bacterial inner membrane. This is reminiscent of ATP production via the electron transport system present within the inner membrane of mitochondria in eukaryotic cells.

In anaerobic bacteria, the removal of halogen atoms by reductive dehalogenases drives the dehalorespiration process by helping to form a proton gradient on either side of the bacterial inner membrane (Fig. 7). Membrane-bound hydrogenases and/or formate hydrogenases produce H^+ on the periplasmic side of the inner membrane, while on the cytoplasmic side, H^+ is utilized by a soluble or membrane-associated reductive dehalogenase. The presence of the reductive dehalogenase at the end of the electron transport chain allows its halogenated substrate to serve as a terminal electron acceptor. The incorporation of a proton and the electrons results in the release of the halogen atom. Thus, the dehalogenation pathway performs two important tasks. The first is the formation of a proton motive force (PMF), resulting in H^+ movement from the periplasmic to cytoplasmic sides of the membrane through an ATP-generating ATPase. The second is the dehalogenation of toxic halogenated compounds resulting in their complete or partial biodegradation. This pathway has been termed the chemiosmotic model of dehalorespiration and was first proposed by Mohn and Tiedje (1990) based on studies with *Desulfomonile tiedjei*, which possesses the 3-CB reductive dehalogenase.

3.1 3-Chlorobenzoate Dehalogenase

Desulfomonile tiedjei is a non-motile bacterium, isolated from an enrichment culture, that utilizes 3-chlorobenzoate (3-CB) as a terminal

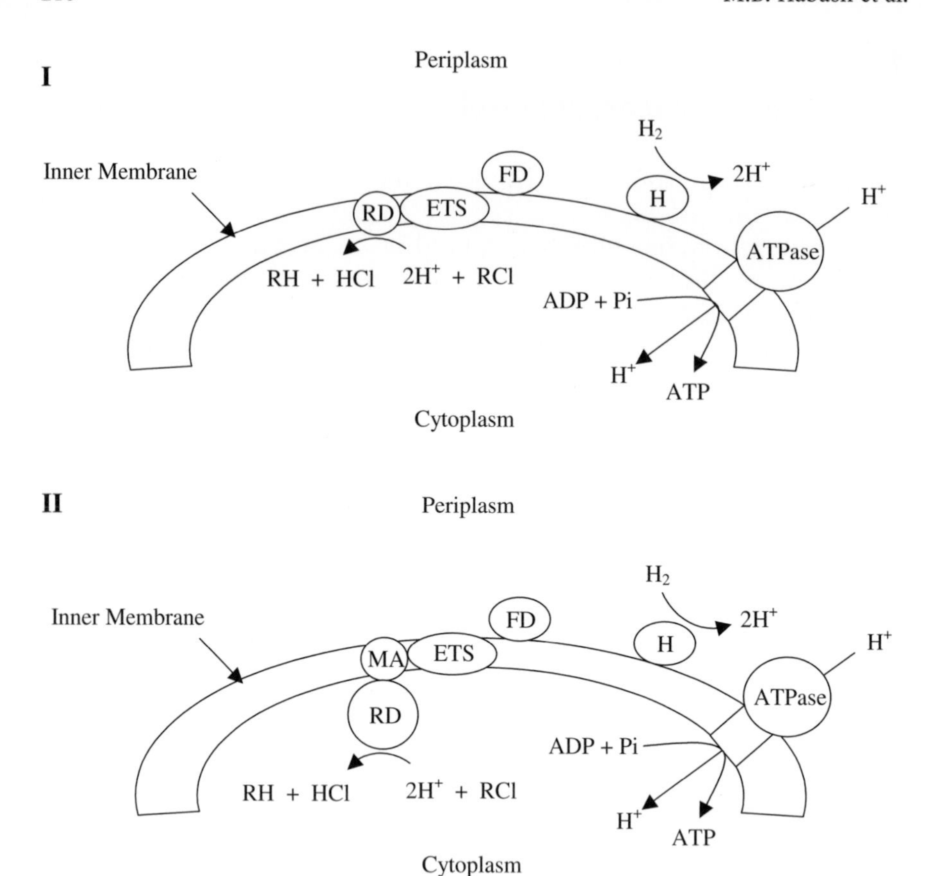

Fig. 7. Proposed dehalorespiratory pathways in anaerobic bacteria that are capable of growth on halogenated alkyl and aryl compounds (RCl). An electron transport system (ETS) with the reductive dehalogenase as the terminal enzyme utilizes the halogenated compounds as electron acceptors. The *RD* can be associated with the inner cell membrane as (1) an integral membrane protein, and (2) through attachment by a membrane anchor protein (MA). Other components include the H^+ producing hydrogenase (H) and/or formate dehydrogenase (FD) present on the periplasmic side of the inner membrane. An ATPase translocates H^+ from the periplasmic to cytoplasmic side of the inner membrane. This reaction is coupled to ATP production

electron acceptor (Mohn and Tiedje 1990). Although not all of the components of the complete electron transport pathway have been identified, studies have identified several components that include a hydrogenase, formate dehydrogenase, a novel cytochrome c, a membrane-bound quinone of unknown type and the terminal 3-CB reductive dehalogenase (Louie et al. 1997; Louie and Mohn 1999). Evidence has shown that a scalar mechanism, i.e. where protons are not translocated, establishes a PMF which results in ATP production. It has also been suggested that a translocation process takes place. However, this has not yet been identified (Louie and Mohn 1999).

An important component of this pathway is the inducible terminal 3-CB reductive dehalogenase, which was the first reductive dehalogenase to be purified from an anaerobic bacterium (Ni et al. 1995). This enzyme is comprised of two subunits (64 and 37kDa) and is believed to be an integral membrane protein (Ni et al. 1995). Unlike other reductive dehalogenases, 3-CB dehalogenase possesses a heme group that acts as a redox center and is not adversely affected by oxygen (Ni et al. 1995). In addition to 3-CB, pure cultures of *D. tiedjei* appear to co-metabolically dehalogenate tetrachloroethene (PCE) (Townsend and Suflita 1996) and pentachlorophenol (PCP) (Mohn and Kennedy 1992). The ability to dehalogenate PCE and PCP was induced in the presence of 3-CB, while neither TCE nor PCP alone induced dehalogenation of PCE and PCP by *D. tiedjei* (Mohn and Kennedy 1992; Townsend and Suflita 1996). Furthermore, it has been shown that reductive dehalogenation occurs at the location where the halogen atom is removed using cell suspensions of *D. tiedjei* in D_2O. This indicates that a partial ring reduction, which would allow incorporation of a proton at a location other than the site of halogen removal, does not happen. To date, no structural or mechanistic data have been presented for the purified enzyme.

Other anaerobic bacteria have been discovered that are capable of dehalorespiration. Some of these include various *Desulfitobacterium* sp. (Utkin et al. 1995; Christiansen and Ahring 1996; Sanford et al. 1996; Miller et al. 1997; Holliger et al. 1998; Gerritse et al. 1999; van de Pas et al. 2001b; Suyama et al. 2002), *Dehalobacter restrictus* (Holliger et al. 1998), *Clostridium bifermentans* DPH-1 (Okeke et al. 2001), *Dehalospirillum multivorans* (Neumann et al. 1994), and *Dehalococcoides ethenogenes* 195 (Maymo-Gatell et al. 1999). The common feature amongst these various anaerobic microorganisms is the presence of reductive dehalogenases that are the terminal enzymes in an electron transport chain involved in dehalorespiration. The inducible reductive dehalogenases from these anaerobic organisms remove halogen atoms from four main substrates:PCE, TCE, 3-chloro-4-hydroxybenzoate, and 3-chloro-4-hydroxyphenylacetate.

The reductive dehalogenases involved in dehalorespiration share many common characteristics. These enzymes have MW ranging from 48 to 65kDa, with optimal pH and temperature values ranging from 7.2 to 8.2 and 37 to 52°C, respectively. They are all inhibited by oxygen and require an electron donor, such as methyl viologen, for activity. The physiological electron donors are not known. These enzymes possess identical redox centers, comprising of a cobalamin-containing corrinoid factor and iron-sulfur clusters. In addition, these enzymes are typically membrane-associated either as integral membrane proteins or peripherally through a separate anchoring protein that contains two or three membrane-spanning regions. This anchor, known as PceB or CprB, is found either upstream or downstream from the gene coding for the reductive dehalogenase and is co-transcribed with the dehalogenase.

Sequence analyses have revealed important conserved regions in all reductive dehalogenase-encoding genes. Two areas coding for the iron-sulfur clusters of the reductive dehalogenases are highly conserved. Typically, one iron-sulfur cluster is CXXCXXCXXXCP), while the other is in the form of Fe_4S_4 (represented by the consensus sequence GXXCXXCXXXCP). Many of the sequences contain the twin Arg signal sequence (RRXFXK), which plays a role in protein maturation and translocation of mainly periplasmic proteins that bind different redox cofactors (Berks 1996; Berks et al. 2000). Sequence analysis also indicated that the reductive dehalogenases from this group of microorganisms do not possess the consensus sequence for a corrinoid-binding motif (DXHXXG). This perhaps reflects the fact that the corrinoid binds non-covalently to the enzyme (Neumann et al. 1998).

3.2 3-Chloro-4-Hydroxyphenylacetate Reductive Dehalogenase

3.2.1 General Characterization

Separate reductive dehalogenases from *Desulfitobacterium dehalogenans*, *Desulfitobacterium hafniense*, and *Desulfitobacterium* sp. PCE1 have been identified that are induced by and can degrade 3-chloro-4-hydroxyphenylacetate (Cl-OHPA). The enzyme from *Desulfitobacterium* sp. PCE1 will be described in the PCE/TCE reductive dehalogenase section. One of the best-studied reductive dehalogenases in this group is from *Desulfitobacterium dehalogenans*, a strain first isolated from a freshwater sediment sample by enrichment on 2,4 dichlorophenol (2,4-DCP) (Kohring et al. 1989a,b). The strain was capable of 2,4-DCP and Cl-OHPA dechlorination utilizing Cl-OHPA as an electron acceptor. In addition, *D. dehalogenans* can dehalogenate a variety of penta-, tetra-, tri-, and di-chlorophenols when grown with either Cl-OHPA or 2,4 DCP as inducers of reductive dehalogenation (Utkin et al. 1995). Experiments involving whole and broken cells with methyl viologen, an electron donor unable to penetrate the cell membrane, indicated the dehalogenating activity was present in the cytoplasm (van de Pas et al. 1999). The ability of *D. dehalogenans* to reductively dehalogenate *ortho*-chlorophenols is linked to a 48-kDa membrane-associated *ortho*-chlorophenol reductive dehalogenase (o-CP RD) (van de Pas et al. 1999). The substrate range of the o-CP RD includes Cl-OHPA, PCP, several dichlorophenols (2,3-, 2,4-, and 2,6-DCP), and 2-chlorophenol (van de Pas et al. 1999). *D. dehalogenans* can reductively dechlorinate the *para*-hydroxylated PCB derivatives 4-hydroxylated PCB and 4,4′-dihydroxylated PCB (Wiegel et al. 1999). However, it is not known if o-CP RD is specifically responsible for this activity. No measurable activity has been observed against tetrachloroethene or trichloroethene (van de Pas et al. 1999).

The energy-producing pathway in *D. dehalogenans* possesses many proteins with functions similar to the 3-CB dehalorespiratory pathway of *D. tiedjei*. The o-CP reductive dehalogenase is likely to be the terminal reductase involved in the energy-producing electron transport system that utilizes o-CP as an electron acceptor. However, only a fraction of the theoretically possible ATP (El Fantroussi et al. 1998) is produced, indicating an inefficient energy-producing pathway (van de Pas et al. 2001a). Other components of the electron transport chain of *D. dehalogenans* have been examined using random Tn916 insertion to create various dehalorespiring mutants (Smidt et al. 1999). H_2-hydrogenase and formate dehydrogenase mutants were found to be incapable of dehalorespiration, this is probably due to the disruption of an electron transport system required for dehalorespiration by *D. dehalogenans* (Smidt et al. 1999).

A second Cl-OHPA reductive dehalogenase has been purified from *Desulfitobacterium hafniense*, a strain isolated from an enrichment culture obtained from a trichlorophenol-converting microbial consortium from municipal sludge (Christiansen and Ahring 1996). The purified enzyme was similar to the o-CP reductive dehalogenase from *D. dehalogenans* in several ways. The enzyme can reductively dehalogenate Cl-OHPA, but not PCE (Christiansen et al. 1998). The 47-kDa membrane-associated dehalogenase contains a corrinoid factor as well as iron-sulfur clusters (Christiansen et al. 1998). One difference between the reductive dehalogenase from *D. hafniense* and all other reductive dehalogenases is the calculated values of iron and sulfur present in the enzyme. The *D. hafniense* reductive dehalogenase was determined to have 12 iron and 12 sulfur atoms per subunit (Christiansen et al. 1998). In contrast, all other reductive dehalogenases, with the exception of the 3-CB reductive dehalogenase from *D. tiedjei* that does not possess any FeS clusters, have seven to eight iron and eight sulfur atoms per subunit. The authors suggest this may indicate the presence of three Fe_4S_4 clusters in the *D. hafniense* reductive dehalogenase (Christiansen et al. 1998). However, based on the sequence of the amino acid sequence of the o-CP reductive dehalogenase, CprA, the *D. hafniense* enzyme contains the conserved FeS sequences found in other reductive dehalogenases, suggesting that it may contain only two FeS clusters (Smidt et al. 2000b). This apparent discrepancy remains to be investigated.

3.2.2 Molecular Characterization

Currently, the *D. hafniense* genome is being sequenced. An examination of the primary sequences of the o-CP reductive dehalogenases (CprA) from *D. hafniense*, *D. dehalogenans* and *Desulfitobacterium* PCE1 show nearly identical sequences, sharing approximately 99% identity (Smidt et al. 2000b).

The entire o-chlorophenol gene cluster from *D. dehalogenans* has been characterized. The o-CP RD, encoded by the *cprA* gene, is responsible for dehalogenating activity towards ortho-chlorophenols (o-CP) (van de Pas et al. 1999). An examination of the CprA primary sequence reveals features typical of this class of reductive dehalogenase, including two iron-sulfur clusters, one Fe_4-S_4 ferredoxin-like center and one Fe_3-S_4 truncated iron-sulfur binding cluster binding motif, and a cobalamin-containing center per monomer (van de Pas et al. 1999). Also, characteristic of a periplasmic redox center containing proteins, a twin Arg-type signal sequence is present in CprA (van de Pas et al. 1999). The upstream of *cprA* is the *cprB* gene coding for the small hydrophobic anchor protein (van de Pas et al. 1999).

Several other genes, along with *cprA* and *cprB* have been found that are likely to be involved in regulation of gene transcription and protein folding. *cprC* and *cprK* code for potential transcriptional regulators, while *cprD* and *cprE* code for proteins which are very similar to the GroEL chaperonines that aid in protein folding and prevention of protein aggregation (Smidt et al. 2000a). The *cprT* codes for a protein with significant similarity to a propyl peptidyl isomerase that catalyzes the *cis-trans* isomerization of proline residues. This activity aids in protein folding (Smidt et al. 2000a). A comparison of the dehalorespiratory pathway of *D. dehalogenans* to others (Smidt et al. 2000a) indicates a similarity to the proposed respiratory complexes found in denitrifying bacteria, which use an electron transport system for the reduction of nitrate to nitrogen gas (Moura and Moura 2001).

3.3 3-Chloro-4-Hydroxybenzoate Reductive Dehalogenase

An integral membrane dehalogenase induced by and capable of 3-chloro-4-hydroxybenzoate (Cl-OHCB) reductive dehalogenation has been found in *Desulfitobacterium chlororespirans*. This microorganism, isolated from an enrichment culture originating from compost soils, can dechlorinate 2,3-dichlorophenol (DCP) (Sanford et al. 1996) and is capable of growth on 2,3- and 2,6-DCP, 2,4,6-tribromophenol, Cl-OHCB, and Cl-OHPA, which serve as electron acceptors (Sanford et al. 1996). Dehalogenation is carried out by a 50-kDa membrane-bound protein with many characteristics similar to the other reductive dehalogenases involved in dehalorespiration (Krasotkina et al. 2001). This includes oxygen sensitivity, possible post-translational modification (since methionine was not the first amino acid), and the presence of the conserved sequence NYVPG, which is present in all published sequences of reductive dehalogenases isolated from *Desulfitobacterium* strains (Krasotkina et al. 2001). Though no role is postulated for this conserved sequence. Other similarities involve the redox reaction centers present in the enzyme, including two FeS clusters and one corrinoid (Krasotkina et al. 2001). An important characteristic that distinguishes the *D. chlororespirans* reductive dehalogenase from others is its ability to dehalogenate hydroxylated PCBs. This is currently the only known enzyme with this capability (Krasotkina et al. 2001).

The proposed catalytic mechanism of this reductive dehalogenase involves the donation of a proton by the solvent to aid in the removal of the chlorine atom from the substrate. Solvent kinetic experiments using $^{2}H_2O$ indicated that a partially rate-limiting step is an intramolecular proton transfer, with the proton originating from the surrounding solvent (Krasotkina et al. 2001). In addition, substrates with a hydroxyl group ortho to the chlorine are necessary for dehalogenase activity (Krasotkina et al. 2001). Competitive inhibitors to the enzyme also require this ortho-hydroxyl configuration to be effective. The presence of another functional group meta to the chlorine was necessary for recognition of the substrate by the enzyme. It was shown that 2-chlorophenol was not a substrate or inhibitor, however, the addition of a meta-substituted carboxyl or acetyl group, such as with Cl-OHCB and Cl-OHPA, resulted in dehalogenation of the substrate (Krasotkina et al. 2001). It is not certain why the ortho hydroxyl and meta-substituent in the substrate are required for dehalogenase activity. It has been suggested that they may be involved in binding to the substrate pocket of the enzyme (Krasotkina et al. 2001). Further structural studies with these reductive dehalogenases may help answer this question.

3.4
Tetrachloroethene and Trichloroethene Reductive Dehalogenases

Many genera of anaerobic bacteria have been isolated with the ability to dehalogenate tetrachloroethene (PCE) and/or trichloroethene (TCE). These include several *Desulfitobacterium* species including strain PCE-S (Miller et al. 1997; Miller et al. 1998), PCE1 (van de Pas et al. 2001b), *D. frappieri* TCE1 (Gerritse et al. 1999), and strain Y51 (Suyama et al. 2002). All of these microorganisms are capable of converting PCE to TCE and then to *cis*-1,2-dichloroethene (DCE) except for strain PCE1 which only converts PCE to TCE.

3.4.1
General Characteristics

The PCE reductive dehalogenases from *Desulfitobacterium* sp. PCE-S and *D. frappieri* TCE1 are very similar, with MW of 65 and 59kDa, respectively (Miller et al. 1998; van de Pas et al. 2001b). Furthermore, their N-terminal sequences are highly homologous (van de Pas et al. 2001b). The dehalogenase from strain PCR-S has been shown to bind to the membrane facing the cytoplasmic side (Miller et al. 1997). It is likely that the TCE1 dehalogenase is also membrane-bound. Another characteristic of these dehalogenases is their narrow chlorinated substrate range. Both enzymes only dehalogenate PCE and TCE, as shown in assays with cell extracts (Miller et al. 1997; Gerritse et al. 1999) and purified enzymes (Miller et al. 1998, van de Pas et al. 2001b). This is similar to other PCE/TCE reductive dehalogenases, such as those from strain PCE1 (van de Pas et al. 2001b) and *Dehalobacter restrictus* (Schumacher et al. 1997; Holliger et al. 1998).

Desulfitobacterium sp. PCE1 is an interesting strain which is able to dechlorinate both PCE and chlorophenols using two separately inducible enzymes (van de Pas et al. 2001b). As described earlier, *Desulfomonile tiedjei* is also capable of chlorophenol and PCE/TCE dehalogenation. The 3-CB reductive dehalogenase in *D. tiedjei* is involved in energy conservation during the utilization of 3-CB (Ni et al. 1995). Its ability to dehalogenate PCE/TCE has been shown to be due to co-metabolism, rather than dehalorespiration or dechlorination for carbon metabolism (Townsend and Suflita 1996).

Induction of strain PCE1 with either PCE or Cl-OHPA results in different protein profiles and substrate ranges, suggesting different degradation pathways for each compound (van de Pas et al. 2001b). Cell extracts of strain PCE1 grown on PCE exhibit activity towards PCE and

TCE, but not chlorophenols. Furthermore, growth on PCE induces the production of a 48kDa enzyme that degrades PCE to TCE (van de Pas et al. 2001b). Degradation of TCE does occur, albeit at a rate corresponding to approximately 10% of the rate of PCE degradation. In comparison, the PCE reductive dehalogenase of strain TCE1 has nearly identical rates of degradation for PCE and TCE (van de Pas et al. 2001b). When grown on Cl-OHPA, strain PCE1 shows activity towards Cl-OHPA and various DCPs (2,3-, 2,4-, 2,5-, and 2,6-) (van de Pas et al. 2001b). Growth of strain PCE1 on Cl-OHPA induces the production of a 48kDa chlorophenol reductive dehalogenase which does not dehalogenate PCE or TCE, while the PCE reductive dehalogenase does not dehalogenate chlorophenols, indicating that two different enzymes are produced for these two reactions (van de Pas et al. 2001b).

A relatively recent addition to the group of *Desulfitobacterium* reductive dehalogenases is from *Desulfitobacterium* sp. strain Y51. Similar to other PCE/TCE reductive dehalogenases, the purified 58-kDa PCE reductive dehalogenase (PceA) is a corrinoid-containing protein (Suyama et al. 2002). However, unlike other reductive dehalogenases, the mature enzyme from strain Y51 was localized to the periplasm and not in association with the inner membrane. The unprocessed enzyme is localized to the cytoplasm and has an MW of 67kDa (Suyama et al. 2002). In comparison, PCE dehalogenase from *D. multivorans* has been purified from the cytosol, while PCE dehalogenases from strains *Desulfitobacterium* sp. PCE-S, *D. restrictus*, and *D. ethenogenes* have been found to be associated with the membrane fraction. The mature enzyme from strain Y51 is able to dehalogenate PCE, TCE, hexa- and penta-chloroethanes (Suyama et al. 2002). Lower chlorinated ethenes and ethanes (trichloroethanes) and chlorobenzoates were not dehalogenated. Suyama et al. (2002) suggested growth of strain Y51 on PCE or TCE as the sole energy-generating electron acceptor was a result of dehalorespiration. However, no mechanism was proposed to account for the fact that this enzyme is present on the periplasmic side of the inner membrane and how its activity can result in energy conservation. This differs from other reductive dehalogenases, which are integral membrane proteins or membrane-associated but present on the cytoplasmic side of the inner membrane. The orientation of the dehalogenase on the cytoplasmic side of the inner membrane likely results in the development of an electrochemical gradient via a scalar mechanism, which does not involve proton translocation from one side of the membrane to the other to generate a PMF. Thus, for dehalorespiration to occur, strain Y51 would potentially require a mechanism of translocating H^+ to establish a chemiosmotic gradient due to its reductive dehalogenase being present in the periplasm.

Other than *Desulfitobacterium* species, PCE reductive dehalogenases have been described in other bacterial species. We will describe three examples found in species of *Dehalobacter*, *Dehalospirillum* and *Dehalococcoides*.

Dehalobacter restrictus was first isolated from a Rhine river sediment sample mixed with granular sludge (Holliger et al. 1998). Similar to those of *Desulfitobacterium* sp., the enzymes from *D. restrictus* have a narrow substrate range and can dehalogenate PCE and TCE only. The *D. restrictus* PCE reductive dehalogenase is a 60kDa membrane-bound protein containing a cobaltous corrinoid and two Fe_4S_4 clusters. The cobalamin in the enzyme has a relatively high redox potential, as compared to other microbial corrinoid enzymes, resulting in an easily reducible cobalamin, i.e., it readily accepts electrons (Schumacher et al. 1997). It is likely that other PCE reductive dehalogenases with similar corrinoid factors may also be easily reduced. The iron-sulfur clusters were found to be of the Fe_4-S_4 cubane type and had a relatively low redox potential (Schumacher et al. 1997). Thus, the FeS clusters were considered to be electron shuttles rather than locations of electron storage. Although the physiologically relevant electron donor has not been identified, the proposed electron transfer scheme within the *D. restrictus* reductive dehalogenase is from electron donor to cubane 2 to cubane 1 (the iron-sulfur clusters) to the cobalamin and finally to the electron acceptor, PCE (Schumacher et al. 1997). Biochemical analyses of the *D. restrictus* reductive dehalogenase have yet to be published.

Dehalospirillum multivorans, isolated from activated sludge, is capable of dechlorinating PCE to *cis*-1,2-DCE via TCE with the production of energy (Scholz-Muramatsu et al. 1995). The monomeric enzyme with an MW of ~57.5kDa has been purified (Neumann et al. 1996). This reductive dehalogenase is similar to others in several respects, including the presence of FeS clusters, a corrinoid factor and oxygen sensitivity (Neumann et al. 1996). In contrast to other reductive dehalogenases from anaerobic bacteria, the PCE reductive dehalogenase from *D. multivorans* was localized to the cytoplasm (Miller et al. 1996). The substrate range of the purified enzyme is very narrow, being active with PCE, TCE, various chloropropenes and, to a lesser extent, tetraiodoethene (Neumann et al. 1996; Neumann et al. 2002). Interestingly, the purified enzyme can dechlorinate chloropropenes, although *D. multivorans* is unable to grow on these compounds as electron acceptors. The enzyme is not active against a variety of dichloroethenes, trichloroethanes, trichloroacetone, hexachloroethane, 3-CB, 4-chlorophenol, and 3,4-dichlorobenzoate (Neumann et al. 1996).

The identification of a non-dechlorinating strain of *D. multivorans* (strain N) has helped in understanding the essential role of the corrinoid factor during reductive dehalogenation. The non-dechlorinating strain N was isolated from the same site as the dechlorinating strain K. Strain N possesses the reductive dehalogenase gene sequence *pceA* and the anchor protein gene sequence *pceB*, although *pceA* is transcribed and translated to a much lesser extent than in strain K (Siebert et al. 2002). Furthermore, the expressed reductive dehalogenase in strain N lacked the corrinoid factor found in the strain K dehalogenase and this was believed to be responsible for the inability of the strain N enzyme to dechlorinate PCE and TCE (Siebert et al. 2002). The authors suggest that the corrinoid factor may have a regulatory role in the transcription and/or translation of *pceA* (Siebert et al. 2002). The corrinoid factor from *D. multivorans* contains a cobalt center. However, its characteristics are quite different from the more common cobalt-containing corrinoid vitamin B_{12} (Neumann et al. 2002). The precise nature of the *D. multivorans* corrinoid factor remains to be elucidated.

A common feature of the reductive dehalogenases from *Desulfitobacterium*, *Dehalobacter* and *Dehalospirillum* strains is that the enzymes catalyze the dechlorination of PCE to TCE and then to *cis*-1,2-DCE, which is toxic. The accumulation of *cis*-1,2-DCE inhibits the dehalogenases through an as yet unknown mechanism. Thus, another enzyme or microorganism is required to eliminate this toxic metabolite.

Dehalococcoides ethanogenes 195 was the first microorganism found to contain a novel catabolic pathway capable of degrading PCE and TCE to ethene, an environmentally harmless compound (Maymo-Gatell et al. 1997). *Dehalococcoides ethenogenes* 195 utilizes the chloroalkenes for growth via dehalorespiration.This microorganism possesses two enzymes with Co(I) corrinoid cofactor, a PCE reductive dehalogenase and a TCE reductive dehalogenase (Magnuson et al. 1998). PCE reductive dehalogenase, believed to be constitutively expressed, catalyzes the transformation of PCE to TCE (Magnuson et al. 1998; Maymo-Gatell et al. 1999). TCE reductive dehalogenase, a 57-kDa peripheral membrane protein, catalyzes the degradation of TCE to ethene, via DCE and VC (Magnuson et al. 1998; Maymo-Gatell et al. 1999). TCE reductive dehalogenase has been shown to dechlorinate other haloalkenes and haloalkanes. However, it prefers two carbon compounds containing two or three halogens, preferentially on non-adjacent carbons (Magnuson et al. 2000).

3.4.2
Molecular Characterization

Sequence analysis of the *pceA* gene products from *D. spirillum* and *Desulfitobacterium* sp. strain Y51 revealed traits common to other reductive dehalogenases and provided clues to some of their characteristics. These include the presence of the twin Arg translocation consensus sequence (RRXFXK), a ferredoxin-like cluster (CXXCXXCXXXCP) and a second iron-sulfur cluster (GXXCXXCXXCS). Both dehalogenases lack the consensus sequence for a corrinoid-binding site (DXHXXG). For both organisms, the *pceA* gene is upstream of the *pceB* (Suyama et al. 2002), which codes for the small hydrophobic protein thought to anchor PceA to the membrane. In *D. spirillum*, the C-terminus of *pceA* and the N-terminus of *pceB* overlap by 4 bp, indicating that the two proteins are co-transcribed (Neumann et al. 1998). In strain Y51, two ORFs upstream of the *pceA* gene were identified. One (ORF1) has no similarity to sequences present in the database analyzed, while the second ORF (ORF0) was partially sequenced at the C-terminus and it showed similarity to several transposases (Suyama et al. 2002).

The *tceA* gene encoding TCE reductive dehalogenase from *D. ethenogenes* 195 has been cloned and sequenced, along with the *tceB* gene, which is downstream of *tceA*. The two genes are believed to be co-transcribed (Magnuson et al. 2000).The 10-kDa TceB contains three membrane-spanning regions and is believed to be the membrane anchor for TceA. TceA is believed to be a new subclass of oxidoreductase containing cobalamin and iron-sulfur clusters as co-factors (Magnuson et al. 2000). The two other proteins from *Dehalospirillum multivorans* and *Desulfitobacterium dehalogenans* that have been characterized have a similar pattern in the placement of the genes, i.e., the gene for the catabolic enzyme (*pceA* and *cprA*) is followed downstream by a smaller gene (*pceB* and *cprB*) coding for the small integral membrane protein (van de Pas et al. 1999). PceA from *D. multivorans* and TceA from *D. ethenogenes* 195 share a low level of sequence homology despite both being able to dehalogenate TCE (Magnuson et al. 2000). The primary sequence of TceA contains two iron-sulfur clusters of the Fe_4-S_4 variety, while the sequences of PceA and CprA contain one Fe_4-S_4 and one Fe_4-S_3 cluster (Magnuson et al. 2000). However, the role of the different iron-sulfur clusters has not been examined.

4 Conclusions

Bacterial reductive dehalogenases are a major class of enzymes that perform the crucial function of removing halogen substituents from many chlorinated organic compounds. This activity initiates the biodegradation of many xenobiotic and otherwise recalcitrant and toxic organic compounds in the environment, resulting in their detoxification and in some instance their ultimate mineralization. In this chapter, we have presented many of the general biochemical and molecular characteristics of bacterial reductive dehalogenases. While many reductive dehalogenases have been purified, their genes cloned, sequenced and characterized, much remains unknown about these enzymes. For example, the nature of the corrinoid factors in the reductive dehalogenases involved in dehalorespiration is not known. The structure-function relationships of these enzymes are also not well understood, despite the identification of some of the putatively important motifs from sequence comparisons. Between the two highly homologous tetrachlorohydroquinone reductive dehalogenases (PcpCs) from *Sphingobium chlorophenolicum* ATCC 39723 and *Sphingomonas* sp. UG30, 13 amino acid differences are noted. Of these, five are conservative and, thus, not expected to drastically affect activity. This leaves eight amino acid changes each with a potentially dramatic effect on enzyme activity. This poses such questions as the following. How did these changes happen so quickly, assuming the evolution of PcpC occurred after the introduction of PCP in the environment approximately 60 years ago? Also, why are there such differences in activity if both enzymes evolved in a similar fashion to catalyze the same reaction, i.e., the conversion of TCHQ to DCHQ? Further studies will likely resolve some of these issues.

In addition to the many studies already published, a better understanding of how reductive dehalogenases work can be determined by structure–function and mutational studies. To date, none of the reductive dehalogenases have been crystallized and, therefore, detailed structural information is not available. Of the microorganisms in which reductive dehalogenation has been found to occur, the genes coding for many of the reductive dehalogenases have been cloned and sequenced. These studies have shown that many of the reductive dehalgoenases involved in chlorophenol, TCE, and PCE dehalogenation from anaerobic bacteria are based on a similar foundation, involving similar redox centers and signal sequences. From aerobic bacteria, PcpC and DCM reductive dehalogenases have been placed as members of the GST superfamily. Molecular studies have helped classify these enzymes into specific GST subgroups. With the availability of cloned genes, the study of bacterial reductive dehalogenases is poised for significant advancement. Future analyses of whole genomes and DNA fragments coding for the reductive dehalogenases and related sequences will yield novel insight into the functioning, accessory proteins, and evolution of the reductive dehalogenases. Currently, the *R. palustris* genome has been completely sequenced and the sequencing of the genomes from *D. hafniense* and *D. ethenogenes* 195 is partially complete. We expect the study of bacterial reductive dehalogenases to broaden as new genera, species, and strains of anaerobic and aerobic microorganisms are discovered that produce not only reductive dehalogenases, but also various other dehalogenating enzymes. The basic knowledge gained from studying these enzymes may help us understand the evolutionary relationship between dehalogenases. It may also have the potential to be used profitably for the remediation of environments contaminated with toxic halogenated organic compounds.

Acknowledgements. Research in the authors' laboratories on bacterial dehalogenases was supported by a group Strategic grant from the Natural Sciences and Engineering Research Council (NSERC) of Canada. Research by JTT and HL was also supported by individual NSERC Discovery grants. We also thank the Canada Foundation for Innovation and Ontario Challenge Fund for providing laboratory and equipment support to our programs.

References

Anandarajah K, Kiefer Jr PM, Donohoe BS, Copley SD (2000) Recruitment of a double bond isomerase to serve as a reductive dehalogenase during biodegradation of pentachlorophenol. Biochemistry 39:5303–5311

Armstrong RN (1998) Mechanistic imperatives for the evolution of glutathione transferases. Curr Opin Chem Biol 2:618–623

Assaf-Anid N, Nies L, Vogel TM (1992) Reductive dechlorination of a polychlorinated biphenyl congener and hexachlorobenzene by vitamin B_{12}. Appl Environ Microbiol 58:1057–1060

Bader R, Leisinger T (1994) Isolation and characterization of the *Methylophilus* sp. strain DM 11 gene encoding dichloromethane dehalogenase/glutathione S-transferase. J Bacteriol 176:3466–3473

Beaudette LA, Cassidy MB, Habash MB, Lee H, Trevors JT, Staddon WJ (2002) Microbial dehalogenation reactions in microorganisms. In: Burns RG, Dick RP (eds) Enzymes in the environment: activity, ecology, and applications. Marcel Dekker, New York, pp 447–494

Berks BC (1996) A common export pathway for proteins binding complex redox cofactors? Mol Microbiol 22:393–404

Berks BC, Sargent F, Palmer T (2000) The Tat protein export pathway. Mol Microbiol 35:260–274

Board PG, Baker RT, Chelvanayagam G, Jermiin LS (1997) Zeta, a novel class of glutathione transferases in a range of species from plants to humans. Biochem J 328:929–935

Cai M, Xun L (2002) Organization and regulation of pentachlorophenol-degrading genes in *Sphingobium chlorophenolicum* ATCC 39723. J Bacteriol 184:4672–4680

Christiansen N, Ahring BK (1996) *Desulfitobacterium hafniense* sp. nov., an anaerobic, reductively dechlorinating bacterium. Int J Syst Bacteriol 46:1010–1015

Christiansen N, Ahring BK, Wohlfarth G, Diekert G (1998) Purification and characterization of the 3-chloro-4-hydroxy-phenylacetate reductive dehalogenase of *Desulfitobacterium hafniense.* FEBS Lett 436:159–162

Copley SD (1998) Microbial dehalogenases: enzymes recruited to convert xenobiotic substrates. Curr Opin Chem Biol 2:613–617

Copley SD (2000) Evolution of a metabolic pathway for degradation of a toxic xenobiotic: the patchwork approach. Trends Biochem Sci 25:261–265

Crawford RL, Ederer MM (1999) Phylogeny of *Sphingomonas* species that degrade pentachlorophenol. J Ind Microbiol Biotechnol 23:320–325

Egland PG, Gibson J, Harwood CS (2001) Reductive, coenzyme A-mediated pathway for 3-chlorobenzoate degradation in the phototrophic bacterium *Rhodopseudomonas palustris*. Appl Environ Microbiol 67:1396–1399

El Fantroussi S, Naveau H, Agathos SN (1998) Anaerobic dechlorinating bacteria. Biotechnol Prog 14:167–188

Evans GJ, Ferguson GP, Booth IR, Vuilleumier S (2000) Growth inhibition of *Escherichia coli* by dichloromethane in cells expressing dichloromethane dehalogenase/glutathione S-transferase. Microbiology 146 (Pt 11):2967–2975

Fetzner S (1998) Bacterial dehalogenation. Appl Microbiol Biotechnol 50:633–657

Gerritse J, Drzyzga O, Kloetstra G, Keijmel M, Wiersum LP, Hutson R, Collins MD, Gottschal JC (1999) Influence of different electron donors and acceptors on dehalorespiration of tetrachloroethene by *Desulfitobacterium frappieri* TCE1. Appl Environ Microbiol 65:5212–5221

Gisi D, Willi L, Traber H, Leisinger T, Vuilleumier S (1998) Effects of bacterial host and dichloromethane dehalogenase on the competitiveness of methylotrophic bacteria growing with dichloromethane. Appl Environ Microbiol 64:1194–1202

Gisi D, Leisinger T, Vuilleumier S (1999) Enzyme-mediated dichloromethane toxicity and mutagenicity of bacterial and mammalian dichloromethane-active glutathione S-transferases. Arch Toxicol 73:71–79

Gribble GW (1994) The abundant natural sources and uses of chlorinated chemicals. Am J Public Health 84:1183

Gribble GW (1996) Naturally occurring organohalogen compounds-a comprehensive survey. Fortschr Chem Org Naturst 68:1–423

Habash MB, Beaudette LA, Cassidy MB, Leung KT, Hoang TA, Vogel HJ, Trevors JT, Lee H (2002) Characterization of tetrachlorohydroquinone reductive dehalogenase from *Sphingomonas* sp. UG30. Biochem Biophys Res Commun 299:634–640

Häggblöm MM, Nohynek LJ, Salkinoja-Salonen MS (1988) Degradation and O-methylation of chlorinated phenolic compounds by *Rhodococcus* and *Mycobacterium* strains. Appl Environ Microbiol 54:3043–3052

Häggblöm MM, Janke D, Salkinoja-Salonen MS (1989) Hydroxylation and dechlorination of tetrachlorohydroquinone by *Rhodococcus* sp. strain CP-2 cell extracts. Appl Environ Microbiol 55:516–519

Holliger C, Schraa G (1994) Physiological meaning and potential for application of reductive dechlorination by anaerobic bacteria. FEMS Microbiol Rev 15:297–305

Holliger C, Hahn D, Harmsen H, Ludwig W, Schumacher W, Tindall B, Vazquez F, Weiss N, Zehnder AJ (1998) *Dehalobacter restrictus* gen. nov. and sp. nov., a strictly anaerobic bacterium that reductively dechlorinates tetra- and trichloroethene in an anaerobic respiration. Arch Microbiol 169:313–321

Kayser MF, Stumpp MT, Vuilleumier S (2000) DNA polymerase I is essential for growth of *Methylobacterium dichloromethanicum* DM 4 with dichloromethane. J Bacteriol 182:5433–5439

Kayser MF, Vuilleumier S (2001) Dehalogenation of dichloromethane by dichloromethane dehalogenase/glutathione S-transferase leads to formation of DNA adducts. J Bacteriol 183:5209–5212

Kayser MF, Ucurum Z, Vuilleumier S (2002) Dichloromethane metabolism and C1 utilization genes in *Methylobacterium* strains. Microbiology 148:1915–1922

Kiefer PM, Jr., Copley SD (2002) Characterization of the initial steps in the reductive dehalogenation catalyzed by tetrachlorohydroquinone dehalogenase. Biochemistry 41:1315–1322

Kiefer PM Jr, McCarthy DL, Copley SD (2002) The reaction catalyzed by tetrachlorohydroquinone dehalogenase does not involve nucleophilic aromatic substitution. Biochemistry 41:1308–1314

Kohler-Staub D, Leisinger T (1985) Dichloromethane dehalogenase of *Hyphomicrobium* sp. strain DM 2. J Bacteriol 162:676–681

Kohler-Staub D, Hartmans S, Gaelli R, Suter F, Leisinger T (1986) Evidence for identical dichloromethane dehalogenases in different methylotrophic bacteria. J Gen Microbiol 132:2837–2843

Kohring GW, Rogers JE, Wiegel J (1989a) Anaerobic biodegradation of 2,4-dichlorophenol in freshwater lake sediments at different temperatures. Appl Environ Microbiol 55:348–353

Kohring GW, Zhang XM, Wiegel J (1989b) Anaerobic dechlorination of 2,4-dichlorophenol in freshwater sediments in the presence of sulfate. Appl Environ Microbiol 55:2735–2737

Krasotkina J, Walters T, Maruya KA, Ragsdale SW (2001) Characterization of the B_{12} and iron-sulfur-containing reductive dehalogenase from *Desulfitobacterium chlororespirans*. J Biol Chem 276:40991–40997

Kumari R, Subudhi S, Suar M, Dhingra G, Raina V, Dogra C, Lal S, van der Meer JR, Holliger C, Lal R (2002) Cloning and characterization of *lin* genes responsible for the degradation of hexachlorocyclohexane isomers by *Sphingomonas paucimobilis* strain B90. Appl Environ Microbiol 68:6021–6028

La Roche SD, Leisinger T (1990) Sequence analysis and expression of the bacterial dichloromethane dehalogenase structural gene, a member of the glutathione S-transferase supergene family. J Bacteriol 172:164–171

La Roche SD, Leisinger T (1991) Identification of *dcmR*, the regulatory gene governing expression of dichloromethane dehalogenase in *Methylobacterium* sp. strain DM 4. J Bacteriol 173:6714–6721

Leisinger T, Bader R, Hermann R, Schmid-Appert M, Vuilleumier S (1994) Microbes, enzymes and genes involved in dichloromethane utilization. Biodegradation 5:237–248

Leisinger T, Braus-Stromeyer SA (1995) Bacterial growth with chlorinated methanes. Environ Health Perspect 103 Suppl 5:33–36

Leung K, Cassidy MB, Shaw KW, Lee H, Trevors JT, Lohmeier-Vogel EM, Vogel HJ (1997) Pentachlorophenol biodegradation by *Pseudomonas* spp. UG25 and UG30. World J Microbiol Biotech 13:305–313

Louie TM, Ni S, Xun L, Mohn WW (1997) Purification, characterization and gene sequence analysis of a novel cytochrome c co-induced with reductive dechlorination activity in *Desulfomonile tiedjei* DCB-1. Arch Microbiol 168:520–527

Louie TM, Mohn WW (1999) Evidence for a chemiosmotic model of dehalorespiration in *Desulfomonile tiedjei* DCB-1. J Bacteriol 181:40–46

Magnuson JK, Stern RV, Gossett JM, Zinder SH, Burris DR (1998) Reductive dechlorination of tetrachloroethene to ethene by a two-component enzyme pathway. Appl Environ Microbiol 64:1270–1275

Magnuson JK, Romine MF, Burris DR, Kingsley MT (2000) Trichloroethene reductive dehalogenase from *Dehalococcoides ethenogenes*: sequence of *tceA* and substrate range characterization. Appl Environ Microbiol 66:5141–5147

Marsh A, Ferguson DM (1997) Knowledge-based modeling of a bacterial dichloromethane dehalogenase. Proteins 28:217–226
Masai E, Katayama Y, Kubota S, Kawai S, Yamasaki M, Morohoshi N (1993) A bacterial enzyme degrading the model lignin compound beta-etherase is a member of the glutathione-S-transferase superfamily. FEBS Lett 323:135–140
Maymo-Gatell X, Chien Y-T, Gossett JM, Zinder SH (1997) Isolation of a bacterium that reductively dechlorinates tetrachloroethene to ethene. Science 276:1568–1571
Maymo-Gatell X, Anguish T, Zinder SH (1999) Reductive dechlorination of chlorinated ethenes and 1,2-dichloroethane by "*Dehalococcoides ethenogenes*" 195. Appl Environ Microbiol 65:3108–3113
McCarthy DL, Navarrete S, Willett WS, Babbitt PC, Copley SD (1996) Exploration of the relationship between tetrachlorohydroquinone dehalogenase and the glutathione S-transferase superfamily. Biochemistry 35:14634–14642
McCarthy DL, Claude AA, Copley SD (1997) In vivo levels of chlorinated hydroquinones in a pentachlorophenol-degrading bacterium. Appl Environ Microbiol 63:1883–1888
McGrath JE, Harfoot CG (1997) Reductive dehalogenation of halocarboxylic acids by the phototrophic genera *Rhodospirillum* and *Rhodopseudomonas*. Appl Environ Microbiol 63:3333–3335
Miller E, Wohlfarth G, Diekert G (1996) Studies on tetrachloroethene respiration in *Dehalospirillum multivorans*. Arch Microbiol 166:379–387
Miller E, Wohlfarth G, Diekert G (1997) Comparative studies on tetrachloroethene reductive dechlorination mediated by *Desulfitobacterium* sp. strain PCE-S. Arch Microbiol 168:513–519
Miller E, Wohlfarth G, Diekert G (1998) Purification and characterization of the tetrachloroethene reductive dehalogenase of strain PCE-S. Arch Microbiol 169:497–502
Miyauchi K, Suh SK, Nagata Y, Takagi M (1998) Cloning and sequencing of a 2,5-dichlorohydroquinone reductive dehalogenase gene whose product is involved in degradation of gamma-hexachlorocyclohexane by *Sphingomonas paucimobilis*. J Bacteriol 180:1354–1359
Mohn WW, Tiedje JM (1990) Strain DCB-1 conserves energy for growth from reductive dechlorination coupled to formate oxidation. Arch Microbiol 153:267–271
Mohn WW, Tiedje JM (1992) Microbial reductive dehalogenation. Microbiol Rev 56: 482–507
Mohn WW, Kennedy KJ (1992) Reductive dehalogenation of chlorophenols by *Desulfomonile tiedjei* DCB-1. Appl Environ Microbiol 58:1367–1370
Moura I, Moura JJ (2001) Structural aspects of denitrifying enzymes. Curr Opin Chem Biol 5:168–175
Neumann A, Scholz-Muramatsu H, Dickert G (1994) Tetrachloroethene metabolism of *Dehalospirillum multivorans*. Arch Microbiol 162:295–301
Neumann A, Wohlfarth G, Diekert G (1996) Purification and characterization of tetrachloroethene reductive dehalogenase from *Dehalospirillum multivorans*. J Biol Chem 271:16515–16519
Neumann A, Wohlfarth G, Diekert G (1998) Tetrachloroethene dehalogenase from *Dehalospirillum multivorans*: cloning, sequencing of the encoding genes, and expression of the *pceA* gene in *Escherichia coli*. J Bacteriol 180:4140–4145
Neumann A, Siebert A, Trescher T, Reinhardt S, Wohlfarth G, Diekert G (2002) Tetrachloroethene reductive dehalogenase of *Dehalospirillum multivorans*: substrate specificity of the native enzyme and its corrinoid cofactor. Arch Microbiol 177:420–426
Ni S, Fredrickson JK, Xun L (1995) Purification and characterization of a novel 3-chlorobenzoate-reductive dehalogenase from the cytoplasmic membrane of *Desulfomonile tiedjei* DCB-1. J Bacteriol 177:5135–5139

Oakley AJ, Bello ML, Battistoni A, Ricci G, Rossjohn J, Villar HO, Parker MW (1997) The structures of human glutathione transferase P1-1 in complex with glutathione and various inhibitors at high resolution. J Mol Biol 274:84–100

Okeke BC, Chang YC, Hatsu M, Suzuki T, Takamizawa K (2001) Purification, cloning, and sequencing of an enzyme mediating the reductive dechlorination of tetrachloroethylene (PCE) from *Clostridium bifermentans* DPH-1. Can J Microbiol 47: 448–456

Polekhina G, Board PG, Blackburn AC, Parker MW (2001) Crystal structure of maleylacetoacetate isomerase/glutathione transferase zeta reveals the molecular basis for its remarkable catalytic promiscuity. Biochemistry 40:1567–1576

Romanov V, Hausinger RP (1996) NADPH-dependent reductive *ortho* dehalogenation of 2,4-dichlorobenzoic acid in *Corynebacterium sepedonicum* KZ-4 and *Coryneform bacterium* strain NTB-1 via 2,4-dichlorobenzoyl coenzyme A. J Bacteriol 178: 2656–2661

Saber DL, Crawford RL (1985) Isolation and characterization of *Flavobacterium* strains that degrade pentachlorophenol. Appl Environ Microbiol 50:1512–1518

Sanford RA, Cole JR, Loffler FE, Tiedje JM (1996) Characterization of *Desulfitobacterium chlororespirans* sp. nov., which grows by coupling the oxidation of lactate to the reductive dechlorination of 3-chloro-4-hydroxybenzoate. Appl Environ Microbiol 62:3800–3808

Schenk T, Muller R, Morsberger F, Otto MK, Lingens F (1989) Enzymatic degradation of pentachlorophenol by extracts from *Arthrobacter* sp. strain ATCC 33790. J Bacteriol 171:5487–5491

Schmid-Appert M, Zoller K, Traber H, Vuilleumier S, Leisinger T (1997) Association of newly discovered IS elements with the dichloromethane utilization genes of methylotrophic bacteria. Microbiology 143:2557–2567

Scholtz R, Wackett LP, Egli C, Cook AM, Leisinger T (1988) Dichloromethane dehalogenase with improved catalytic activity isolated from a fast-growing dichloromethane-utilizing bacterium. J Bacteriol 170:5698–5704

Scholz-Muramatsu H, Neumann A, Messner M, Moore E, Diekert G (1995) Isolation and characterization of *Dehalospirillum multivorans* gen. nov., sp. nov., a tetrachloroethene-utilizing strictly anaerobic bacterium. Arch Microbiol 163:48–56

Schumacher W, Holliger C, Zehnder AJ, Hagen WR (1997) Redox chemistry of cobalamin and iron-sulfur cofactors in the tetrachloroethene reductase of *Dehalobacter restrictus*. FEBS Lett 409:421–425

Seltzer S (1973) Purification and properties of maleylacetone *cis-trans* isomerase from *Vibrio* 01. J Biol Chem 248:215–222

Sheehan D, Meade G, Foley VM, Dowd CA (2001) Structure, function and evolution of glutathione transferases: implications for classification of non-mammalian members of an ancient enzyme superfamily. Biochem J 360:1–16

Siebert A, Neumann A, Schubert T, Diekert G (2002) A non-dechlorinating strain of *Dehalospirillum multivorans*: evidence for a key role of the corrinoid cofactor in the synthesis of an active tetrachloroethene dehalogenase. Arch Microbiol 178:443–449

Smidt H, Song D, van Der Oost J, de Vos WM (1999) Random transposition by Tn916 in *Desulfitobacterium dehalogenans* allows for isolation and characterization of halorespiration-deficient mutants. J Bacteriol 181:6882–6888

Smidt H, van Leest M, van der Oost J, de Vos WM (2000a) Transcriptional regulation of the cpr gene cluster in ortho-chlorophenol-respiring *Desulfitobacterium dehalogenans*. J Bacteriol 182:5683–5691

Smidt H, Akkermans AD, van der Oost J, de Vos WM (2000b) Halorespiring bacteria-molecular characterization and detection. Enzyme Microb Technol 27:812–820

Suyama A, Yamashita M, Yoshino S, Furukawa K (2002) Molecular characterization of the PceA reductive dehalogenase of *Desulfitobacterium* sp. strain Y51. J Bacteriol 184:3419–3425

Townsend GT, Suflita JM (1996) Characterization of chloroethylene dehalogenation by cell extracts of *Desulfomonile tiedjei* and its relationship to chlorobenzoate dehalogenation. Appl Environ Microbiol 62:2850–2853

Uotila JS, Salkinoja-Salonen MS, Apajalahti JHA (1991) Dechlorination of pentachlorophenol by membrane bound enzymes of *Rhodococcus chlorophenolicus* PCP-1. Biodegradation 2:25–31

Utkin I, Dalton DD, Wiegel J (1995) Specificity of reductive dehalogenation of substituted ortho-chlorophenols by *Desulfitobacterium dehalogenans* JW/IU-DC1. Appl Environ Microbiol 61:346–351

van de Pas BA, Smidt H, Hagen WR, van der Oost J, Schraa G, Stams AJ, de Vos WM (1999) Purification and molecular characterization of ortho-chlorophenol reductive dehalogenase, a key enzyme of halorespiration in *Desulfitobacterium dehalogenans.* J Biol Chem 274:20287–20292

van de Pas BA, Jansen S, Dijkema C, Schraa G, de Vos WM, Stams AJ (2001a) Energy yield of respiration on chloroaromatic compounds in *Desulfitobacterium dehalogenans.* Appl Environ Microbiol 67:3958–3963

van de Pas BA, Gerritse J, de Vos WM, Schraa G, Stams AJ (2001b) Two distinct enzyme systems are responsible for tetrachloroethene and chlorophenol reductive dehalogenation in *Desulfitobacterium* strain PCE1. Arch Microbiol 176:165–169

Vuilleumier S, Leisinger T (1996) Protein engineering studies of dichloromethane dehalogenase/glutathione S-transferase from *Methylophilus* sp. strain DM 11. Ser12 but not Tyr6 is required for enzyme activity. Eur J Biochem 239:410–417

Vuilleumier S (1997) Bacterial glutathione S-transferases: What are they good for? J Bacteriol 179:1431–1441

Vuilleumier S, Pagni M (2002) The elusive roles of bacterial glutathione S-transferases: new lessons from genomes. Appl Microbiol Biotechnol 58:138–146

Vuilleumier S, Sorribas H, Leisinger T (1997) Identification of a novel determinant of glutathione affinity in dichloromethane dehalogenases/glutathione S-transferases. Biochem Biophys Res Commun 238:452–456

Vuilleumier S, Ivos N, Dean M, Leisinger T (2001) Sequence variation in dichloromethane dehalogenases/glutathione S-transferases. Microbiology 147: 611–619

Wheeler JB, Stourman NV, Thier R, Dommermuth A, Vuilleumier S, Rose JA, Armstrong RN, Guengerich FP (2001) Conjugation of haloalkanes by bacterial and mammalian glutathione transferases: mono- and dihalomethanes. Chem Res Toxicol 14:1118–1127

Wiegel J, Zhang X, Wu Q (1999) Anaerobic dehalogenation of hydroxylated polychlorinated biphenyls by *Desulfitobacterium dehalogenans.* Appl Environ Microbiol 65: 2217–2221

Willett WS, Copley SD (1996) Identification and localization of a stable sulfenic acid in peroxide-treated tetrachlorohydroquinone dehalogenase using electrospray mass spectrometry. Chem Biol 3:851–857

Xun L (1992) Glutathione is the reducing agent for the reductive dehalogenation of tetrachloro-p-hydroquinone by extracts from a *Flavobacterium* sp. Biochem Biophys Res Commun 182:361–366

Xun L, Topp E, Orser CS (1992) Purification and characterization of a tetrachloro-p-hydroquinone reductive dehalogenase from a *Flavobacterium* sp. J Bacteriol 174:8003–8007

10 Engineering of Improved Biocatalysts in Bioremediation

Wilfred Chen[1] and Ashok Mulchandani[2]

1 Introduction

Over the past few decades enormous quantities of industrial pollutants have been released into the environment. A large number of them, particularly those structurally related to natural compounds, are readily degraded or removed by microorganisms found in soil and water. However, superimposed on the rich variety of pollutants present in the environment are an increasing number of novel industrial compounds not found extensively in nature. A significant amount of these industrial chemicals are released into the environment deliberately, to function as pesticides or to act as wood preservatives or electrical insulators. Many of these chemicals, produced on a large scale as part of the normal activities of industrialized societies, are considered hazardous to humans, plants and animals. These xenobiotic compounds are usually removed slowly and tend to accumulate in the environment (Gibson and Parales 2000). The local concentration of these contaminants depends on the rate at which the compound is released, its stability and mobility in the environment, and its rate of biological or non-biological removal (Harayama 1997; Ellis 2000; Janssen et al. 2001). Some of these chemicals have long half-lives in the soil, e.g. the environmental persistence of the insecticide DDT ranges between 3–10 years, chlordane ranges from 2–4 years and HCH persists for up to 11 years. Due to the high degree of toxicity, their accumulation can cause severe environmental problems.

However, as pollutant treatment by conventional methods such as incineration or landfill is often not only technically challenging but also

[1] Department of Chemical and Environmental Engineering, University of California, Riverside, California 92521, USA, e-mail: wilfred@engr.ucr.edu, Tel: +1-909-7872473, Fax: +1-909-7875696
[2] Department of Chemical and Environmental Engineering, University of California, Riverside, California 92521, USA

Soil Biology, Volume 2
Biodegradation and Bioremediation
(ed. by. A. Singh and O. P. Ward)

very expensive, increasing consideration has been placed on the development of alternative, economical and reliable biological treatments. Microorganisms have been shown to excel at using organic substances as sources of nutrients and energy. The vast diversity of potential substrates for growth led to the evolution of enzymes capable of transforming many unrelated natural organic compounds by many different catalytic mechanisms. Although simple organic compounds are biodegradable by a variety of degradative pathways, many of the industrial chemicals contain substitutions, which make them more resistant to bacterial attacks. Although natural microorganisms collectively exhibit remarkable evolutionary capabilities to adapt to a wide range of chemicals, natural evolution occurs at a relatively slow rate, particularly when the acquisition of multiple catalytic activities is necessary. In such cases, the acceleration of these events via genetic engineering/metabolic engineering (Bailey 1996) may be helpful since specific genetic alterations can be carefully designed and controlled.

Recent advances in molecular biology have opened up new avenues to move toward the goal of engineering microbes or enzymes to function as "designer biocatalysts" (Lee et al. 1994; Wackett et al. 1994), in which certain desirable traits from different organisms are brought together with the aim of performing specific bioremediation. In this review, we will focus on the advances of the past few years with emphasis on new approaches to engineering improved microbes or enzymes for bioremediation.

2 Engineering Microbes for Improved Bioremediation

2.1 Bioadsorbents for Heavy Metal Removal

The discharge of heavy metals from agricultural, industrial, and military operations has serious adverse effects on the environment (Nriagu and Pacyna 1989). Conventional technologies are often inadequate in reducing heavy metal concentrations in wastewater to acceptable regulatory standards. Recent research has focused on the development of novel bioadsorbents with increased affinity, capacity, and selectivity for target metals.

Eukaryotes limit the concentrations of reactive free metal ions by intracellular sequestration. Glutathione (GSH), GSH-related phytochelatins (PCs) and cysteine-rich metallothioneins (MTs) (Stillman et al. 1992) are the main metal sequestering peptides used to immobilize metal ions. Earlier attempts to produce MTs in *E. coli* to increase their metal-binding capability were successful in some cases (Pazirandeh et al. 1995, Romeyer et al. 1988). Similarly, expression of PC synthase in *E. coli* resulted in a marked accumulation of intracellular PC and cadmium content (Sauge-Merle 2003). However, expression of such cysteine-rich proteins is not devoid of problems because of the potential interference with the redox pathways (Bardwell 1994). More importantly, intracellular accumulation of heavy metals may prevent the reuse of the bioadsorbents because the accumulated metals cannot be easily released (Gadd and White 1993). One clever solution to bypass intracellular accumulation is to express MTs on the cell surface. Sousa et al. (1996) demonstrated this possibility by inserting MTs into the permissive site 153 of the LamB sequence. Expression of the hybrid proteins on the cell surface multiplied the natural Cd^{2+} accumulation ability by more than 20-fold.

In addition to naturally occurring peptides, the de novo design of metal-binding peptides is an attractive alternative as they offer the potential of improved affinity and selectivity for heavy metals. Peptides with an abundance of cysteine or histidine residues are known to bind heavy metals with high affinity. Sousa and coworkers (1996) first applied this approach for improved heavy metal sequestration by inserting hexahistidine clusters onto the outer membrane LamB protein. Strains with surface-exposed histidines accumulated greater than 11-fold more Cd^{2+} than cells expressing the LamB protein without the insert. Novel metal-binding peptides could also be selected from a phage display library (Mejare et al. 1998). The peptide His-Ser-Gln-Lys-Val-Phe, which exhibits the strongest affinity for Cd^2, was cloned into *E. coli* as a fusion to the cell surface exposed area of the outer-membrane protein OmpA. Cells expressing this peptide showed increased survival in growth medium containing toxic levels of $CdCl_2$, demonstrating the binding of Cd^{2+} by the surface-exposed peptide. Similarly, cells expressing other histidine-(Gly-His-His-Pro-His-Gly) or cysteine-(Gly-Cys-Gly-Cys-Pro-Cys) rich peptides, which bind mercury with much higher affinity than other divalent heavy metal, have been anchored on the surface of *E. coli* (Kotrba et al. 1999). The display of these peptides again increased the bioaccumulation of Cd^{2+} although to a lesser extent than the MTs. Peptide sequences have also been selected from a fimbriae display system, which conferred on recombinant *E. coli* the ability to adhere to different metal oxides (Schembri et al. 1999).

Unlike MTs, which can be expressed easily as fusion proteins, the presence of a γ bond between glutamic acid and cysteine in PCs indicates that these peptides must be synthesized enzymatically. One promising alternative has recently been developed by creating synthetic phytochelatins (ECn) with a repetitive metal-binding motif (Glu-Cys)$_n$Gly and were shown to have improved Cd^{2+} binding capability over that of MTs (Bae et al. 2000). The measured Hg^{2+} and Cd^{2+} binding stoichiometry of 20 and 10, respectively, were significantly higher than the typical values reported for MTs (Bae et al. 2001). As a result, the Cd^{2+} binding capability of cells expressing EC20 was almost twice the amount obtained using MTs.

In addition to short metal-binding peptides, another group of useful metal-binding moieties is typically associated with metal resistance. The highly specific nature of these resistance mechanisms is the result of a cleverly designed genetic circuit that is tightly controlled by a specific metalloregulatory protein. To provide sensitive resistance, these metalloregulatory proteins also possess high affinity in the submicromolar range. Examples are MerR and ArsR, which are regulatory proteins used for controlling the expression of enzymes responsible for mercury and arsenic detoxification, respectively (O'Halloran et al. 1989; Diorio et al. 1995). The affinity of these proteins for mercury and arsenate, respectively, is typically in the 10^{-15} M range, on par with those of metallothioneins and phytochelatins. However, the clear advantage of these proteins is their specificity. The binding affinity of MerR and ArsR protein is at least 4–5 orders of magnitude higher for mercury and arsensate, respectively, than other heavy metals (Bontidean et al. 1998). The high affinity and selectivity of MerR toward mercury was exploited for the construction of microbial biosorbents specific for mercury removal (Bae et al. 2003). Whole cell sorbents were constructed with MerR genetically engineered onto the surface of *E. coli* using an ice nucleation protein anchor. The presence of surface-exposed MerR on the engineered strains enabled a six-fold higher Hg^{2+} biosorption as compared to the wild-type JM109 cells. Hg^{2+} binding via MerR was very specific with no observable decline even in the presence of 100-fold excess Cd^{2+} and Zn^{2+}.

Besides lab-born *E. coli* strains, metal-binding peptides have also been expressed on the surface of soil bacteria that are known to survive in contaminated environments for an extended period. The mouse MT was fused to the autotransporter β-domain of the IgA protease from *Neisseria gonorrhoeae* and displayed on the surface of *Pseudomonas putida* (Valls et al. 2000a) and *Ralstonia metallidurans* CH34 (Valls et al. 2000b), resulting in a threefold increase in the binding of Cd^{2+}. This modest increase in binding was sufficient to improve growth and chlorophyll production of the tobacco plant *Nicotiana betamiana* in contaminated soil (Valls et al. 2000b). A genetically engineered *Moraxella* sp. with surface-expressed EC20 has also been developed with almost a ten-fold improvement in mercury binding (Bae et al. 2002). This higher level of improvement again reflects the improved binding capacity of EC20 over that of MTs.

2.2 Metal Precipitation

In addition to their use as biosorbents, bacteria can be engineered to efficiently immobilize heavy metals through a reduction to lesser bioactive metal species. Typically, toxic metals are precipitated as metal sulfides (Labrenz et al. 2000). However, even low levels (20–200 μM) of free metal ions are toxic to sulfate-reducing bacteria, limiting their use (White and Gadd 1998). One alternative for improving sulfide-dependent metal removal is to engineer the sulfate reduction pathway into more robust bacteria. The first attempt in this direction was the expression of a thiosulfate reductase gene from *Salmonella typhimurium* in *E. coli.* Strains expressing thiosulfate reductase produced significantly more sulfide and more than 98% of the available cadmium was removed (from up to 200 mM solutions) under anaerobiosis (Bang et al. 2000). Further improvement in metal precipitation was achieved by engineering effective sulfate reduction under aerobic conditions. *E. coli* expressing both serine acetyltransferase and cysteine desulfhydrase overproduced cysteine and converted it to sulfide (Wang et al. 2000). The resulting strain was effective in aerobically precipitating cadmium as cadmium sulfide on the cell surface.

Metal precipitation can also be mediated by the liberation of inorganic phosphate from organic phosphate donors. The precipitation process was catalyzed by the formation of inorganic phosphate by the action of an acid phosphatase (named PhoN) (Basnakova et al. 1998; Macaskie et al. 2000). The Introduction of PhoN to other bacteria to enhance metal precipitation has been shown to be effective (Basnakova et al. 1998). Another possibility is the use of bacteria with engineered polyphosphate pathways. The introduction of polyphosphate kinase into *P. aeruginosa* resulted in the accumulation of large quantities of polyphosphate. Exposure to phosphate-limiting conditions resulted in polyphosphate degradation and inorganic phosphate secretion, removing the uranyl from the solution (Keasling et al. 2000).

2.3 Enzymatic Transformation of Metals and Metalloids

Bacteria could transform a wide range of metals such as mercury and arsenic to generate less poisonous species. The mechanism of bacterial resistance to Hg^{2+} is the reduction by mercuric reductase (the product of the merA gene) to the less toxic and volatile Hg^0 species (Misra 1992), providing a means of mercury removal by mobilization to the atmosphere. An *E. coli* variant containing simultaneously the merA and glutathione S-transferase genes was able to tolerate high mercury concentrations (30 mg/l) and to reduce mercury (Cursino et al. 2000). *Deinococcus radiodurans*, the most radiation-resistant organism known, has been recently modified to express merA, enabling the engineered strain to effectively volatilize mercury (Brim et al. 2000). A combined method of chemical leaching and subsequent volatilization of mercury by bacteria has been developed, which removed about 70% of mercury from polluted Minamata Bay sediments (Nakamura et al. 1999). The mobilization of mercury could solve a local problem, but there is public concern that this might eventually contribute to global atmospheric pollution.

2.4 Designing Strains for Enhanced Biodegradation

Many environmental pollutants are readily degraded by naturally occurring microbes. Very often, however, the rate of degradation may not be optimized for practical, large-scale bioremediaton. Genetic engineering of biodegradative pathways offers the potential of expanding the existing capabilities found in nature. Several notable examples will be discussed.

Neurotoxic organophospates are used extensively as pesticides and are among the most toxic compounds known. Organophosphorus hydrolase (OPH), isolated from soil microorganisms, has been shown to degrade these pesticides effectively. However, the use of OPH for detoxification has always been limited by the high cost associated with purification. Whole cell detoxification is more cost effective, however, it is limited by the transport barrier of organophosphates, which lies across the cell membrane. Surface expression of OPH can circumvent transport limitations imposed by cell membranes in much the same way that surface expression of metallothioneins enhanced the metal binding capability of cells. Whole cells expressing OPH on the cell surface degraded parathion and paraoxon are seven-fold faster compared to whole cells expressing OPH intracellularly (Richins et al. 1997). The resulting cellular biocatalysts were also considerably more stable and robust than purified OPHs, retaining 100% activity over a period of 1 month when maintained at 37 °C (Chen and Mulchandani 1998). Immobilization of these novel biocatalysts by physical adsorption onto solid supports provides an attractive means for pesticide detoxification in place of immobilized OPH (Mulchandani et al. 1999). However, a gradual cell detachment from the support reduced the effectiveness of the immobilized-cell system for long-term operation. A significant improvement, in terms of both economics and technology, could be achieved with reversible and specific adhesion to the support.

Specific adhesion of whole cells to cellulosic materials with high affinity has been demonstrated by anchoring the cellulose-binding domain (CBD) from *Cellulomonas fimi* on the surface of *E. coli* (Francisco et al. 1993). This was exploited to enable very strong attachment of the organophosphate-degrading cells to cellulose supports for long-term usage (Wang et al. 2002). Two different surface anchors (Lpp-OmpA and INPNC) were employed to target OPH and CBD onto the cell surface, respectively, in order to minimize direct competition of the same translocation machinery. Whole-cell immobilization with surface-anchored CBD was very specific, forming essentially a monolayer of cells onto different supports as shown by electron micrographs. Immobilized cells degraded paraoxon rapidly and retained almost 100% efficiency over a period of 45 days. This is also the first reported genetic co-immobilization of two functional moieties onto the surface of *E. coli*.

Although the enzymatic hydrolysis of organophosphates such as parathion and methyl parathion reduces the toxicity by nearly 120-fold, the hydrolyzed product, *p*-nitrophenol (PNP), is still considered a priority pollutant by the US EPA. A novel approach was developed to enable the simultaneous degradation of organophosphates and PNP by anchoring OPH on the surface of a native PNP degrader, *Moraxella* sp (Shimazu et al. 2001). The result is a single microorganism that is endowed with the capability to rapidly degrade organophosphate pesticides and PNP. This is also the first report on the functional expression of enzymes on the surface of gram-negative bacteria other than *E. coli*.

Although OPH hydrolyzes a wide range of organophosphates, the effectiveness of hydrolysis varies dramatically. For example, some highly used organophosphorus insecticides such as methyl parathion, chlorpyrifos and diazinon are hydrolyzed 30–1000 times slower than the preferred substrate, paraoxon. Sequential cycles of DNA shuffling and screening were used to “fine tune” and enhance the activity of OPH towards poorly degraded substrates. However, due to the inaccessibility of these pesticides across the cell membrane, OPH variants were displayed on the surface of *E. coli* using the truncated ice-nucleation protein in order to isolate novel enzymes with truly improved substrate specificities (Cho et al. 2002). Two rounds of DNA shuffling and screening were carried out and several improved variants were isolated. One variant 22A11, in particular, hydrolyzes methyl parathion 25-fold faster than the wild type. As a result of the success we achieved with directed evolution of OPH for improved hydrolysis of methyl parathion, this method can be extended in creating other OPH variants with improved activity against poorly degraded nerve agents such as sarin and soman.

The expression of biocatalytic pathways in foreign microorganisms can significantly enhance the efficiency of the process. One such example has been recently reported for the desulfurization of fossil fuels. A number of organisms, such as *Rhodococcus erythropolis*, have been found to remove sulfur from fossil fuel via a sulfur-specific pathway, selectively cleaving the sulfur with ring destruction and therefore maintaining the fuel content (Izumi et al. 1994). *Pseudomonas* strains have been engineered by inserting the catabolic genes, *dszA*, *-B*, and *-C*, responsible for desulfurization into their chromosomes (Gallardo et al. 1997). These recombinants desulfurized dibenzothiophene (DBT) more efficiently than the native *R. erythropolis*. Moreover, by expressing the desulfurization pathway in *Pseudomonas aeruginosa* EGSOX, the ability to carry desulfurization and to produce rhamnolipid, a biosurfactant that increases the aqueous DBT concentration, was effectively combined into a single strain. The design of novel biocatalysts endowed with a desulfurization phenotype offers new insights for the development of commercially viable desulfurization processes.

The expression of detoxifying enzymes in foreign organisms may also confer specific selective advantages for in situ applications. This strategy was recently reported using a wheat rhizosphere system for the detoxification of soil-borne trichloroethylene (TCE) (Yee et al. 1998). The toluene *o*-monooxygenase (Tom) genes of *Burkholderia cepacia* G4 were engineered into *Pseudomonas fluorescens* 2–79. *P. fluorescens* colonizes wheat roots better than other known colonizers, enabling the establishment of a bacterium-plant-soil microcosm. Treatment of TCE-contaminated surface and near-surface soil was demonstrated, with more than 63% of the initial TCE removed within 4 days. The most attractive aspect of this technology is the low cost associated with it, since the only expenses required are for planting.

Deinococcus radiodurans is a soil bacterium that can survive acute exposures to the ionizing radiation of 15,000 Gy without lethality (Daly et al. 1994). A recombinant *D. radiodurans* strain, expressing the toluene dioxygenase, was shown to effectively oxidize toluene, chlorobenzene and TCE in a highly irradiating environment (Lange et al. 1998). The recombinant strains were also tolerant to the solvent effects of toluene and TCE at levels exceeding those of many radioactive waste sites. The prospect of using this strategy to treat a variety of organic wastes in the presence of radionuclides is very promising.

The construction of a hybrid strain which is capable of mineralizing components of a benzene, toluene and *p*-xylene mixture simultaneously was attempted by redesigning the metabolic pathway of *Pseudomonas putida* (Lee et al. 1994). Genetic and biochemical analyses of the *tod* and the *tol* pathways revealed that dihydrodiols formed from benzene, toluene and *p*-xylene by toluene dioxygenase in the *tod* pathway could be channeled into the *tol* pathway by the action of *cis-p*-toluate-dihydrodiol dehydrogenase, leading to the complete mineralization of a benzene, toluene and *p*-xylene mixture. Consequently, a hybrid strain was constructed by cloning *todC1C2BA* genes, encoding toluene dioxygenase on RSF1010 and introducing the resulting plasmid into *P. putida* mt-2. The hybrid strain of *P. putida* TB105 was found to mineralize a benzene, toluene and *p*-xylene mixture without the accumulation of any metabolic intermediate.

3 Protein Engineering for Improved Bioremediation

Many wild-type microorganisms are capable of transforming or, in some cases, completely mineralizing man-made compounds through biodegradative pathways, which were evolved for naturally occurring compounds of a similar structure. The ever-increasing volume of information on the structure and function of enzymes and pathways involved in biodegradation of recalcitrant pollutants offers opportunities for improving enzymes or entire pathways by genetic engineering. Control mechanisms and enzyme properties can be tailored by site-directed mutagenesis, which is often guided by computer-assisted modeling of the three-dimensional protein structures. For example, site-directed approaches have been recently applied to engineer the active-site topology of cytochrome P450 (Stevenson et al. 1998), to enlarge the binding pocket of haloalkane dehalogenase (Holloway et al. 1998) and to influence the substrate range of aromatic dioxygenases (Bruhlmann and Chen 1999).

Haloalkane dehalogenase is an enzyme responsible for the substitution of a terminal chlorine atom by a hydroxyl group. This enzyme has been shown to dechlorinate a wide range of substrates (Verschueren et al. 1993; Stevenson et al. 1998). Using structural information as a guide, the bulky amino acids lining the catalytic cavity were replaced with alanine thereby increasing active-site volume (Verschueren et al. 1993) and producing an enzyme several-fold more active in its ability to dechlorinate dichlorohexane. However, no mutant tested could utilize more bulky substrates, such as TCE (Holloway et al. 1998).

Cytochrome P450 belongs to a class of enzymes that are useful for a wide range of oxidation reactions. These enzymes are particularly useful for bioremediation because of the possibility to transform different hydrocarbons. Cytochrome $P450_{cam}$ is one example that uses camphor as the natural substrate and has a low level of activity towards other small hydrophobic molecules such as styrene, ethylbenzene and some small alkanes. By engineering its active site volume and topology, Stevenson et al. (1998) enhanced the activity, substrate specificity and regioselectivity of $P450_{cam}$ for hexane and 3-methylpentane (3-MP). Reducing the active site by the Y96A-V247L double mutation resulted in a four-fold higher activity for the oxidation of hexane over 3-MP, thus, increasing the active site volume by Y96A-Y247A double mutations generated an enzyme with a two-fold preference for 3-MP over hexane. Another successful example was demonstrated with the oxidation of polycyclic aromatic hydrocarbons (PAHs) (Appel et al. 2001). Carmichael and Wong (2001) found that two hydrophobic substitutions in CYP102 (Arg47Leu/Tyr51Phe) increased the activity of this enzyme up to 40-fold against PAHs such as phenanthrene, fluoranthene and pyrene. If these mutations were combined with the active site mutations Phe87Ala or Ala264Gly, PAH oxidation increased significantly with simultaneous enhancement of NADPH oxidation.

The biphenyl dioxygenases of strains *Burkholderia cepacia* LB400 (LB400) and *Pseudomonas pseudoalcaligenes* KF707 (KF707) are structurally very similar but exhibit different specificities for polychlorinated biphenyls (PCBs). Guided by sequence comparison, Mondello and coworkers (1997) altered several amino acid residues within regions III and IV of the large subunit of LB400 biphenyl dioxygenase to mimic those in KF707. Similarly, Kimura and coworkers (1997) produced chimeric enzymes from the same wild-type dioxygenases by exchanging restriction fragments. With both approaches, variant enzymes were obtained with the capability to hydroxylate double *ortho*- and double *para*-substituted PCBs, thus combining the substrate range of both parental enzymes. The construction of hybrid nitrotoluene dioxygenases was also helpful in identifying the C-terminal region of the large subunit of this enzyme as critical for substrate specificity (Parales et al. 1998). The recent resolution of the crystal structure of naphthalene dioxygenase (Kauppi et al. 1998) will undoubtedly be very helpful for tailoring aromatic dioxygenases by rational design.

Even with increasing information on the structural and functional aspects of biodegradative enzymes, rational design approaches can fail due to unexpected influences exerted by the substitution of one or more amino acid residues. It is known that mutations far from the active site can modulate catalytic activity or substrate recognition but are difficult to predict a priori. Rational (site-directed) approaches are also restrictive because they allow the exploration of only a very limited sequence space at a time. Therefore, irrational approaches such as DNA shuffling

(Stemmer 1994) can be a preferable alternative to direct the evolution of enzymes or pathways with highly specialized traits. Similar to natural selection, in which multiple environmental forces selected enzymes to meet a variety of challenges, irrational approaches do not require prior extensive structural or biochemical data. When combined with focused selection or screening, irrational approaches offer a useful alternative for generating both the desired improvements and a database for future rational approaches to protein design.

Perhaps the most powerful and promising utility of DNA shuffling is in the cross-breeding of genes between diverse classes of species because of the extended sequence space that can be explored (Crameri et al. 1998). In two independent studies, the substrate range of biphenyl dioxygenases toward PCBs have been successfully extended using directed evolution (Kumamaru et al. 1998; Bruhlmann and Chen 1999). Variants were obtained by random shuffling of DNA segments between the large subunit of two wild type biphenyl dixoxygenases. Several variants had extended substrate ranges for PCBs exceeding those of the two parental enzymes. Molecular evolution is probably the most useful way for evolving enzymes with extended substrate specificities for PCBs and other recalcitrant compounds, since microbial degradation of xenobiotics is usually by co-metabolism and does not exert a selective pressure on bacteria.

Two non-heme monooxygenases, which are capable of hydroxylating aromatic compounds have recently been engineered by directed evolution (Meyer et al. 2002; Canada et al. 2002). The following compound, 2-hydroxybiphenyl 3-monooxygenase (HpbA) catalyzes the regioselective *ortho*-hydroxylation of a wide range of 2-substituted phenols to the corresponding catechols (Held et al. 1999). Meyer et al. (2002) used random mutagenesis to generate HpbA mutants for improved monooxygenase activity on various 2-substituted phenols. Mutants were isolated with improved activities as well as improved coupling efficiency. Toluene *ortho*-monooxygenase (TOM) converts toluene to methylcatechol in a two-step process; it also oxidizes naphthalene and trichloroethylene, making TOM a potential biocatalyst for bioremediation (Luu et al. 1995). Using error prone DNA shuffling and spectrophotometric screens that detect naphthol or chloride, Canada et al. (2002) improved TOM's ability to hydroxylate naphthalene and degrade chlorinated compounds.

The optimization of an entire biodegradative pathway is more likely to be achieved by a directed evolution process than by rational design. This was recently demonstrated by the modification of an arsenic resistance operon using DNA shuffling (Crameri et al. 1997). Cells expressing the optimized operon grew in up to 0.5 M arsenate, a 40-fold increase in resistance. Moreover, a 12-fold increase in the activity of one of the gene products (arsC) was observed in the absence of any physical modification to the gene itself. The authors speculated that modifications to other genes in the operon effected the function of the *arsC* gene product. Such unexpected but exciting results are more likely to be realized using irrational approaches.

Irrational strategies have also been employed to amplify homologues of biodegradative enzymes and incorporate them into recombinant enzymes without isolation (or, in theory, knowledge of) the host microorganisms. This approach was recently demonstrated for the modification of catachol 2,3-dioxygenase (C23O) (Okuta et al. 1998). Degenerate primers were used to amplify the central segment of C23O present in a consortia of bacteria derived from soil and sea water samples. A second round of PCR incorporated the amplified central domains into 5′ and 3′ "arms" of the *nahH* gene. Such an approach allows rapid exploitation of the natural sequence diversity already present in the environment for creation of novel hybrid enzymes or pathways with desired features.

4
Conclusions

Engineering microbes or enzymes for better bioremediation, until recently, have been successful mostly for a defined group of related problems. However, this has gradually been replaced by either rational or irrational approaches that can now be applicable to a wider range of pollutants. Although the ability to predictively design microbes or enzymes for any given remediation remains an overwhelming task, the increasing understanding of fundamental mechanistic principles, generated from both basic research or directed evolution, will likely lead to the emergence of novel solutions for improved bioremediation.

Acknowledgement. The financial supports from NSF and U.S. EPA are gratefully acknowledged.

References

Appel D, Lutz-Wahl S, Fischer P, Schwaneberg U, Schmid RD (2001) A P450 BM-3 mutant hydroxylates alkanes, cycloalkanes, arenas and heteroarenes. J Biotechnol 88:167–171

Bae W, Chen W, Mulchandani A, Mehra R (2000) Enhanced bioaccumulation of heavy metals by bacterial cells displaying synthetic phytochelatins. Biotechnol Bioeng 70:518–524

Bae W, Mehra R, Mulchandani A, Chen W (2001) Genetic engineering of *Escherichia coli* for enhanced uptake and bioaccumulation of mercury. Appl Environ Microbiol 67:5335–5338

Bae W, Mulchandani A, Chen W (2002) Cell surface display of synthetic phytochelatins using ice nucleation protein for enhanced heavy metal bioaccumulation. J Inorg Biochem 88:223–227

Bae W, Wu C, Kostal J, Mulchandai A, Chen W (2003) Enhanced mercury biosorption by bacterial cells with surface-displayed MerR. Appl Environ Microbiol 69:3176–3180

Bailey JE (1996) Metabolic engineering. Adv Mol Cell Biol 15A:289–296

Bang SW, Clark DS, Keasling JD (2000) Engineering hydrogen sulfide production and cadmium removal by expression of the thiosulfate reductase gene (phsABC) from *Salmonella enterica serovar typhimurium* in *Escherichia coli*. Appl Environ Microbiol 66:3939–3944

Bardwell JCA (1994) Building bridges: Disulphide bond formation in the cell. Mol Microbiol 14:199–205

Basnakova G, Stephens ER, Thaller MC, Rossolini GM, Macaskie LE (1998) The use of Escherichia coli bearing a phoN gene for the removal of uranium and nickel from aqueous flows. Appl Microbiol Biotechnol 50:266–272

Bontidean I, Berggren C, Johansson G, Csorgi E, Mattiasson B, Lloyd JR, Jakeman KJ, Brown NL (1998) Detection of heavy metal ions at femtomolar levels using protein-based biosensors. Anal Chem 70:4162–4169

Brim H, McFarlan SC, Fredrickson JK, Minton KW, Zhai M, Wackett LP, Daly MJ (2000) Engineering *Deinococcus radiodurans* for metal remediation in radioactive mixed waste environments. Nat Biotechnol 18:85–90

Brühlmann F, Chen W (1999) Tuning biphenyl dioxygenase for extended substrate specificity. Biotechnol Bioeng 63:544–551

Canada KA, Iwashita S, Shim H, Wood TK (2002) Directed evolution of toluene ortho monooxygenase for enhanced 1-naphthol synthesis and chlorinated ethene degradation. J Bacteriol 184:344–349

Carmichael AB, Wong LL (2001) Protein engineering of *Bacillus megaterium* CYP102. The oxidation of polycyclic aromatic hydrocarbons. Eur J Biochem 268:3117–3125

Crameri A, Dawes G, Rodriguez E Jr, Silver S, Stemmer WPC (1997) Molecular evolution of an arsenate detoxification pathway by DNA shuffling. Nat Biotechnol 15:436–438

Crameri A, Raillard S-A, Bermudez E, Stemmer WPC (1998) DNA shuffling of a family of genes from diverse species accelerates directed evolution. Nature 391:288–291

Chen W, Mulchandani A (1998) The use of live biocatalysts for pesticides detoxification. Trends Biotechnol 16:71–76

Cho CHM, Mulchandani A, Chen W (2002) Bacterial cell surface display of organophosphorus hydrolase for selective screening of improved hydrolysis of organophosphate nerve agents. Appl Environ Microbiol 68:2026–2030

Cursino L, Mattos SV, Azevedo V, Galarza F, Bucker DH, Chartone-Souza E, Nascimento A (2000) Capacity of mercury volatilization by mer (from *Escherichia coli*) and glutathione S-transferase (from *Schistosoma mansoni*) genes cloned in *Escherichia coli*. Sci Total Environ 261:109–113

Daly MJ, Ouyang L, Minton KW (1994) In vitro damage and *recA*-dependent repair of plasmid and chromosomal DNA in the radioresistant bacterium *Dinococcus radiodurans*. J Bacteriol 176:3508–3517

Diorio C, Cai J, Marmor J Shinder R, DuBow MS (1995) An *Escherichia coli* chromosomal *ars* operon homolog is functional in arsenic detoxification and is conserved in gram-negative bacteria. J Bacteriol 177:2050–2056

Ellis BML (2000) Environmental biotechnology informatics. Curr Opin Biotechnol 11: 232–235

Francisco JA, Stathopoulos C, Warren RAJ, Kilburn DG, Georgiou G (1993) Specific adhesion and hydrolysis of cellulose by intact *Escherichia coli* expressing surface anchored cellulase or cellulose binding domains. Bio/Technology 11:491–495

Gadd GM, White C (1993) Microbial treatment of metal pollution: a working biotechnology? Trends Biotechnol 11:353–359

Gallardo ME, Ferrandez A, de Lorenzo V, Garcia JL, Diaz E (1997) Designing recombinant *Pseudomonas* strains to enhance biodesulfurization. J Bacteriol 179:7156–7160

Gibson DT, Parales RE (2000) Aromatic hydrocarbons dioxygenases in environmental biotechnology. Curr Opin Biotechnol 11:236–243

Haryama S (1998) Artificial evolution by DNA shuffling. Trends Biotechnol 16:76–82

Held M, Schmid A, Kohler HP, Suske W, Witholt B, Wubbolts MG (1999) An integrated process for the production of toxic catechols from toxic phenols based on a designer biocatalyst. Biotechnol Bioeng 62:641–648

Holloway P, Knoke KL, Trevors JT, Lee H (1998) Alteration of the substrate range of haloalkane dehalogenase by site-directed mutagenesis. Biotechnol Bioeng 59:520–523

Janssen DB, Oppentocht JE, Poelarends G (2001).Microbial dehalogenation. Curr Opin Biotechnol 12:254–258

Izumi Y, Ohshiro T, Ogino H, Hine Y, Shimao M (1994) Selective desulfurization of dibenzothiophene by *Rhodococcus erythropolis*. Appl Environ Microbiol 60:223–226

Kauppi B, Lee K, Carredano E, Parales RE, Gibson DT, Eklund H, Ramaswamy S (1998) Structure of an aromatic-ring-dioxygenase-naphthalene 1,2-dioxygenase. Structure 6:571–586

Keasling JD, Van Dien SJ, Trelstad N, Renninger N, McMahon K (2000) Application of polyphosphate metabolism to environmental and biotechnological problems. Biochemistry 65:324–331

Kimura N, Nishi A, Goto M, Furukawa K (1997) Functional analysis of a variety of chimeric dioxygenases constructed from two biphenyl dioxygenases that are similar structurally but different functionally. J Bacteriol 179:3996–3943

Kotrba P, Doleckova L, de Lorenzo V, Ruml T (1999) Enhanced bioaccumulation of heavy metal ions by bacterial cells due to surface display of short metal binding peptides. Appl Environ Microbiol 65:1092–1098

Kumamaru T, Suenaga H, Mitsuoka M, Watanabe M, Furukawa K (1998) Enhanced degradation of polychlorinated biphenyls by directed evolution of biphenyl dioxygenase. Nat Biotechnol 16:663–666

Labrenz M, Druschel GK, Thomsen-Ebert T, Gilbert B, Welch SA, Kemner KM, Logan GA, Summons RE, De Stasio G, Bond PL, Lai B, Kelly SD, Banfield JF (2000) Formation of sphalerite (ZnS) deposits in natural biofilms of sulfate-reducing bacteria. Science 290:1744–1747

Lange CC, Wackett LP, Minton KW, Daly MJ (1998) Engineering a recombinant *Deinococcus radiodurans* for organopollutant degradation in radioactive mixed waste environments. Nat Biotechnol 16:929–933

Lee JY, Roh JR, Kim HS (1994) Metabolic engineering of *Pseudomonas putida* for the simultaneous biodegradation of benzene, toluene, and *p*-xylene mixture Biotechnol Bioeng 4:1146–1152

Luu PP, Yung CW, Sun AK, Wood TK (1995) Monitoring trichloroethylene mineralization by *Pseudomonas cepacia* G4 PR1. Appl Microbiol Biotechnol 44:259–264

Macaskie LE, Bonthrone KM, Yong P, Goddard DT (2000) Enzymically mediated bioprecipitation of uranium by a *Citrobacter* sp.: a concerted role for exocellular lipopolysaccharide and associated phosphatase in biomineral formation. Microbiology 146:1855–1867

Mejare M, Ljung S, Bulow L (1998) Selection of cadmium specific hexapeptides and their expression as OmpA fusion proteins in *Escherichia coli.* Protein Eng 11: 489–494

Meyer A, Schmid A, Held M, Westphal AH, Rothlisberger M, Kohler HP, van Berkel WJ, Witholt B (2002) Changing the substrate reactivity of 2-hydroxybiphenyl 3-monooxygenase from *Pseudomonas azelaica* HBP1 by directed evolution. J Biol Chem 277:5575–5582

Misra TK (1992) Bacterial resistances to inorganic mercury salts and organomercurials. Plasmid 27:4–16

Mondello FJ, Turcich MP, Lobos JH, Erickson BD (1997) Identification and modification of biphenyl dioxygenase sequences that determine the specificity of polychlorinated biphenyl degradation. Appl Environ Microbiol 63:3096–3103

Mulchandani A, Kaneva I, Chen W (1999) Detoxification of organophosphorus nerve agents by immobilized *Escherichia coli* with surface-expressed organophosphorus hydrolase. Biotechnol Bioeng 63:216–223

Nakamura K, Hagimine M, Sakai M, Furukawa K (1999) Removal of mercury from mercury-contaminated sediments using a combined method of chemical leaching and volatilization of mercury by bacteria. Biodegradation 10:443–447

Nriagu JO, Pacyna JM (1989) Quantitative assessment of worldwide contamination of air, water and soils by trace metals. Nature 333:34–139

O'Halloran TV, Frantz B, Shin MK, Ralston DM, Wright JG (1989) The MerR heavy metal receptor mediates positive activation in a topologically novel transcription complex. Cell 56:119–129

Okuta A, Ohnishi K, Harayama S (1998) PCR isolation of catechol 2,3-dioxygenase gene fragments from environmental samples and their assembly into functional genes. Gene 212:221–228

Pazirandeh M, Chrisey LA, Mauro JM, Campbell JR, Gaber BP (1995) Expression of the *Neurospora crassa* metallothionein gene in *Escherichia coli* and its effect on heavy-metal uptake. Appl Microbiol Biotechnol 43:1112–1117

Parales JV, Parales RE, Resnick SM, Gibson DT (1998) Enzyme specificity of 2-nitrotoluene 2,3-dioxygenase from *Pseudomonas* sp. strain JS42 is determined by the C-terminal region of the alpha subunit of the oxygenase component. J Bacteriol 180:1194–1199

Richins R, Kaneva I, Mulchandani A, Chen W (1997) Biodegradation of Organophosphorus pesticides by surface-expressed organophosphorus hydrolase. Nat Biotechnol 15:984–987

Romeyer FM, Jacobs FA, Brousseau R (1988) Expression of a *Neuospora crassa* metallothionein and its variants in *Escherichia coli.* Appl Environ Microbiol 56:2748–2754

Sauge-Merle S, Cuine S, Carrier P, Lecomte-Pradines C, Luu DT, Peltier G (2003) Appl Environ Microbiol 69:490–494

Schembri MA, Kjaergaard K, Klemm P (1999) Bioaccumulation of heavy metals by fimbrial designer adhesins. FEMS Microbiol Lett 170:363–371

Shimazu M, Mulchanani A, Chen W (2001) Simultaneous degradation of organophosphorus pesticides and *p*-nitrophenol by a genetically engineered *Moraxella* sp. with surface-expressed organophosphorus hydrolase. Biotechnol Bioeng 76:318–324

Sousa C, Cebolla A, de Lorenzo V (1996) Enhanced metalloadsorption of bacterial cells displaying poly-His peptides. Nature Biotechnol 14:1017–1020

Sousa C, Kotrba P, Ruml T, Cebolla A, de Lorenzo V (1998) Metalloadsorption by *Escherichia coli* cells displaying yeast and mammalian metallothioneins anchored to the outer membrane protein LamB. J Bacteriol 180:2280–2284
Stemmer WPC (1994) Rapid evolution of a protein in vitro by DNA shuffling. Nature 370:389–391
Stevenson J-A, Bearpark JK, Wong LL (1998) Engineering molecular recognition in alkane oxidation catalyzed by cytochrome $P450_{cam}$. New J Chem 551–552
Stillman MJ, Shaw III FC, Suzuki KT (1992) Metallothioneins, VCH, Weinheim
Valls M, Atrian S, de Lorenzo V, Fernandez LA (2000b) Engineering a mouse metallothionein on the cell surface of *Ralstonia eutropha* CH34 for immobilization of heavy metals in soil. Nature Biotechol 18:661–665
Valls M, de Lorenzo V, Gonzalez-Duarte R, Atrian S (2000a) Engineering outer-membrane proteins in *Pseudomonas putida* for enhanced heavy-metal bioadsorption. J Inorg Biochem 79:219–223
Verschueren KHG, Selje F, Rozeboom HJ, Kalk KH, Dijkstra BW (1993) Crystallographic analysis of the catalytic mechanism of haloalkane dehalogenase. Nature 363: 693–698
Wackett LP, Sadowsky MJ, Newman LM, Hur HG, Li S (1994) Metabolism of polyhalogenated compounds by a genetically engineered bacterium. Nature 368:627–629
Wang CL, Maratukulam PDL, AM, Clark DS, Keasling JD (2000) Metabolic engineering of an aerobic sulfate reduction pathway and its application to precipitation of cadmium on the cell surface. Appl Environ Microbiol 66:4497–4502
Wang AA, Mulchandani A, Chen W (2002) Specific adhesion to cellulose and hydroysis of organophosphate nerve agents by a genetically engineered *E. coli* with surface-expressed cellulose-binding domain and organophosphorus hydrolase. Appl Environ Microbiol 68:1684–1689
White C, Gadd GM (1998) Accumulation and effects of cadmium on sulphate-reducing bacterial biofilms. Microbiology 144:1407–1415
Yee DC, Maynard JA, Wood TK (1998) Rhizoremediation of trichloroethylene by a recombinant root-colonizing *Pseudomonas fluorescens* strain expressing toluene *ortho*-monooxygenase constitutively. Appl Environ Microb 64:112–118

11 Combined Biological and Abiological Degradation of Xenobiotic Compounds

R.L. Crawford,[1] T.F. Hess,[2] and A. Paszczynski[1]

1 Introduction

In this chapter, we will discuss approaches to bioremediation of recalcitrant xenobiotic chemicals, specifically approaches that employ a combination of chemical (or physical) and biological steps to increase the efficacy of contaminant destruction. Varieties of chemical agents and processes have been applied toward this goal. Prominent among these procedures is the use of pretreatments with strong oxidizing agents such as ozone, Fenton's reagent ($Fe^{2+} + H_2O_2$), potassium permanganate ($KMnO_4$), or ferrate (K_2FeO_4) to convert recalcitrant molecules to oxidized products that are more amenable to biodegradation than the parent contaminant. Other chemical and physical agents, however, also have been employed in this manner, including powerful reductants such as zero-valent iron (Fe^0) and the use of ultrasound or electric fields. Studies of combined chemical-biological treatment processes have been performed using pure microbial cultures, mixed microbial systems under rigorously controlled laboratory conditions' or actual wastewaters or contaminated soils. We will discuss these different investigations in separate sections of our review. We will begin with a brief description of the primary chemical and physical agents used in combined physiochemical–biological treatment schemes.

[1] Environmental Biotechnology Institute, Food Research Center 103, University of Idaho, Moscow, Idaho 83844-1052, USA, e-mail: crawford@uidaho.edu, Tel: +1-208-8856580, Fax: +1-208-8855741
[2] Biological and Agricultural Engineering Department, 401 Engineering and Physics, University of Idaho, Moscow, Idaho 83844-0904, USA

Soil Biology, Volume 2
Biodegradation and Bioremediation
(ed. by. A. Singh and O. P. Ward)

2
Chemical Agents for Pretreatment of Recalcitrant Contaminants

2.1
Fenton's Reagent

The most commonly used chemical pretreatment used in chemical–biological treatment sequences is Fenton's reagent. This approach takes advantage of Fenton's (1894) discovery of the strong oxidizing power of mixtures of hydrogen peroxide and ferrous iron solutions. Haber and Weiss (1934) later identified the oxidizing element in Fenton's reagent as hydroxyl radical ($OH^{\bullet}$). The hydroxyl radical is a nonspecific, exceptionally strong oxidant that reacts with most organic and biological molecules at near diffusion-controlled rates ($>10^9 M^{-1} s^{-1}$) (Dorfman and Adams 1973; Hoigne et al. 1989). The classic Fenton's reaction involves the addition of dilute hydrogen peroxide to a degassed, acidic ferrous iron solution, which generates hydroxyl radicals as shown in Eq. (1). The degradation of organic chemicals by hydroxyl radicals then proceeds via hydroxylation, hydrogen atom abstraction, or dimerization (Walling 1975).

$$H_2O_2 + Fe^{2+} \rightarrow Fe^{3+} + OH^- + OH^{\bullet} \quad (1)$$

Some environmental applications of Fenton's reagent involve reaction modifications, including the use of high concentrations of hydrogen peroxide, the substitution of different catalysts such as ferric iron or naturally occurring iron oxides, and the use of phosphate-buffered media and metal-chelating agents (Lu 2000). These conditions, although not as stoichiometrically efficient as the standard Fenton's reactions, are often necessary to treat industrial waste streams and contaminants in soils and groundwater (Tyre et al. 1991).

2.2
Ozone

Ozone (O_3) is a natural resonance-stabilized form of oxygen (Fig. 1). It is produced industrially in very large quantities and is thus readily available for use in bioremediation processes. Ozone, like oxygen, is an excellent oxidant, though it is a more powerful oxidant than oxygen because it does not require a catalyst such as a metal ion to initiate a reaction. The reduction of ozone results in the release of molecular oxygen (O_2)

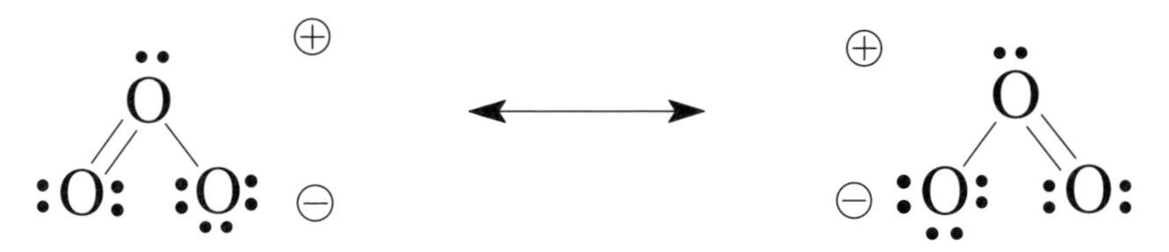

Fig. 1. Structure of ozone

and the formation of an oxygen atom with a –2 oxidation state, usually in the form of water. Thus, no problematic by-products are produced, as is the case with some other oxidants (see below). Ozone is already in use as a stand-alone wastewater treatment technology in many industries and is often investigated for this type of use (Beltran-Heredia et al. 2000; Freire et al. 2001). The compound is a good candidate for use in combined chemical–biological treatment processes because it is readily available and easily deliverable to environments such as water and even to groundwater. Ozone reacts rapidly with organic compounds that have multiple bonds (e.g., C=C, C—N′ and N=N) and with ions such as S^{2-} to form oxyanions such as SO_3^{2-} and SO_4^{2-}. It does not react so rapidly with singly bonded organic structures such as C—C or C—O. Thus, the use of ozone in a combined chemical-biological process must take into account the structure of the target contaminant(s) and the presence of potential interfering oxidizable ions.

2.3 Permanganate

The well-known chemical oxidizing agent potassium permanganate ($KMnO_4$) has been used for in situ treatment of contaminants in groundwater. Targeted contaminants have often been dissolved or dense non-aqueous phase liquid (DNAPL) forms of solvents such as trichloroethylene (TCE) and perchloroethylene (PCE) (Nelson et al. 2001; MacKinnon and Thomson 2002; Huang et al. 2002a). Thus, permanganate should be useful as a pretreatment for recalcitrant xenobiotic compounds in much the same way as peroxide (Huang 2002b), though it produces as a by-product precipitated hydrous MnO_2 that forms during oxidation processes (MacKinnon and Thomson 2002). Permanganate may react too slowly with some contaminants to be of practical value (Damm et al. 2002).

2.4
Ferrate (FeVI)

Iron has a set of oxidation states between +2 and +6, with the higher oxidation states being very strong, potent oxidants. The reduction potential of FeO_4^{2-} (Fe^{6+}) has been determined to be $E^0 = +2.20\,V$ under acidic conditions and $E^0 = +0.72\,V$ under alkaline conditions (Wood 1958). Thus, the use of iron in the +5 or +6 oxidation state has great potential as a pretreatment approach for recalcitrant xenobiotic molecules. Ferrate (FeVI) is known to be highly active in water as a biocide (Kazama 1994) and is drawing considerable interest for treatment of organic contaminants in water. The crimson-colored Fe(VI), as potassium ferrate K_2FeO_4, can be prepared and stored (dry) according to the method of Thompson et al. (1951), though there are still needs for more effective industrial-scale methods for preparation of this reagent (Jiang and Lloyd 2002).

2.5
Zero-Valent Iron (Fe^0)

Presently, there is great interest in the use of zero-valent iron held in situ within subsurface barriers to intercept and degrade or remove contaminants within aquifers through various reductive processes that convert the contaminants to less problematic products (Ghauch 2001; Morrison et al. 2002). The primary removal processes include: (1) sorption and precipitation, (2) chemical reaction, and (3) associated biologically mediated reactions (Scherer et al. 2000). Iron metal, especially its colloidal form, is a potent reductant as it gives up electrons and produces hydrogen during aerobic corrosion in water to form iron hydroxide rusts ($Fe_2O_3 \cdot X\ H_2O$). The formation of rust is thought to begin with the oxidation of Fe^0 to Fe^{2+} as shown in Eq. (2):

$$Fe^0 \rightarrow Fe^{+2} + 2e^- \tag{2}$$

In the presence of water, Fe^{2+} is further oxidized to Fe^{3+} (Eq. 3):

$$Fe^{+2} \rightarrow Fe^{+3} + 1e^- \tag{3}$$

The electrons provided from these oxidations then reduce oxygen as shown in Eq. (4):

$$O_2 + 2H_2O + 4e^- \rightarrow 4OH^- \tag{4}$$

Fe^{3+} then reacts with oxygen to form the rust product. Thus, an overall reaction for rust formation can be written as follows:

$$4Fe^{2+} + O_2 + 4e^- + 2 \times H_2O_2 \rightarrow Fe_2O_3 \times H_2O + 8H^+ \qquad (5)$$

Therefore, iron metal in water promotes oxygen consumption and the creation of anaerobic conditions. In oxygen-limited environments, the reductive equivalents from the iron metal oxidation can form hydrogen and/or reduce organic contaminants (the basis for pretreatment technologies).

2.6 Chlorine

Chlorine is a commonly used chemical agent that can be used as a pretreatment prior to biological degradation. Chlorine is a very strong oxidizing agent and is used commercially as a bleaching agent and as a disinfectant. When pure chlorine is added to water, it forms hypochlorous acid and hydrochloric acid (Eq. 6).

$$Cl_2 + H_2O \rightarrow HOCl + HCl \qquad (6)$$

Hypochlorous acid further dissociates to hypochlorite ion (OCl^-), as shown in Eq. (7):

$$HOCl \rightarrow OCl^- + H^+ \qquad (7)$$

This dissociation is reversible and pH dependent. For example, at pH 7.0, about 70% of the equilibrium is in the form of HOCl, with 30% as OCl^-. The HOCl acts as a strong oxidizing agent and can be used as a pretreatment for recalcitrant contaminants. As HOCl is consumed, the pool of OCl^- replenishes the HOCl in a pH-dependent equilibrium, as shown in Eqs. (8) and (9):

$$Cl_2 + H_2O \rightarrow HOCl + HCl \qquad (8)$$

$$HOCl \rightarrow OCl^- + H^+ \qquad (9)$$

The use of chlorine has the disadvantage of often creating chlorinated oxidation products that may themselves be difficult to biologically degrade. For example, chlorine can react with naturally occurring organic matter in water to produce chlorinated byproducts such as trihalomethane.

2.7 Other Pretreatment Approaches

Other methods, besides the use of strong chemical agents, have been investigated as pretreatments to increase the biodegradability of recalcitrant hazardous compounds. For example, Tiehm et al. (2001) employed ultrasound as a pretreatment to improve the biodegradability of chlorobenzene and the disinfectant 2,4-dichlorophenol. Other approaches involve the use of an oxidizing PbO_2 anode (Longhi et al. 2001) and photolytic pretreatments using sunlight (Maki et al. 2001) or UV radiation in the presence of TiO_2 (Maillacheruvu et al. 2001). A recent investigation indicates that a combination of pulsed-electric discharge and bioremediation may be an effective treatment strategy for certain compounds such as TCE and PCE. This involves the use of repetitive (0.1–1 kHz), short pulse (approximately 100 ns), low voltage (40–80 kV) discharges prior to biotreatment (Yee et al. 1998).

In the following sections, we will discuss some of the experiments performed using the above processes and reagents to promote combined chemical-biological remediation processes.

3 Fenton-Like Systems for Soil and Water

3.1 Combined Remediation Technologies

The application of combined technologies for the destruction of hazardous wastes has been the subject of recent attention (Carberry and Benzing 1991; Ravikumar and Gurol 1991; Koyama et al. 1994; Scott and Ollis 1995). In these studies, the researchers investigated sequential processes using abiotic reactions as a pretreatment step followed by a separate biological reaction. Such technologies were developed to overcome the biorecalcitrance of a particular compound inherent with stand-alone biological processes. Our own research into sequential, coupled technologies has indicated that the use of either TiO_2-mediated photocatalysis or Fenton-like reactions followed by biological degradation is an effective technology for the mineralization of TNT in aqueous solution (Hess et al. 1998; Hess and Schrader 2002) and soil. The investigation of coexisting abiotic and biotic transformation processes has also received attention recently (Büyüksönmez et al. 1998a, b, 1999; Howsawkeng 2001). Coexisting reactions may have both economic and process advantages when applied to industrial pollution prevention schemes or to in situ remediation of hazardous wastes.

3.2
Combined Remediation Technologies for TNT Destruction

Sequential, coupled abiotic/biotic processes have recently been shown at a proof-of-concept level to be effective for the ultimate destruction of TNT. Using TiO_2-mediated photocatalysis as a pretreatment process followed by fungal-mediated oxidation of the photo-products resulted in an approximate 32% overall TNT mineralization extent, more than double the 14% mineralization by fungal treatment alone and ten times the 3% mineralization by photocatalytic treatment alone (Hess et al. 1998). Optimization of the coupled process time showed that 6 hours of photo-pretreatment resulted in the greatest extent of TNT mineralization (Fig. 2).

Further work with another combined abiotic/biotic technology (Fenton-like reactions together with nonspecific bacterial cultures) also

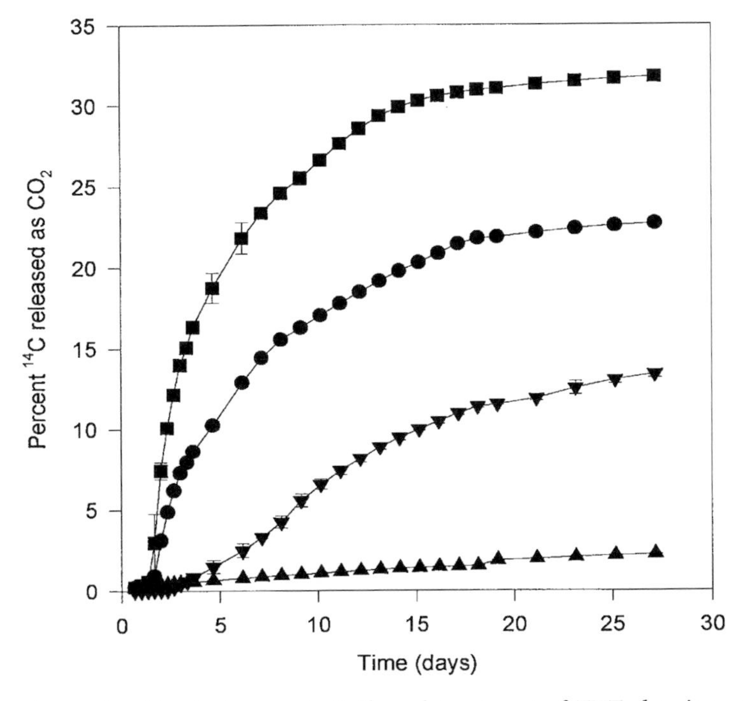

Fig. 2. Combined photocatalytic/fungal treatment of TNT showing an approximate 30% improvement in mineralization extent using the combined process over photocatalysis alone (100 mg/l initial TNT concentration). Data are from fungal degradation of: 6-h (*filled squares*) and 2-h (*filled circles*) photocatalytic pretreatment products, an uninoculated aliquot of 6-h photocatalysis products (*filled inverted triangles*), and [U-^{14}C]-TNT alone (*filled triangles*) (Adapted from Hess et al. 1996)

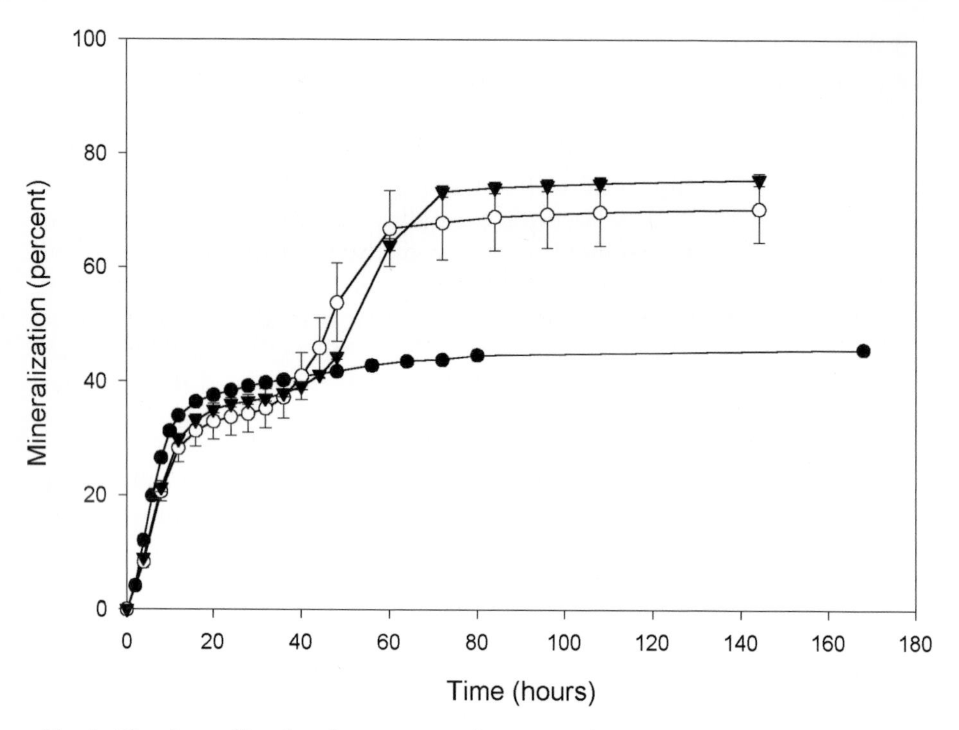

Fig. 3. Kinetic profiles showing an approximate 33% increase in TNT mineralization extent (aqueous solution) in combined abiotic/biotic reactions compared to Fenton reactions alone. Abiotic Fenton reactions alone (*filled circles*) and coupled abiotic/biotic reactions with two concentrations of biomass (*filled inverted triangles* = 467 mg/l; *open circles* = 93 mg/l) (Adapted from Hess and Schrader 2002)

showed improved TNT mineralization in a coupled process over that of either the abiotic or biotic process alone (Hess and Schrader 2002). The TNT mineralization extent (78%) in aqueous solution due to the optimized, coupled abiotic/biotic process at acidic pH was an approximate 33% increase over mineralization from Fenton-like treatment alone (45%; Fig. 3). A similar increase in TNT mineralization was shown for neutral pH. An accumulation of carboxylic acids in solution (Fenton reaction products of TNT degradation) with subsequent microbial mineralization was shown as the mechanism for improved bacterial mineralization of organic carbon in the system. In soil solution experiments (Hess and Schrader 2002), the use of the coupled abiotic/biotic process resulted in improved TNT mineralization over that of the Fenton reaction alone (Fig. 4), although the type of soil and the number of hydrogen peroxide applications greatly influenced overall TNT mineralization extent.

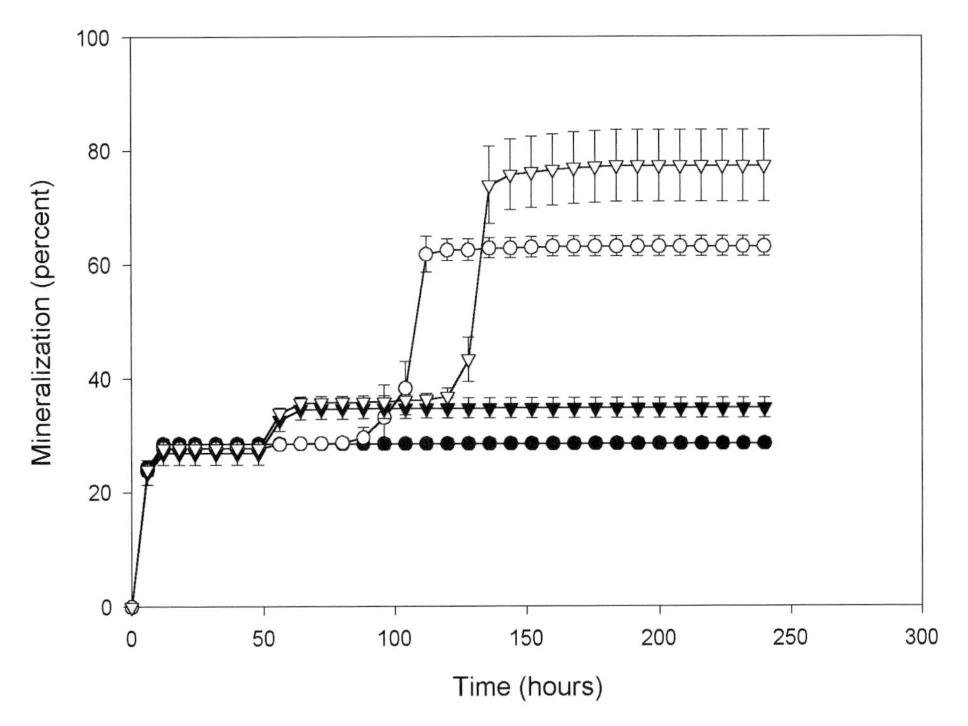

Fig. 4. Kinetic profiles showing an approximate 40% increase in TNT mineralization extent (soil solution) in combined abiotic/biotic reactions compared to Fenton reactions alone. Profiles shown include: (1) abiotic Fenton reaction alone (*filled circles*-one application of H_2O_2); (2) abiotic Fenton reaction alone (*filled inverted triangles*-two applications of H_2O_2); (3) coupled abiotic/biotic reaction (*open circles*-one application of H_2O_2); (4) coupled abiotic/biotic reaction (*open inverted triangles*-two applications of H_2O_2)

Coexisting abiotic-biotic reactions may have application for in situ soil and groundwater remediation and could possibly occur during such efforts. Hydrogen peroxide, which is miscible and dissociates to oxygen and water, has been commonly used as an oxygen source for in situ bioremediation (Pardieck et al. 1992; Brown et al. 1994; Norris and Dowd 1994). However, in some instances, due to rapid dissociation of the hydrogen peroxide, oxygen transfer was found to be limited to the area in the close vicinity of the injection well. This quick dissociation was later attributed to microbial enzymatic activity and iron and copper salts present in the soil matrix (Spain et al. 1989; Brown and Norris 1994; Brown et al. 1994). Data from Tyre et al. (1991) and Watts et al. (1993) suggest that naturally occurring iron oxyhydroxides can serve as effective Fenton catalysts. Fenton reactions, then, were likely taking place in the in situ bioremediation applications where hydrogen peroxide was

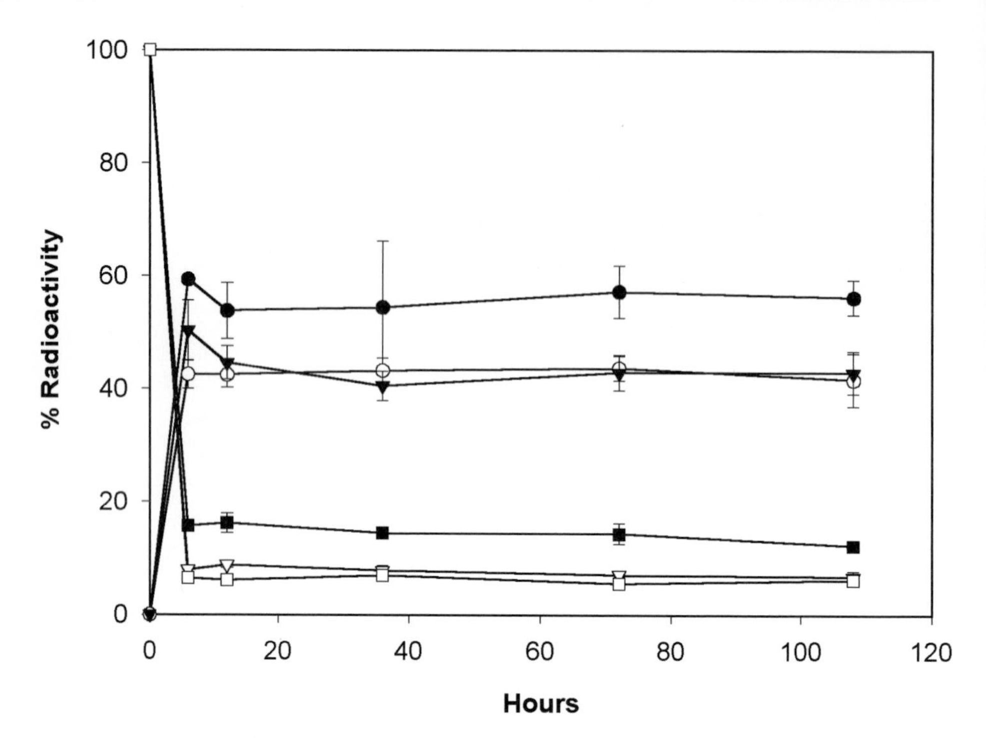

Fig. 5. Kinetic profiles showing an approximate 12% increase in PCE mineralization extent (aqueous solution) in coexistent abiotic/biotic reactions compared to Fenton reactions alone (Adapted from Büyüksönmez et al. 1998b)

being used to increase dissolved oxygen concentrations. Additionally, interest in abiotic remediation technologies has led to recent commercialization of modified Fenton reaction systems for use in situ (Andrews et al. 1997). Both abiotic and bioremediation technologies may ultimately benefit from coexisting abiotic and biotic reactions.

Coexisting reactions using an aerobic, biodegradative organism in the presence of modified Fenton's reactions have recently been shown to exist (Büyüksönmez et al. 1998b, 1999; Howsawkeng et al. 2001). Data obtained in our laboratories revealed that the overall mineralization of PCE was enhanced by the addition of microorganisms degrading dichloroacetic acid (DCAA), a Fenton-oxidation product of PCE degradation (Fig. 5). Howsawkeng et al. (2001) provided additional evidence for simultaneous abiotic/biotic reactions by showing that PCE destruction via hydroxyl radical attack could coexist with microbial oxalate destruction. In both studies, maximal destruction of the target compound (PCE) occurred at intermediate organism concentration due to reaction quenching by the microbial enzymes catalase and peroxidase. While the overall combined abiotic and biotic processes still need to be optimized for practical use, it appears that the concept of coexisting reactions is possible.

Although effective in remediation, the elements of Fenton reactions, such as hydrogen peroxide and hydroxyl radicals, are highly toxic to living organisms and must be taken into consideration during process optimization. The toxicity of the reactive oxygen species derives from their ability to oxidize a large number of cellular constituents. Toxicity mechanisms include DNA disruption (Ananthaswamy and Eisentrark 1977; Storz et al. 1987), oxidation of proteins and amino acids (Brot et al. 1981; Wolff et al. 1986), and lipid peroxidation of membrane fatty acids (Mead 1976). While the toxicity of hydrogen peroxide to microorganisms has been the subject of many investigations (Ananthaswamy and Eisentrark 1977; Hassan and Fridovich 1978; Winquist et al. 1984; Buchmeier et al. 1995; Izawa et al. 1996; Fiorenza and Ward 1997), only recently has there been an attempt to quantify the toxic effects of modified Fenton reaction species on similar biological systems. Using a peroxide-acclimated pure culture of bacteria, Büyüksönmez et al. (1998a) showed that cells do survive hydroxyl radical attack during degradative processes. The mechanism of cellular protection against hydroxyl radicals was found to be similar to that against hydrogen peroxide: an increase in catalase and peroxidase activity.

4 Other Treatment Systems

4.1 Ferrate

Ferrate (Fe^{6+}) is a stronger oxidant than most of the alternative oxidizing agents being employed for pretreatments of recalcitrant environmental contaminants (Bielski 1991). Its decomposition products, when it is applied in water or moist soils, are principally iron hydroxide rusts, which themselves are innocuous. Thus, ferrate has great potential for use in combined chemical–biological systems. One limitation on its large-scale use has been the challenge of producing it in bulk quantities. The primary synthetic procedures used to date involve the use of concentrated and dangerous solutions of NaOH and hypochlorite (Thompson et al. 1951). Fortunately, researchers are focusing on less troublesome ferrate synthesis techniques. For example, Lee et al. (2002) have recently produced ferrate electrochemically in an acidic aqueous medium for the first time.

Rush et al. (1995) examined the reactions of potassium ferrate, K_2FeO_4, with the simple aromatic model compound phenol in aqueous solution between pH 5.5 and 10. They observed a process that is second order in reaction rate for both reactants and gave primarily a mixture of hydroxylated products, including paraquinone, and biphenols that indicate that oxidation of phenol occurs by both one-electron and two-electron pathways. The two-electron oxidant, producing both *para*- and *ortho*-hydroxylated phenols, was considered to be ferrate(V), which is itself produced by the initial one-electron reduction of ferrate(VI). Since hydroxylation reactions generally activate aromatic rings and make them more susceptible to biodegradation, this work indicates a great potential for use of ferrate in combined chemical–biological treatment systems.

Prior work indicates that ferrate might be useful in the pretreatment of toxic metal-contaminated soil or water to lower overall system toxicity and thereby allow more effective biodegradation of any organic co-contaminants that are present. For example, Bartzatt et al. (1992) used K_2FeO_4 to remove toxic metals from an aqueous solution. The toxic material present in water was precipitated from solution and was readily removed with precipitated potassium ferrate-derived rusts.

Chao (1996) showed that ferrate (VI) applied to sludges, oxidizes ammonia to nitrate and sulfide to sulfate. In theory such a process might be used to shift the overall microbial community metabolism of a reducing environment rich in sulfides and ammonia from the inefficient processes dominated by sulfate reduction toward the use of nitrate as a respiratory electron acceptor, thus improving biodegradative potential and microbial diversity in general. Other examples of chemicals highly susceptible to oxidation by ferrate include nitrosamines (Bartzatt and Nagel 1991).

4.2 Permanganate

Thus far, little research has focused on the possible *sequential* use of permanganate and biological processes for bioremediation applications. However, some work has been performed looking at permanganate as the primary degradative agent. Work conducted thus far suggests that permanganate in combination with a following biological degradation step might be very useful for developing remediation strategies in some environmental contamination situations.

Huang et al. (2002a, b) examined the kinetics, reaction pathways, and product distribution of oxidation of tetrachloroethylene (perchloroethylene, PCE) by potassium permanganate ($KMnO_4$). Reactions were performed in reactors in phosphate-buffered solutions under constant pH and isothermal conditions with complete mixing and no headspace. Though the PCE–$KMnO_4$ reaction was thought to proceed through various pathways to ultimately yield CO_2, readily biodegradable intermediates observed included oxalic acid, formic acid, and glycolic acid. Column studies showed complete destruction of TCE in a sandy matrix.

Damm et al. (2002) investigated the reaction kinetics for the oxidation of methyl-*tert*-butyl ether (MTBE) by permanganate. Batch tests demonstrated that the oxidation of MTBE by permanganate is second order overall and first order individually with respect to permanganate and MTBE. However, the slow rates of MTBE oxidation by permanganate were suggested to limit the applicability of this process for rapid MTBE cleanup strategies. However, the authors felt that permanganate oxidation of MTBE has potential for passive oxidation risk management strategies. This would certainly be the case if MTBE transformation products were confirmed to be biodegradable.

MacKinnon and Thomson (2002) undertook a laboratory investigation using a physical model of silica sand overlying a silica flour base. Their system was designed to assess in situ chemical oxidation using permanganate as an oxidant to reduce the mass of a PCE dense nonaqueous phase liquid (DNAPL) pool. They found that permanganate will remove substantial mass from a DNAPL pool; however, the precipitation of hydrous MnO_2 that occurs during the oxidation process may limit performance.

Nelson et al. (2001) injected aqueous $KMnO_4$ solutions into an unconfined sand aquifer contaminated by a DNAPL containing PCE. Monitoring involved depth-specific, multilevel groundwater samplers and continuous cores. Aquifer cores exhibited dark brown to black bands of manganese oxide reaction products in sand layers where DNAPL was originally present and had been exposed to the injected permanganate. The manganese oxide coatings were uniformly distributed over the mineral grains and were about $1\,\mu m$ thick. Plugging of pores by precipitates of manganese oxides was of concern; however, in this case, the decrease in porosity through the formation caused by hydrous MnO_2 precipitates was negligible.

Clearly, more work needs to be performed to examine the effectiveness of permanganate in combined chemical–biological systems for contaminants other than chlorinated solvents. This approach may be of considerable unrealized potential.

4.3 Zero-Valent Iron

Metallic iron can be an excellent pretreatment agent to support the ultimate bioremediation of troublesome or recalcitrant organic or inorganic compounds. For example, there is a need at many locations around the world to remove nitrate from groundwater. Zero-valent iron can be used to convert nitrate to ammonia, which then can be removed by microbial assimilation. The reduction of nitrate to ammonia requires eight electrons/nitrate, as shown in Eq. (10). Iron metal can facilitate the reduction (Choe et al. 2000). In fact, nitrate reduction to ammonia by zero-valent iron is a spontaneous process occurring over a broad pH range. Reduction of nitrate by zero-valent iron consumes four molar equivalents of iron per mole of nitrate (Eqs. 11 and 12).

$$NO_3^- + 10H^+ + 8e^- \rightarrow NH_4^+ + 3H_2O \qquad (10)$$

$$Fe \rightarrow Fe^{2+} + 2e^- \qquad (11)$$

$$NO_3^- + 10H^+ + 4Fe \rightarrow NH_4^+ + 3H_2O + 4Fe^{2+} \qquad (12)$$

Importantly, the hydrogen produced during anaerobic iron corrosion in water can support the growth of anaerobic bacterial communities that include microorganisms such as methanogens, homoacetogens, sulfate reducers, and denitrifiers. Metallic iron barriers placed within natural environments have a diverse microflora (Gu et al. 2002). Since many hydrogen-utilizing and other microorganisms associated with iron barriers can mediate cometabolic reductive dechlorination reactions (Mazur and Jones 2001), the combined chemical and biological process of iron-metal corrosion can result in the removal of chlorinated contaminants, such as TCE and PCE that are found in many groundwater systems (Fetzner 1998). However, in some instances reduction products can be more hazardous than the initial contaminant, as in the transformation of TCE to vinyl chloride (Lampron et al. 2001). The potential for this type of transformation must always be monitored. Certain anaerobic bacteria associated with zero-valent iron barriers are respiratory dehalogenators using hydrogen or other reduced substrates as electron donors for dechlorination (Drzyzga and Gottschal 2002; Neumann et al. 2002).

An accumulation of reports indicates that zero-valent iron is a good pretreatment to promote the ultimate biodegradation of recalcitrant compounds. For example, azo dyes are a group of chemicals that are largely resistant to aerobic biodegradation and persist in wastewater treatment processes. Perey et al. (2002) proposed that zero-valent iron can be used to reduce the azo bond in these dyes, cleaving the dye molecule into products that are more amenable to mineralization by bacteria in biological treatment processes such as activated sludge. Hozalski et al. (2001) employed batch experiments to investigate reactions of Fe^0 with four trihaloacetic acids: trichloroacetic acid, tribromoacetic acid, chlorodibromoacetic acid, and bromodichloroacetic acid. All four compounds readily reacted with Fe^0, and reaction products indicated sequential hydrogenolysis to compounds amenable to biodegradation.

Morrison et al. (2002) used zero-valent iron treatment for the passive removal of contamination from groundwater at a uranium mill tailings site. Zero-valent iron that had been powdered, bound with aluminosilicate, and molded into plates was used as a reactive material in one treatment cell. Another cell used granular Fe^0 and steel wool. The treatment cells significantly reduced concentrations of As, Mn, Mo, Se, U, V, and Zn in groundwater as it passed through the matrixes. Aside from metals and radionuclide treatment, this process might also be employed to detoxify water prior to further biotreatment of other contaminants.

Lampron et al. (2001) examined trichloroethene (TCE) transformation and transformation product distribution in an aqueous medium containing zero-valent iron. They investigated these transformations in the presence of an anaerobic mixed culture to assess the potential role of microorganisms in reactions promoting permeable iron barriers. The presence of the culture increased the rate of TCE disappearance and changed the product distribution, suggesting the probable development of dechlorinating microbial populations in metallic iron barriers. However, the alteration of TCE reduction pathways appeared to be responsible for the production of the carcinogen vinyl chloride, which would have significant negative impact on the performance of a Fe^0 barrier at an actual site.

Ghauch and Suptil (2000) and Ghauch (2001) studied the reduction of several agricultural chemicals (*s*-triazines, benomyl, picloram, and dicamba) in batch systems by investigating the lowering of concentrations of these compounds upon contact with zero-valent iron powder. Aqueous buffered solutions of the compounds were added to the system followed by the iron powder. Declines in the concentrations of these molecules were monitored over time, showing the complete disappearance of picloram after 20 min of contact. Benomyl and dicamba disappeared after 25 and 40 min, respectively. The half-lives of the *s*-triazine pesticides were 7–10 min when they were treated separately and 10–14 min when they were treated together under the same conditions. In other work, Ghauch et al. (2001) demonstrated that the reductive degradation of carbaryl with iron powder as the source of electrons was effective, with a half-life of several minutes. Speculations on the intermediates produced by reduction of these various pesticides and herbicides indicate a likelihood of their biodegradability in a succeeding treatment step, should that be deemed desirable. For example, the same primary reduction product was produced from all the *s*-triazines. It was identified as 4,6-(diamino)-*s*-triazine. This is likely to be a biodegradable molecule, but studies of this question are still needed. Nonetheless, it appears that treatment of pesticide-herbicide mixtures with zero-valent iron should lead to products that can be degraded in a succeeding biological system. This is a process strategy that could be useful as a means for treatment of agricultural chemical residues in the environment.

5 Brown-Rot Fungi: Nature's Example of Organisms Employing Fenton-Like Reactions

Wood-degrading, saprophytic higher Basidiomycetes use biological means to generate free radicals, which initiate depolymerization of this ubiquitous plant material. Koenigs (1974a, b) first demonstrated rapid loss of wood weight due to depolymerization of cellulose and hemicelluloses when exposed to low concentrations of ferrous iron and hydrogen peroxide. He suggested that iron and peroxide are substrates for production of highly reactive, reduced oxygen species, including $OH^{\bullet}$ by these fungi. Using this system, he was able to simulate brown-rot decay of wood. Koenigs (1974a, b) also detected hydrogen peroxide in several fungal cultures, which was later confirmed as being a very common product in white-rot cultures as well (Forney et al. 1982; Zhao et al. 1996).

5.1 Brown-Rot Decay Mechanisms

Several subsequent review articles and research reports have attempted to explain the biochemical mechanisms underlying brown-rot decay using Koenigs' original idea (Highley 1980; Ander and Marzullo 1997; Enoki et al. 1997; Goodell et al. 1997; Hyde and Wood 1997; Shimada et al. 1997). Experimental data only partially support current models of brown-rot decay. All models assume that such decay involves a Fenton-type catalytic system that produces hydroxyl radicals that attack wood components, emphasizing different electron donors or iron chelators. However, large amounts of scientific data indicate that in biological systems that include fungi the classical Fenton reaction is only part of the radical-generating system. In living systems in which Fenton-like reactions occur, these reactions can be defined as reductions of peroxides by transition metal complexes (Liochev 1999; Mau 2001). These reactions are important for all living systems (including humans) that dwell in oxygen-rich environments, and they are the cause of many undesirable reactions, such as mutations, protein damage, and cancer induction.

The brown-rot mechanism of wood proposed by Enoki et al. (1992, 1997) involves low-molecular-weight extracellular substances that the authors isolated from brown- and white-rot fungi. These substances, never fully characterized by the authors, have one-electron reducing capability. They can reduce molecular oxygen to $OH^{\bullet}$ and ferric to ferrous iron. The model requires an extracellular reductant. Enoki et al. used NADH, ascorbate, and *N,N′*-dimethyl-4-nitrosoaniline to demonstrate that an isolated low-molecular-weight substance will work in vivo (Hirano et al. 1995, 1997). The problem with this model is that there is no evidence that fungal cells release this type of compound, and there is lack of chemical characterization of the isolated catalyst(s).

Recently, Tanaka et al. (1999a, b) have reported that the catalytic substances are peptides in the molecular weight range of 1000–5000 Da produced by white-rot fungi; they also isolated similar compounds from soft-rot fungi (Tanaka et al. 2000). These peptides have been proposed to catalyze a redox reaction between molecular oxygen and an electron donor to produce hydroxyl radicals and other reduced oxygen species in fungal culture.

Goodells' group reported the isolation of extracellular, low-molecular-weight biochelators (Gt chelators) from culture filtrates of the brown-rot fungus *Gloeophyllum trabeum* (Fekete et al. 1989; Jellison et al. 1991). This study suggested an alternative pathway for brown-rot decay. Strong positive Chrome Azural-S (CAS) assay results indicated that low-molecular-weight, phenolate-type biochelators produced by *G. trabeum* can chelate transition metals and can also participate in redox cycling (Goodell et al. 1997). Although the mechanism of one-electron oxidation of the chelators in the presence of iron or manganese ions is not completely understood, the isolation of two methoxylated benzoquinones (4,5-dimethoxy-1,2-benzenediol and 2,5-dimethoxy-1,4-benzenediol) from *G. trabeum* culture filtrate by our group suggests the involvement of quinone/hydroquinone redox reactions in brown-rot decay (Paszczynski et al. 1999). Kerem et al. (1999) confirmed our observation of the presence of 2,5-dimethoxy-1,4-benzenediol in the culture medium of *G. trabeum*, and later Jensen et al. (2001) confirmed the presence of a second quinone, 4,5-dimethoxy-1,2-benzenediol. Their results helped us understand how quinones can contribute to radical generation in extracellular brown-rot wood degradation. Xu and Goodell (2001) investigated the changes in molecular weight distribution of cellulose with changes in the concentration and ratio of Fenton chemistry reactants as mediated by Gt chelator and by changes in pH. In addition, cellulose-iron affinity has been determined using inductively coupled plasma (ICP) spectroscopy.

Hyde and Wood (1997) proposed yet another mechanism for the production of hydroxyl radical by the brown-rot fungus *Coniophora puteana*. Their model suggested reduction of Fe(III) by cellobiose dehydrogenase (CDH) within the cells, diffusion of the produced Fe(II) away from the hyphae, formation of an Fe(II)-oxalate complex, and finally, Fenton-based hydroxyl radical formation at a "safe" distance from the hyphae. Near the fungal cell where pH is about 2, ferrous iron is uncomplexed and reacts very slowly with oxygen or with its reduction products. As iron diffuses from the fungal cell and pH increases (a pH gradient is created by a decreasing concentration of oxalic acid as distance from hyphae increases), oxalate chelates and activates the iron. The very slow interaction of CDH with Fe(III) is a weak point of this model. Eriksson et al. (1990) have questioned the production of CDH by brown-rot fungi growing on natural substrates, whereas Schmidthalter and Canevascini (1993) did purify CDH from *C. puteana*. However, according to Ander and Marzullo (1997), *C. puteana* might not be a typical brown-rot fungus.

5.2 Secretion of Organic Acids

Research has shown that brown- and white-rot fungi secrete organic acids, particularly oxalic acid (Espejo and Aguilar 1991; Dutton et al. 1993; Dutton and Evans 1996; Jordan et al. 1996). Both groups of fungi can also rapidly degrade oxalic acid (Akamatsu et al. 1990; Dutton et al. 1994). Shimada et al. (1997) suggested roles for oxalic acid in both brown-rot and white-rot decay postulating that brown-rot fungi might use oxalic acid as a proton donor for enzymatic and nonenzymatic hydrolysis of polysaccharides and as a chelator for an Fe(II)/H_2O_2 system, generating the hydroxyl radical. On the other hand, in white-rot decay oxalic acid may also play multiple roles, including acting as an inhibitor of fungal peroxidases, an electron donor in multiple redox reactions, and a source of formate radicals (Liochev 1999). The importance of fungal-produced oxalic acid in the speciation of metals in the biosphere and in biogeochemical cycles has recently been reviewed (Sayer et al. 1997; Gadd 1999; Gharieb and Gadd 1999). Interestingly, the fungal release of metal-complexing organic acids has been implicated as a factor in the impoverishment of Norway spruce roots causing spruce needle yellowing (Devevre et al. 1996).

Our recent research indicates that the brown-rot fungus *Gloeophyllum trabeum* secretes oxalic acid and also 4,5-dimethoxycatechol and 2,5-dimethoxyhydroquinone into its growth medium (Paszczynski et al. 1999). Oxalic acid production and its role in the stabilization of metal ions is a very well-known phenomenon in fungal systems (Gadd 1999). However, less is known about the catalytic properties of oxalate:metal(s) complexes. We found that eight species belonging to the genus *Gloeophyllum* produce oxalic acid, and five of them produce methoxyhydroquinones (Newcombe et al. 2002). We elucidated structures of Fe(III):oxalate and Fe(II):oxalate and postulated their involvement in redox reactions in fungal Fenton-like systems. These small organic molecules, together with transition metal ions and hydrogen peroxide, play roles in Fenton-like extracellular reactions that are believed to be major mechanisms of initial wood attack by brown-rot fungi.

Other researchers have shown that in in vitro reactions, this system can produce hydrogen peroxide and depolymerize polyethylene glycol (Kerem et al. 1999; Jensen et al. 2001). Recently, Jensen et al. (2002) have purified and characterized NADH:quinone oxidoreductase from *G. trabeum*. This intracellular enzyme could be responsible for redox transfer from fungal cytoplasm to the extracellular Fenton-like' wood-degrading reactants. We also found that methoxylated quinones would not only promote Fenton-like reactions but also efficiently scavenge hydroxyl radicals (Paszczynski, unpubl. data). Recent observations by Zhu et al. (2002), using electron-spin resonance (ESR) spin trapping to study the reactions of quinones with hydrogen peroxide, have shown that substituted quinones are capable of the metal-independent production of $OH^{\bullet}$. This finding suggests that brown-rot wood decay may not require a transition metal and could be simpler than we thought originally.

5.3 Fenton-Like Reactions for Degradation of Man-Made Chemicals

There is strong evidence that brown-rot fungi use Fenton-like reactions from the study of degradation products of the man-made chemicals Enrofloxacin (ENR) and Ciprofloxacin (CIP). These are fluoroquinolone antibacterial drugs widely used in the treatment of serious infections in humans and animals (Fig. 6). These and other similar antibiotics will find a way to contaminate soils and water supplies, particularly in areas of large animal operations.

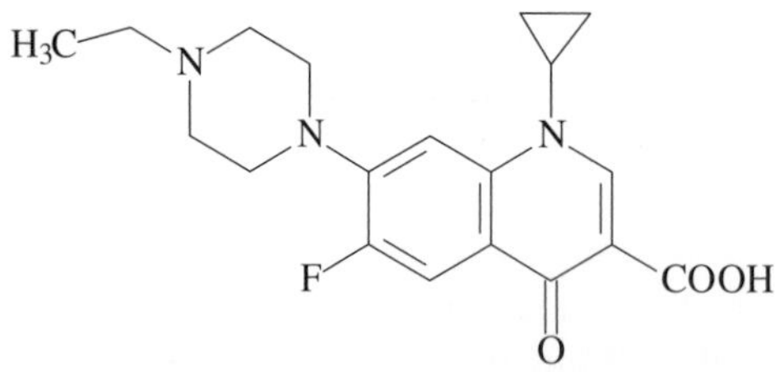

Fig. 6. Structures of Enrofloxacin and Ciprofloxacin

Martens et al. (1996) reported fungal degradation/mineralization of an ENR. The brown-rot fungus *Gleophyllum striatum* mineralized 53% of a ^{14}C-labeled ENR in eight weeks of growth. The rate decreased many times when the antibiotic was adsorbed to soils, but the use of a low concentration of surfactant might make the antibiotic more available in soil systems (Dombrovskaya and Kostyshin 1996). Several other brown-rot fungi also mineralized the compound but to a lesser extent than *G. striatum*. Mineralization of ^{14}C-ENR labeled at either C2 or C4 or in the piperazinyl moiety was subsequently examined using the same brown-rot fungus (Wetzstein et al. 1997). Four principal routes of degradation were proposed, which may reflect an initial attack by hydroxyl radicals at different sites of the drug. Five of the 11 fungal metabolites produced from the antibiotic had structures identical to those observed when the same compound was degraded by Fenton's reagent.

Degradation of CIP by Basidiomycetes fungi was also studied by monitoring $^{14}CO_2$ production from ^{14}C-CIP in liquid cultures (Wetzstein et al. 1999). Sixteen species of fungi isolated from wood, soil, humus, or animal dung produced up to 35% $^{14}CO_2$ during eight weeks of incubation. Despite some low rates of drug mineralization, all species tested reduced the antibacterial activity of CIP in supernatants. A culture of *G. striatum* was used to identify the metabolites formed from CIP. After eight weeks, mycelium mineralized 17% of C-4 and 10% of the piperazinyl moiety, although more than half of 10 ppm of CIP had been transformed after 90 hours. The structures of 11 metabolites were elucidated using HPLC, mass spectrometry, and proton NMR. They fell into four categories: monohydroxylated congeners, dihydroxylated congeners, an isatintype compound, and metabolites indicating both elimination and degradation of the piperazinyl moiety. A metabolic scheme previously described for ENR degradation was confirmed and extended. A new type of metabolite, 6-defluoro-6-hydroxydeethylene-CIP, provided confirmatory evidence for the proposed sequences of degradation. Again, reported results suggested sequential hydroxylation of CIP and its congeners by hydroxyl radicals generated by *G. striatum* culture. The findings also revealed widespread potential for ENR and CIP degradation among Basidiomycetes from different environments. Studies of the degradation of these two chemicals strongly supported the hypothesis that brown-rot fungi are able to produce hydroxyl radicals and to use them to degrade anthropogenic chemicals in the biosphere.

5.4 Additional Evidence for Fungal Fenton-Like Reactions

Using a similar approach and methodology, Fahr et al. (1999) produced more evidence that brown-rot fungi employ Fenton-type reactions. They investigated the degradation of chlorinated phenols, phthalic hydrazide, and $OH^{\bullet}$ scavenging compounds. Wheat straw cultures of the brown-rot fungi *G. striatum* and *G. trabeum* mineralized 2,4-dichlorophenol and pentachlorophenol. After six weeks of growth, up to 54 and 27% $^{14}CO_2$ were liberated from uniformly labeled substrates. Under identical conditions from the same substrates, *Trametes versicolor* (a white-rot species) evolved 42 and 43% $^{14}CO_2$ and expressed high activities of laccase, manganese peroxidase, and lignin peroxidase. No such activity could be detected in straw or liquid cultures of *Gloeophyllum*. Moreover, *G. striatum* degraded both chlorophenols most efficiently under idophasic conditions, i.e., on defined mineral medium lacking sources of carbon, nitrogen, and phosphate.

Later, Schlosser et al. (2000) investigated the degradation pathway of 2,4-dichlorophenol by two strains of *G. striatum*. Two metabolites identified, 4-chlorocatechol and 3,5-dichlorocatechol, were indicative of $OH^{\bullet}$ involvement in the pathway. Accordingly, in vitro Fenton's reagents formed identical metabolites from 2,4-dichlorophenol. These metabolites were identified by comparing HPLC retention times and mass spectra with those of chemically synthesized standards. Under similar conditions, 3-hydroxyphthalic hydrazide was generated from phthalic hydrazide, a reaction assumed to indicate $OH^{\bullet}$ formation. An absence of iron or presence of mannitol (the hydroxyl radical scavenger) in the culture strongly inhibited both reactions. These results provided more evidence of a Fenton-like degradation mechanism operating in the brown-rot fungus *G. striatum*. To date, data indicate that biological Fenton-like reactions require organic molecules (methoxylated quinones) as carriers of electrons from inside cells to the extracellular environment. Methoxylated hydroquinone will reduce oxygen to H_2O_2 and Fe(III) to Fe(II) providing the essential components of a biological Fenton reaction.

6
Conclusions

In this chapter, we have examined examples of chemical-biological remediation of some man-made pollutants. In most cases, pollutant treatment involves the addition of electron donors or acceptors to the contaminated sample (or site) in order to promote redox reaction of the contaminant, thereby making its reduction or oxidation product more susceptible to biological degradation by indigenous microorganisms. The agents that are most frequently used decompose to compounds already present in the environment. Recent reports describing the use of iron(VI) and iron(0) are particularly interesting, and we predict that the application of iron to various oxidation stages in the chemical-biological treatment of pollutants will play an important role in the future.

Similarly, Fenton-like reactions will be an important future tool in pollutant treatment. We are only just beginning to understand the details of this reaction. Because the saprophytic brown-rot fungi employ this reaction in lignocellulose degradation, it may be possible to utilize this group of fungi for in situ degradation of man-made pollutants.

References

Akamatsu Y, Ma DB, Higuchi T, Shimada M (1990) A novel enzymatic decarboxylation of oxalic acid by the lignin peroxidase system of white rot fungus *Phanerochaete chrysosporium*. FEBS Lett 269:261–263

Ananthaswamy HN, Eisenstark A (1977) Repair of hydrogen peroxide-induced single-strand breaks in *Escherichia coli* deoxyribonucleic acid. J Bacteriol 130:187–191

Ander P, Marzullo L (1997) Sugar oxidoreductases and veratryl alcohol oxidase as related to lignin degradation. J Biotechnol 53:115–131

Andrews T, Zervas D, Greenburg RS (1997) Oxidizing agent can finish cleanup where other systems taper off. Soil Groundwater Cleanup July:39–42

Bartzatt R, Nagel D (1991) Removal of nitrosamines from waste water by potassium ferrate oxidation. Arch Environ Health 46:313–315

Bartzatt R, Cano M, Johnson L, Nagel D (1992) Removal of toxic metals and nonmetals from contaminated water. J Toxicol Environ Health 35:205–210

Beltran-Heredia J, Torregrosa J, Dominguez JR, Garcia J (2000) Aerobic biological treatment of black table olive washing wastewaters: effect of an ozonation stage. Proc Biochem 35:1183–1190

Bielski BH (1991) Studies of hypervalent iron. Free Rad Res Commun 12–13:469–477

Brot N, Weissbach L, Wearth J, Weissnach H (1981) Enzymatic reduction of protein-bound methionine sulfoxide. Proc Natl Acad Sci USA 78:2155–2158

Brown RA, Norris RD (1994) The evolution of a technology: hydrogen peroxide in in situ bioremediation. In: Hinchee RE, Alleman BC, Hoeppel RE, Miller RN (eds) Hydrocarbon bioremediation. Lewis, Boca Raton, pp 148–162

Brown RA, Norris RD, Raymond RL (1994) Oxygen transport in contaminated aquifers with hydrogen peroxide. In: Proceedings of the NWWA/API Conference on Petroleum hydrocarbons and organic chemicals in groundwater: prevention, detection and restoration. National Water Well Association, Worthington, pp 441–450

Buchmeier NA, Libby SJ, Xu Y, Loewen PC, Switala J, Guiney DG (1995) DNA repair is more important than catalase for *Salmonella* virulence in mice. J Clin Invest 95: 1047–1053

Büyüksönmez F, Hess TF, Crawford RL, Watts RJ (1998a) Toxic effects of modified Fenton reactions on *Xanthobacter flavus* FB71. Appl Environ Microbiol 64:3759–3764

Büyüksönmez F, Hess TF, Crawford RL, Paszczynski, A, Watts RJ (1998b) Simultaneous abiotic-biotic mineralization of perchloroethylene (PCE). In: Wicramanayake GB, Hinchee RE (eds) Designing and Applying Treatment Technologies: Remediation of Chlorinated and Recalcitrant Compounds. Battelle Press, Columbus, OH, pp 277–282

Büyüksönmez F, Hess TF, Crawford RL, Paszczynski A, Watts RJ (1999) Optimization of simultaneous chemical and biological mineralization of perchloroethylene (PCE). Appl Environ Microbiol 65:2784–2788

Carberry JB, Benzing TM (1991) Peroxide pre-oxidation of recalcitrant toxic waste to enhance biodegradation. Water Sci Technol 23:367–376

Chao AC (1996) Quality improvement of biosolids by ferrate(VI) oxidation of offensive odour compounds. Water Sci Technol 33:119–130

Choe S, Chang YY, Hwang KY, Khim J (2000) Kinetics of reductive denitrification by nanoscale zero-valent iron. Chemosphere 41:1307–1311

Damm JH, Hardacre C, Kalin RM, Walsh KP (2002) Kinetics of the oxidation of methyl *tert*-butyl ether (MTBE) by potassium permanganate. Water Res 36:3638–3646

Devevre D, Garbaye J, Botton B (1996) Release of complexing organic acids by rhizosphere fungi as a factor in Norway spruce yellowing in acidic soil. Mycol Res 100: 1367–1374

Dombrovskaya EN, Kostyshin SS (1996) Effects of surfactants of different tonic nature on the ligninolytic enzyme complexes of the white-rot fungi *Pleurotus floridae* and *Phellinus igniarius*. Biochemistry 61:215–220

Dorfman LM, Adams GE (1973) Document NSRDS-NBS-46. National Bureau of Standards, Washington, DC

Drzyzga O, Gottschal JC (2002) Tetrachloroethene dehalorespiration and growth of *Desulfitobacterium frappieri* TCE1 in strict dependence on the activity of *Desulfovibrio fructosivorans*. Appl Environ Microbiol 68:642–649

Dutton MV, Evans CS (1996) Oxalate production by fungi: Its role in pathogenicity and ecology in the soil environment. Can J Microbiol 42:881–895

Dutton MV, Evans CS, Atkey PT, Wood DA (1993) Oxalate production by Basidiomycetes, Including the White-Rot Species *Coriolus versicolor* and *Phanerochaete chrysosporium*. Appl Microbiol Biotechnol 39:5–10

Dutton MV, Kathiara M, Gallagher IM, Evans CS (1994) Purification and characterization of oxalate decarboxylase from *Coriolus versicolor*. FEMS Microbiol Lett 116: 321–326

Enoki A, Hirano T, Tanaka H (1992) Extracellular substance from brown-rot basidiomycete *Gloeophyllum trabeum* that produces and reduces hydrogen peroxide. Mater Org 27:247–261

Enoki A, Itakura S, Tanaka H (1997) The involvement of extracellular substances for reducing molecular oxygen to hydroxyl radical and ferric iron to ferrous iron in wood degradation by wood decay fungi. J Biotechnol 53:265–272

Eriksson KEL, Blanchette RA, Ander P (1990) Lignin degradation by brown-rot, soft-rot and other fungi. In: Timell TE (ed) Microbial and enzymatic degradation

of wood and wood components. Springer, Berlin Heidelberg New York, pp 312–319
Espejo E, Aguilar A (1991) Production and degradation of oxalic acid by brown rot fungi. Appl Environ Microbiol 57:1980–1986
Fahr K, Wetzstein HG, Grey R, Schlosser D (1999) Degradation of 2,4-dichlorophenol and pentachlorophenol by two brown rot fungi. FEMS Microbiol Lett 175:127–132
Fekete F, Chandhoke V, Jellison J (1989) Iron-binding compounds produced by wood-decaying basidiomycetes. Appl Environ Microbiol 55:2720–2722
Fenton HJH (1894) Oxidation of tartaric acid in presence of iron. J Chem Soc 65: 899–910
Fetzner S (1998) Bacterial dehalogenation. Appl Microbiol Biotechnol 50:633–657
Fiorenza S, Ward CH (1997) Microbial adaptation to hydrogen peroxide and biodegradation of aromatic hydrocarbons. J Ind Microbiol Biotechnol 18:140–151
Forney L, Reddy C, Packrats H (1982) Ultrastructural localization of hydrogen peroxide production in ligninolytic *Phanerochaete chrysosporium* cells. Appl Environ Microbiol 44:732–736
Freire RS, Kubota LT, Duran N (2001) Remediation and toxicity removal from Kraft E1 paper mill effluent by ozonization. Environ Technol 22:897–904
Gadd GM (1999) Fungal production of citric and oxalic acid: importance in metal speciation, physiology and biogeochemical processes. Adv Microb Physiol 41:47–92
Gharieb MM, Gadd GM (1999) Influence of nitrogen source on the solubilization of natural gypsum ($CaSO_42H_2O$) and the formation of calcium oxalate by different oxalic and citric acid-producing fungi. Mycol Res 103:473–481
Ghauch A (2001) Degradation of benomyl, picloram, and dicamba in a conical apparatus by zero-valent iron powder. Chemosphere 43:1109–1117
Ghauch A, Suptil J (2000) Remediation of *s*-triazines contaminated water in a laboratory scale apparatus using zero-valent iron powder. Chemosphere 41:1835–1843
Ghauch A, Gallet C, Charef A, Rima J, Martin-Bouyer M (2001) Reductive degradation of carbaryl in water by zero-valent iron. Chemosphere 42:419–424
Goodell B, Jellison J, Liu J, Daniel G, Paszczynski A, Fekete F, Krishnamurthy S, Jun L, Xu G (1997) Low molecular weight chelators and phenolic compounds isolated from wood decay fungi and their role in the fungal biodegradation of wood. J Biotechnol 53:133–162
Gu B, Watson DB, Wu L, Phillips DH, White DC, Zhou J (2002) Microbiological characteristics in a zero-valent iron reactive barrier. Environ Monit Assess 77:293–309
Haber F, Weiss J (1934) The catalytic decomposition of hydrogen peroxide by iron salts. Proc R Soc Lond 147A:332–351
Hassan HM, Fridovich I (1978) Regulation of the synthesis of catalase and peroxidase in *Escherichia coli.* J Biol Chem 253:6445–6450
Hess TF, Schrader PS (2002) Coupled abiotic-biotic mineralization of 2,4,6-trinitrotoluene (TNT). J Environ Qual 31:736–744
Hess TF, Lewis TA, Crawford RL, Katamneni S, Wells JH, Watts, RJ (1998) Combined photocatalytic and fungal treatment for the destruction of 2,4,6-trinitrotoluene (TNT). Water Res 32:1481–1491
Highley T (1980) Degradation of cellulose by *Postia placenta* in the presence of compounds that affect hydrogen peroxide. Mater Org 15:93–113
Hirano T, Tanaka H, Enoki A (1995) Extracellular substance from the brown-rot basidiomycete *Tyromyces palustris* that reduces molecular oxygen to hydroxyl radicals and ferric iron to ferrous iron. Mokuzai Gakkaishi 41:334–341
Hirano T, Tanaka H, Enoki A (1997) Relationship between production of hydroxyl radicals and degradation of wood by the brown-rot fungus, *Tyromyces palustris*. Holzforschung 51:389–395

Hoigne J, Faust BC, Haag WR, Scully FE, Zepp RG (1989) Aquatic humic substances as source and sink of photochemically produced transient reactants. In: Suffet I, McCarty PL (eds) Aquatic humic substances: influence on fate and treatment of pollutants. American Chemical Society, Washington, DC, pp 363–381

Howsawkeng J, Watts RJ, Washington DL, Hess TH, Crawford RL (2001) Evidence for simultaneous abiotic-biotic oxidations in a microbial-Fenton's system. Environ Sci Technol 35:2961–2966

Hozalski RM, Zhang L, Arnold WA (2001) Reduction of haloacetic acids by Fe(0): implications for treatment and fate. Environ Sci Technol 35:2258–2263

Huang K-C, Hoag GE, Chheda P, Woody BA, Dobbs GM (2002a) Chemical oxidation of trichloroethylene with potassium permanganate in porous medium. Adv Environ Res 7:217–230

Huang KC, Hoag GE, Chheda P, Woody BA, Dobbs GM (2002b) Kinetics and mechanism of oxidation of tetrachloroethylene with permanganate. Chemosphere 46: 815–825

Hyde SM, Wood PM (1997) A mechanism for production of hydroxyl radicals by the brown-rot fungus *Coniophora puteana*: Fe(III) reduction by cellobiose dehydrogenase and Fe(II) oxidation at a distance from the hyphae. Microbiology 143: 259–266

Izawa S, Inoue Y, Kimura A (1996) Importance of catalase in the adaptive response to hydrogen peroxide: analysis of acatalasaemic *Saccharomyces cerevisiae*. Biochem J 320:61–67

Jellison J, Chandhoke V, Goodell B, Fekete FA (1991) The isolation and immunolocalization of iron-binding compounds produced by *Gloeophyllum trabeum*. Appl Microbiol Biotechnol 35:805–809

Jensen KA, Houtman CJ, Ryan ZC, Hammel KE (2001) Pathways for extracellular Fenton chemistry in brown rot basidiomycete *Gloeophyllum trabeum*. Appl Environ Microbiol 67:2705–2711

Jensen KA, Ryan ZC, Wymelenberg AV, Cullen D, Hammel DE (2002) An NADH:quinone oxidoreductase active during biodegradation by brow-rot basidiomycete *Gloeophyllum trabeum*. Appl Environ Microbiol 68:2699–2703

Jiang JQ, Lloyd B (2002) Progress in the development and use of ferrate(VI) salt as an oxidant and coagulant for water and wastewater treatment. Water Res 36:1397–1408

Jordan CR, Dashek WV, Highley TL (1996) Detection and quantification of oxalic acid from brown-rot decay fungus *Postia placenta*. Holzforschung 50:312–318

Kazama F (1994) Inactivation of coliphage Q beta by potassium ferrate. FEMS Microbiol Lett 118:345–349

Kerem Z, Jensen KA, Hammel KE (1999) Biodegradative mechanism of the brown rot basidiomycete *Gloeophyllum trabeum*: evidence for an extracellular hydroquinone-driven Fenton reaction. FEBS Lett 446:49–54

Koenigs JW (1974a) Production of hydrogen peroxide by wood-rotting fungi in wood and its correlation to with weight loss, depolymerization and pH changes. Arch Microbiol 99:129–145

Koenigs JW (1974b) Hydrogen peroxide and iron: A proposed system for decomposition of wood by brown-rot basidiomycetes. Wood Fiber 6:66–80

Koyama O, Kamagata Y, Nakamura K (1994) Degradation of chlorinated aromatics by Fenton oxidation and methanogenic digester sludge. Water Res 28:885–899

Lampron KJ, Chiu PC, Cha DK (2001) Reductive dehalogenation of chlorinated ethenes with elemental iron: the role of microorganisms. Water Res 35:3077–3084

Lee J, Tryk DA, Fujishima A, Park SM (2002) Electrochemical generation of ferrate in acidic media at boron-doped diamond electrodes. Chem Commun (Camb) 7: 486–487

Liochev SI (1999) The mechanism of "Fenton-like" reactions and their importance for biological systems. A biologist's view. In: Sigel A, Sigel H (eds) Metals ions in biological systems, vol 36. Marcel Dekker, New York, pp 1–39

Longhi P, Vodopivec B, Fiori G (2001) Electrochemical treatment of olive oil mill wastewater. Ann Chim 91:169–174

Lu MC (2000) Oxidation of chlorophenols with hydrogen peroxide in the presence of goethite. Chemosphere 40:125–130

MacKinnon LK, Thomson NR (2002) Laboratory-scale in situ chemical oxidation of a perchloroethylene pool using permanganate. J Contam Hydrol 56:49–74

Maillacheruvu K, Buck L, Lee E (2001) Biodegradation potential of photocatalyzed surfactant washwater. J Environ Sci Health Part A Tox Hazard Subst Environ Eng 36:883–895

Maki H, Sasaki T, Harayama S (2001) Photo-oxidation of biodegraded crude oil and toxicity of the photo-oxidized products. Chemosphere 44:1145–1151

Martens R, Wetzstein HG, Zadrazil F, Capelari M, Hoffmann P, Schmeer N (1996) Degradation of fluoroquinolone Enrofloxacin by wood-rotting fungi. Appl Environ Microbiol 62:4206–4209

Mau J-L, Chao G-R, Wu K-T (2001) Antioxidant properties of methanolic extracts from silver ear mushrooms. J Agric Food Chem 49:5461–5467

Mazur CS, Jones WJ (2001) Hydrogen concentrations in sulfate-reducing estuarine sediments during PCE dehalogenation. Environ Sci Technol 35:4783–4788

Mead JF (1976) Free radical mechanisms of lipid damage and consequences for cellular membranes. In: Pryor WA (ed) Free radicals in biology. Academic Press, New York, pp 51–68

Morrison SJ, Metzler DR, Dwyer BP (2002) Removal of As, Mn, Mo, Se, U, V and Zn from groundwater by zero-valent iron in a passive treatment cell: reaction progress modeling. J Contam Hydrol 56:99–116

Nelson MD, Parker BL, Al TA, Cherry JA, Loomer D (2001) Geochemical reactions resulting from in situ oxidation of PCE-DNAPL by $KMnO_4$ in a sandy aquifer. Environ Sci Technol 35:1266–1275

Neumann A, Siebert A, Trescher T, Reinhardt S, Wohlfarth G, Diekert G (2002) Tetrachloroethene reductive dehalogenase of *Dehalospirillum multivorans*: substrate specificity of the native enzyme and its corrinoid cofactor. Arch Microbiol 177: 420–426

Newcombe D, Paszczynski A, Gajewska W, Kröger M, Crawford RL, Felis G (2002) Production of small molecular weight catalysts and the mechanism of trinitrotoluene degradation by several *Gloeophyllum* species. Enzyme Microb Technol 30:506–517

Norris RD, Dowd, KD (1994) In situ bioremediation of petroleum hydrocarbon-contaminated soil and groundwater in a low permeability aquifer. In: Flathman PE, Jerger DE, Exner JH (eds) Bioremediation: field experience, Lewis, Boca Raton, pp 457–474

Pardieck DL, Bouwer EJ, Stone AT (1992) Hydrogen peroxide use to increase oxidant capacity for in situ bioremediation of contaminated soils and aquifers: a review. J Contam Hydrol 9:221–242

Paszczynski A, Crawford R, Funk D, Goodell B (1999) De novo synthesis of 4,5-dimethoxycatechol and 2,5-dimethoxyhydroquinone by the brown rot fungus *Gloeophyllum trabeum*. Appl Environ Microbiol 65:674–679

Perey JR, Chiu PC, Huang CP, Cha DK (2002) Zero-valent iron pretreatment for enhancing the biodegradability of Azo dyes. Water Environ Res 74:221–225

Ravikumar JX, Gurol MD (1991) Effectiveness of chemical oxidation to enhance the biodegradation of pentachlorophenol in soils: a laboratory study. In:

Neufeld RD, Casson LW (eds) Hazardous and industrial wastes. Proceedings of the 23rd Mid-Atlantic Industrial Waste Conference. Technomic Publishing, Lancaster, pp 211–221

Rush JD, Cyr JE, Zhao Z, Bielski BH (1995) The oxidation of phenol by ferrate(VI) and ferrate(V). A pulse radiolysis and stopped-flow study. Free Radic Res 22:349–360

Sayer JA, Kierans M, Gadd GM (1997) Solubilization of some naturally occurring metal-bearing minerals, lime scale and lead phosphate by *Aspergillus niger*. FEMS Microbiol Lett 154:29–35

Scherer MM, Richter S, Valentine RL, Alvarez PJ (2000) Chemistry and microbiology of permeable reactive barriers for in situ groundwater clean up. Crit Rev Microbiol 26:221–264

Schlosser D, Fahr K, Karl W, Wetzstein, HG (2000) Hydroxylated metabolites of 2,4-dichlorophenol imply a Fenton-type reaction in *Gloeophyllum striatum*. Appl Environ Microbiol 66:2479–2483

Schmidthalter DR, Canevascini G (1993) Isolation and characterization of cellobiose dehydrogenase from the brown-rot fungus *Coniophora puteana* (Schum ex Fr) Karst. Arch Biochem Biophys 300:559–563

Scott JP, Ollis DF (1995) Integration of chemical and biological oxidation processes for water treatment: review and recommendations. Environ Prog 14:88–103

Shimada M, Akamtsu Y, Tokimatsu T, Mii K, Hattori T (1997) Possible biochemical roles of oxalic acid as a low molecular weight compound involved in brown-rot and white-rot wood decays. J Biotechnol 53:103–113

Spain JC, Milligan JD, Downey DC, Slaughter JK (1989) Excessive bacterial decomposition of H_2O_2 during enhanced biodegradation. Ground Water 27:163–167

Storz G, Christman MF, Sies H, Ames BN (1987) Spontaneous mutagenesis and oxidative damage to DNA in *Salmonela typhimurium*. Proc Natl Acad Sci USA 84: 8917–8921

Tanaka H, Itakura S, Enoki A (1999a) Hydroxyl radical generation by an extracellular low-molecular-weight substance and phenol oxidase activity during wood degradation by the white-rot basidiomycete *Trametes versicolor*. J Biotechnol 75:57–70

Tanaka H, Itakura S, Enoki A (1999b) Hydroxyl radical generation by an extracellular low-molecular-weight substance and phenol oxidase activity during wood degradation by the white-rot basidiomycete *Phanerochaete chrysosporium*. Holzforschung 53:21–28

Tanaka H, Itakura S, Enoki A (2000) Phenol oxidase activity and one-electron oxidation activity in wood degradation by soft-rot deuteromycetes. Holzforschung 54: 463–468

Thompson GW, Ockerman LT, Schreyer JM (1951) Preparation and purification of potassium ferrate IV. J Am Chem Soc 73:1379–1381

Tiehm A, Kohnagel I, Neis U (2001) Removal of chlorinated pollutants by a combination of ultrasound and biodegradation. Water Sci Technol 43:297–303

Tyre BW, Watts RJ, Miller GC (1991) Treatment of four biorefractory contaminants in soils using catalyzed hydrogen peroxide. J Environ Qual 20:832–838

Walling C (1975) Fenton's reagent revised. Acta Chem Res 8:125–131

Watts RJ, Udell MD, Monsen RM (1993) Use of iron minerals in optimizing the peroxide treatment of contaminates soils. Water Environ Res 65:839–844

Wetzstein HG, Schmeer N, Karl W (1997) Degradation of the fluoroquinolone enrofloxacin by the brown rot fungus *Gloeophyllum striatum*: identification of metabolites. Appl Environ Microbiol 63:4272–4281

Wetzstein HG, Stadler M, Tichy HV, Dalhoff A, Karl W (1999) Degradation of ciprofloxacin by basidiomycetes and identification of metabolites generated by the brown rot fungus *Gloeophyllum striatum*. Appl Environ Microbiol 65:1556–1563

Winquist L, Rannug U, Rannug A, Ramel C (1984) Protection from toxic and mutagenic effects of H_2O_2 by catalase induction in *Salmonella typhimurium*. Mutat Res 141: 145–147

Wolff SP, Garner A, Dean RT (1986) Free radicals, lipids and protein degradation. Trends Biochem Sci 11:27–31

Wood RH (1958) The heat, free energy and entropy of the ferrate (VI) ion. J Am Chem Soc 80:2038–2041

Xu G, Goodell B (2001) Mechanisms of wood degradation by brown-rot fungi: chelator-mediated cellulose degradation and binding of iron by cellulose. J Biotechnol 87:43–57

Yee DC, Chauhan S, Yankelevich E, Bystritskii VV, Wood TK (1998) Degradation of perchloroethylene and dichlorophenol by pulsed-electric discharge and bioremediation. Biotechnol Bioeng 59:438–444

Zhao J, Janse BJH (1996) Comparison of H_2O_2-producing enzymes in selected white rot fungi. FEMS Microbiol Lett 139:215–221

Zhu B-Z, Zhao H-T, Kalyanaraman B, Frei B (2002) Metal-independent production of hydroxyl radicals by halogenated quinones and hydrogen peroxide: an ESR spin rapping study. Free Rad Biol Med 32:465–473

12 Methods for Monitoring and Assessment of Bioremediation Processes

Ajay Singh,[1] Ramesh C. Kuhad,[2] Zarook Shareefdeen,[3] and Owen P. Ward[4]

1 Introduction

The cleanup of the contaminated soils is a priority environmental task due to the risks contaminants pose to the groundwater, drinking water and soil fertility. A wide variety of biological, physical and chemical methods have been developed to decontaminate polluted sites. Any successful remediation technology should not simply transfer the contaminants to other environmental compartments. Bioremediation provides a cost-effective and contaminant/substrate-specific treatment technology (Ward et al. 2003). A successful bioremediation approach requires sufficient proof for the detoxification of the contaminants, preferably proven by complete mineralization (Dua et al. 2002). However, the determination of effectiveness and completeness to satisfactory status is one of the major problems. Current monitoring practices require the determination of the disappearance of the contaminants and their degradation products to regulatory levels are monitored followed by toxicity testing, usually on a single organism or species to make sure that there is no product or induced change resulting in any residual toxicity. The problems related to these monitoring approaches and to the assessment of successful bioremediation have been widely recognized and discussed (Höhner et al. 1998; White et al. 1998; van Straalen 2002; Widada et al. 2002a). The microbial community response may prove to be a much more comprehensive indicator of residual toxicity, which is more sensitive than single species toxicity screens, and can be used to complement the disappearance or sequestration of contaminants.

[1] Petrozyme Technologies, 7496 Wellington Road 34, R.R. #3, Guelph, ON N1H 6H9, Canada, e-mail: asingh@petrozyme.com, Tel: +1-519-7672299, Fax: +1-519-7679435
[2] Department of Microbiology, University of Delhi South Campus, New Delhi-110 020, India
[3] BIOREM Technologies, 7496 Wellington Road 34, R.R. #3, Guelph, ON N1H 6H9, Canada
[4] Department of Biology, University of Waterloo, Waterloo, Ontario N2L 3G1, Canada

Soil Biology, Volume 2
Biodegradation and Bioremediation
(ed. by. A. Singh and O. P. Ward)

Traditional culture-dependent methods are generally based on differential morphological, physiological and metabolic traits (Leahy and Colwell 1990; Rosenberg 1992; Atlas and Cerniglia 1995). These include isolation and cultivation on solid media, most probable number (MPN) assays, and more recently, BIOLOG substrate utilization patterns. However, we should also keep in mind that only a small fraction of microorganisms can currently be cultured from environmental samples and, even if a microorganism is cultured, its role in a community and contribution to ecosystem function are not revealed. Attempts to characterize natural microbial communities, impacted upon by environmental contaminants, are further hampered by individual substrate and metabolite interactions. Culture-independent methods for community analysis began with the direct examination of metabolically active microorganisms using differential stains, fluorescence in situ hybridization (FISH) and phospholipid fatty acid analysis (PFLA).

Recent advances in molecular biology have extended our understanding of the metabolic processes related to the microbial transformation of environmental contaminants and their possible role in bioremediation. Physiological responses of microorganisms to the presence of hydrocarbons, including cell surface alterations and adaptive mechanisms for uptake and efflux of these substrates, have been characterized. New molecular and biochemical techniques, such as PCR-DGGE, PFLA and BIOLOG, have enhanced our ability to investigate the dynamics of microbial communities in contaminated ecosystems. Bacterial biosensors have been developed with potential to continuous online monitoring of pollutant concentrations in environmental applications and also offer the possibility to characterize, identify, quantify and determine the biodegradability of target contaminants present in complex environmental samples. This chapter provides an overview of different chemical, microbial, biochemical and molecular methods used for the monitoring and assessment of bioremediation processes.

2 Biodegradation Estimation Methods

As a principal abatement process in the environment, biodegradation is the most important parameter influencing the toxicity, persistence and ultimate fate of pollutants in soils. Biodegradation is dependent upon many factors including temperature, microbial population, degree of acclimation, accessibility of nutrients, cellular transport properties, and chemical partitioning in growth medium. Standard assay procedures using simple consortia are being developed for Environment Canada (Foght et al. 1998, 1999) and the United States Environmental Protection Agency (Haines et al. 2003) in order to test commercially available cultures for bioaugmentation to the contaminated soils. However, third party testing of such products has not proven them to be more effective than autochthonous microbial communities once additional nutrients and sorbents are removed (Venosa et al. 1992; Thouand et al. 1999).

Microbial activity and aerobic metabolism can be quantified by the consumption of molecular oxygen or the evolution of carbon dioxide using respirometry. In addition, respirometric tests can also be applied to assess the degradation potential of petroleum hydrocarbons in soil, nutrient limitations and possible inhibitory effects of heavy metals, toxic compounds and pH on the microbial activities in soil (Graves et al. 1991; Aichinger et al. 1992; Thurmann et al. 1999). Mineralization studies involving measurements of total CO_2 production can provide excellent information on the biodegradability potential of contaminated soil (Balba et al. 1998). Respirometry tests can also be used to evaluate different biological treatment options, the effect of nutrient supplementation and the bioaugmentation of cultures and to confirm active hydrocarbon degradation during a full-scale bioremediation.

Soil microcosm tests can be a useful tool to assess the biodegradation potential of hydrocarbon-contaminated soils and the development of models for predicting the fate of these pollutants. During the test, the concentration of the contaminants and their degradation products can be monitored to obtain useful biodegradation kinetics information, as well as to establish the most suitable bioremediation strategy for the large-scale application. The biodegradation potential can similarly be assessed using slurry bioreactors of various sizes, which offer several advantages, including efficient mixing, aeration and improved substrate availability, to reduce the treatment time significantly (Ward and Singh 2000; Singh et al. 2001). The rates of contaminant degradation and formation of products are monitored using gas chromatography (GC), mass spectroscopy (MS), high performance liquid chromatography (HPLC), thin layer chromatography (TLC) and infrared (IR) absorption etc.

In petroleum-contaminated samples, aromatic hydrocarbons can be analyzed by GC-MS or UV synchronous luminescence spectroscopy; polar compounds may be monitored by UV fluorescence and Fourier-transform infrared spectroscopy (FTIR) and asphaltene by FTIR (Korda et al. 1997). Huessmann (1995) utilized a comprehensive petroleum characterization procedure involving group-type separation analysis, boiling point distribution and hydrocarbon typing by field desorption mass spectroscopy to determine the biodegradation of petroleum hydrocarbons. Since microbes usually degrade branched alkanes and isoprenoid compounds, such as pristane, phytane and hopane, at much slower rates than straight-chain alkanes, the ratio of straight-chain alkanes to these highly branched biomarkers can represent the extent to which microorganisms have degraded the hydrocarbons in a petroleum-contaminated soil.

Environmental regulators are now taking a more risk-based approach to the management of oily wastes. Here, risk is defined as the probability or likelihood that public health may be impacted upon from exposure to chemicals contained in waste management units (US EPA 1999). The mere presence of a petroleum hydrocarbon fraction does not mean that there is necessarily an environmental risk. A compound must be both toxic and mobile (leachable) in the environment to actually pose a risk to the public (Harju et al. 1999). With the land application of oily sludges, the important routes of exposure are the direct exposures to potentially contaminate groundwater. Untreated sludge often contains compounds such as benzene with a low molecular weight and relatively high aqueous solubility. Benzene is both toxic and mobile (leachable) and hence poses a significant environmental risk. Bioreactor treatment is very effective at degrading the toxic, leachable organic compounds in oily sludges. The toxicity characteristic leaching procedure (TCLP) method (US EPA Method #1311) can be used to identify compounds that are toxic and leachable. Often, it is more cost-effective to treat the compounds that are toxic and mobile, rather than the entire fraction of petroleum hydrocarbons.

Solid-phase microextraction (SPME) has been used to monitor the biodegradation of semivolatile hydrocarbons in diesel fuel-contaminated water and soil (Eriksson et al. 1998) and volatile hydrocarbons during bacterial growth on crude oil (Van Hamme and Ward 2000). Although the method requires external calibration with several standard calibration curves, SPME has been proven to be a rapid and accurate method for monitoring volatile and semivolatile hydrocarbons in petroleum biodegradation systems.

Microbial transformations of minerals can be investigated by a variety of surface spectroscopic techniques. In addition to identifying and mapping the distribution of elements on a surface, some surface-sensitive spectroscopic techniques can resolve the oxidation state of an element and provide information on the identity of neighboring atoms and their orientation. In the bioremediation of metal-contaminated and radionuclide-contaminated subsurface sites, spectroscopic methods play an important role in characterizing biotransformation of minerals (Geesey et al. 2002). Advances in surface-sensitive spectroscopic techniques have provided useful analytical tools to identify many new microbiological-mediated biogeochemical processes. Some surface spectroscopic instrumentation is now being modified for use in the field, to permit researchers to evaluate mineral biotransformations under in situ conditions.

Biodegradation of an organic compound in a natural system may be classified as primary (transformation or alteration of molecular integrity), ultimate (complete mineralization) and acceptable (toxicity ameliorated). A considerable amount of research has been performed to develop reliable structure–activity relationships (SAR) that can predict and describe the biodegradability of organic chemicals in the natural environment (Raymond et al. 2001). The correlations between molecular structure, activity and biodegradation have been termed as quantitative structure biodegradability relationships (QSBR). The QSBRs commonly employ simple or multiple regression analyses and the predicted biodegradability is represented by a diversity of parameters, such as biodegradation rates and constants, half-lives, theoretical oxygen demand (ThOD), biochemical oxygen demand (BOD). For a detailed discussion on QSBRs, an excellent review by Raymond et al. (2001) may be consulted.

3
Conventional Plating and Microbial Enumeration

Traditional culture techniques can yield valuable information about microbial interactions in the environment and the methods by which they access hydrocarbons as a substrate. Initial soil analysis for total heterotrophic microbial counts and specific hydrocarbon-degrading microbial counts in the contaminated soil provide useful information on the extent to which the indigenous microbial population has acclimated to the contaminated site conditions and if it is capable of supporting bioremediation (Balba et al. 1998). Conventional plating methods, used in early studies, revealed a broadly distributed and diverse collection of bacteria, yeast and fungi capable of hydrocarbon utilization (Atlas 1981; Sorkhoh et al. 1995; Ijah 1998; Chaîneau et al. 1999; Bouchez-Naîtali 1999; Petrikevich et al. 2003).

Microbial counts are usually determined in representative soil composite samples and a strong correlation between microbial counts and hydrocarbon degradation has been reported (Balba et al. 1998).

However, care must be taken when reporting numbers of hydrocarbon-degrading organisms, as non-hydrocarbon-degrading microorganisms may also be developed on agar plates prepared with solid, liquid or volatile hydrocarbons (Bogardt and Hemmington 1992; Randall and Hemmington 1994). The conventional plate count technique has several limitations, particularly in relation to non-culturable organisms.

In an evaluation of mineral agar plates, with and without toluene-xylene fumes, it was revealed that little selection was provided against non-toluene and xylene degrading bacteria. A rapid MPN test (sheen screen), using tissue culture plates, can be employed for non-volatile hydrocarbons based on the formation of emulsions (Brown and Braddock 1990). A similar assay to screen for hydrocarbon degraders based on a redox indicator has been described (Hanson et al. 1993) and combined with the sheen screen to produce an MPN assay based on both emulsification and respiration (Van Hamme et al. 2000).

Generally based on MPN-assays, dividing communities into physiological types is best served if numerous categories are examined and associated with relevant site characteristics (Telang et al. 1997; Eckford and Fedorak 2002a, b; Magot et al. 2002). Kâmpfer et al. (1993) monitored in situ bioremediation of a waste oil-contaminated site subjected to various bioremediation treatments. Microorganisms were divided into the following classes: methylotrophic, facultative anaerobes, denitrifiers, sulfate reducers, oil-degrading denitrifiers and anaerobic vacuum gas-oil degraders. In a separate study of a crude oil-contaminated aquifer, Bekins et al. (1999) used a similar MPN approach to study ecological succession, microbial nutrient demands and the importance of free-living versus attached populations. MPN determinations of aerobes, denitrifiers, iron reducers, heterotrophic fermenters, sulfate reducers and methanogens were used. In Antarctica, Delille et al. (1997) examined seasonal changes in the functional diversity of ice bacteria over 9 months in uncontaminated, contaminated and treated (Inipol EAP22 fertilizer) plots. The total bacteria (acridine orange), saprophytes, and hydrocarbon-utilizing bacteria (MPN) were assayed. In all cases, changes in total bacterial abundance, reaching a minimum in the winter ($>10^5$ cells ml^{-1}), were caused by seasonal variation. Both saprophytic and oil-degrading bacteria increased with the addition of Inipol.

Other potential indicators of soil contamination assessment are related to the microbial biomass. The bacterial biomass may be characterized by muramic acid, whereas ergosterol has been suggested as an indicator for fungi (Naseby and Lynch 1997; Barajas-Aceves 2002). Barajas-Aceves et al. (2002) determined the ergosterol content of 20 white-rot fungi and found linear correlation of ergosterol content with

Table 1. Biodegradation estimation methods for monitoring and assessment of bioremediation processes

Method	Comments
Gas chromatography (GC)	Commonly used technique to separates complex mixture of hydrocarbons into its components; high sensitivity, accurate identification
High performance liquid chromatography (HPLC)	Used commonly for analysis of various contaminants and their (bio)transformation products
Thin layer chromatography–flame ionization detector (TLC-fiD)	Fast method, a number of samples can be replicated and run in a short period; good sensitivity and easy to use
Respirometry	Quantified by the consumption of molecular oxygen or evolution of carbon dioxide using respirometry
Spectroscopic	For example, X-ray diffraction, Ramans spectroscopy and high resolution electron diffraction; surface-sensitive technique to determine biotransformation of metals, due to dehydrated samples real-time measurement difficult
Fluorescence	Useful in identification of hydrocarbons; commonly used in coastal oil spill monitoring
Total petroleum hydrocarbons (TPH)	Gravimetric; infra-red or GC method; oil and grease analysis
Solid-phase microextraction (SPME)	Determination of volatile and semivolatile compounds during biodegradation of oily sludge
Toxicity characteristics leachate procedure (TCLP)	Measurement of toxic and mobile hydrocarbons in sample leachates
Quantitative structure biodegradability relationships (QSBR)	The QSBRs are correlations between molecular structure, activity and biodegradation; commonly employ simple or multiple regression analyses and predicts biodegradability in terms of biodegradation rates and constants, half-lives, theoretical oxygen demand (ThOD), biochemical oxygen demand (BOD).

fungal biomass C in both polluted and control soil cultures. The study showed that ergosterol could be a useful indicator for fungal biomass in polluted soils and can be applied for monitoring bioremediation processes.

Various methods for the estimation of biodegradation used for monitoring and the assessment of bioremediation processes are shown in Table 1.

4 Biochemical/Physiological Methods

4.1 Phospholipid Fatty Acid Analysis

Phospholipids are essential membrane components of all living cells and make up a relatively constant proportion of the biomass of organisms under natural conditions, hence phospholipid fatty acid (PLFA) patterns can be used to provide insight into biomass and microbial community profiling of bacteria and fungi (Zelles 1997; van Elsas et al. 1998; Kozdroj and van Elsas 2001). Whole cell fatty acid is also used for the identification of microbial species. PLFAs are effective taxonomic and phylogenetic markers, similar to the analysis based on the sequence homology of 16S rDNA. Changes in the community structure can be monitored by the comparison of the relative abundance of certain PLFAs that differ considerably among specific groups of microorganisms. Various biochemical and molecular methods used for the monitoring of bioremediation processes are shown in Table 2.

An increase in monounsaturated fatty acids, lower concentrations of branched-chain fatty acids and a low level of methyl branching on the tenth C atom fatty acid indicated an increase in the number of Gram-negative bacteria, and a decrease in the proportion of Gram-positive bacteria and actinomycetes, respectively (Zelles 1999). A dominance of Gram-negative over Gram-positive bacteria is indicated by an increase in cy17:0 and a decrease in several iso- and anteiso-branched PLFAs. The 18:26ω9 is a reliable fungal marker and 18:2ω6,9 and 20:4 have decreased in response to long-term heavy metal deposition in soil. In soil, contaminated with jet fuel, there was an increase in PLFAs typical of Gram-negative heterotrophs (16:1ω7c, 18:1ω7c) and aerobic actinomycetes (10Me18:0), and a decrease in the proportion of Gram positives as shown by the ratio of iso/anteiso 15:0. Table 3 shows the fatty acid markers for identification of microbial groups.

PLFA analysis is one of the suitable methods for culture-independent characterization of microbial communities in soil. However, there are certain limitations, which require attention when interpreting the results. There is a significant variation in the presence or absence of signature fatty acids across taxa, whereas many fatty acids are common to different microbes. Thus, the quantitative interpretation of particular fatty acids indicating the abundance of specific species may be deceptive sometimes. It is unlikely that microbial PLFA will provide a detailed picture of the community profile, however, combining molecular and

Table 2. Biological methods for monitoring and assessment of bioremediation processes

Method	Comments
Plating	Only a small proportion of microbial community involved is detected, isolates not necessarily reflective of a specific metabolic function
MPN	Metabolic function of interest detected, selective media may limit proportion of community detected
BIOLOG	Overall metabolic activity detected, rapid and easy to use, selective media may limit proportion of community detected, sensitive to inoculum size
Staining for active microbes	Enumerate live microorganisms, no bias from culture media, does not differentiate microorganisms with catabolic activity of interest
PFLA	Changes in signature fatty acid can indicate change in community structure
Soil enzymes	Soil enzymes such as dehydrogenases, lipases etc. can be considered attractive indicators for monitoring impacts of contaminated soils
Immunochemical	Based on interaction of an antigen (pollutant) and an antibody (detector), depending on quality of the antibodies, intact specific organism can be detected, laborious counting unless coupled to flow cytometric methods
Protein banding	No selection pressure if extracted directly, no measurement of community function, difficult to link fingerprints to specific microbial groups
MPN-PCR	Number of target DNA in environmental samples can be detected, only semi-quantitative determinations possible
DNA:DNA hybridization	Abundance of gene fragments can be determined, this relatively fast method can be performed on bacterial colonies or on isolated DNA
fiSH	Spatially visualize specific microorganisms in an environment, not necessarily detecting active microorganisms, laborious technique
RSGP	Quantitative analysis of specific microorganisms in environmental samples, limited to those microorganisms included in the screen
PCR-DGGE	Can identify microorganisms by sequencing resolved bands, bulk changes in community structure detected, differential DNA or RNA extraction from different cells, no information on activity
Microarray	Rapid method for automated determination composition and abundance of species within a microbial community
Probes for specific metabolic genes	Detect genes with function of interest, mRNA detection can reveal information about expression, limited to known genes, activity cannot be inferred from presence of genes alone
Bacterial sensors	Constructed by fusing a reporter gene to a promoter element induced by the target compound, offer the possibility to characterize the biodegradability of specific contaminants present in a complex mixture without pretreatment of the environmental samples

Table 3. Marker phospholipid fatty acids (PLFAs) for selected microbial groups (Adapted from Kozdroj and van Elsas 2001)

Group	Fatty acids
Gram-negative bacteria	OH fatty acids, usually 3 OH Monounsaturated fatty acids, 16:1ω7t; 16:1ωSc; 18:1ω7 Cy 17:0; cy 19:0
Gram-positive bacteria	Iso- and anteiso fatty acids, i15:0; A15:0; i16:0; i17:0; a17:0
Actinomycetes	10Me fatty acids, 10 Me 16:0; 10 Me 17:0; 10 Me 18:0
Fungi	18:1ω,9c; 18:2ω6,9c; 20:4; 21:0; 23: 25:0
Arbuscular fungi	16:1ωSc
Eukaryotic algae and	Polyunsaturated fatty acids, 16:1ω4;
Protozoa	16:3; 18:4ω3; 20:4; 20:5; 22:6

PLFA techniques will allow microbiologists to describe the changes in microbial community structure more accurately.

4.2 Soil Enzyme Assay

Soil enzymes are the catalysts of important degradation processes which include the decomposition of organic matter and the detoxification of xenobiotics. Soil enzymes, such as dehydrogenases, lipases, ureases, acid and alkaline phosphatases, arylsulfatases and catalases, can be considered attractive indicators for monitoring impacts of contaminated soils due to their central role in the soil environment. Although soil enzyme activities have been used as biological indicators of pollution with heavy metals, pesticides and organic contaminants (Margesin et al. 2000), only a little information is available on the potential of soil enzymes as biological indicators of hydrocarbon biodegradation (Rossel et al. 1997; Margesin et al. 1999).

The assay of dehydrogenase activity in contaminated soils and wastewaters has been recognized as a useful indicator of the overall measure of the microbial metabolism and in the assessment of ecotoxicological impacts of environmental contaminants (Rossel et al. 1997; Lee et al. 2000). For example, the presence of high concentrations of toluene or chloroform can strongly inhibit soil dehydrogenase activity, but low concentrations have little or no effect. Dehydrogenase activity is often preferred over cultural methods for enumeration of microbes, as viable cell counts tend to be underestimates due to the lack of homogeneity in distribution (Torstensson 1997). The method has been effectively used in field studies in petroleum-contaminated soils (Mathew et al. 1999) and beach sediments (Mathew and Obbard 2001). However, the presence of nitrate, nitrite and ferric ion may interfere with and inhibit soil dehydrogenase activity.

Brohon et al. (2001) studied the complimentarity of bioassays and microbial activity measurements to evaluate hydrocarbon-contaminated soils. Urease (31% inhibition) and dehydrogenase (50% inhibition) were sensitive to the presence of metals. In the zone containing the highest concentration of metals, the measurement of substrate-induced respiration showed that the soil microflora was stressed, the microbial activities were low and the bioassays revealed a high potential toxicity.

Margesin et al. (1999) evaluated the usefulness of soil lipase activity as a tool to monitor oil biodegradation in soil. A significant negative correlation of residual soil hydrocarbon concentration with soil lipase activity and the number of oil-degrading microorganisms, independent of fertilization, was observed. The study revealed that the induction of soil lipase activity is a valuable indicator of oil biodegradation in naturally attenuated (unfertilized) and bioremediated (fertilized) soils.

There is growing evidence that soil biological activities are sensitive to environmental stresses and hence can be used as a quick method to assess the decontamination of soils in combination with other acceptable methods. However, more studies and data are required before enzymatic assays can be used as the main tool to assess bioremediation.

4.3 BIOLOG

BIOLOG microtiter plate assays can be used as a rapid method for monitoring shifts of the metabolic fingerprints of microbial communities (van Elsas et al. 1998). This method was originally developed for the classification of bacterial isolates, based on their ability to oxidize 95 different carbon sources, and later adapted as a means to characterize functional characteristics of the microbial communities which produce habitat-specific and reproducible patterns of the oxidation of carbon substrates.

BIOLOG substrate utilization patterns have been used to evaluate the functional diversity of microbial communities in continuous flow-through cultures treating C_{16}-contaminated intertidal sediments (Berthe-Corti and Bruns 1999). Measurements of C_{16} degradation, product formation, oxygen consumption, total heterotrophs, MPN determinations of nitrate reducers, sulfate reducers and C_{16}- utilizing bacteria were combined with BIOLOG data. It was observed that substrate utilization became more limited, especially at low dissolved oxygen (0.4%) levels, whereas the parameters of C_{16} degradation, protein production, and oxygen consumption increased with dilution independently of dissolved oxygen. Overall, it appeared that the level of dissolved oxygen (80 or 0.4%) appeared to dictate the structure of the microbial community.

Substrate utilization patterns have been shown to vary depending on both the composition and density of the inoculum used. Since bacterial growth occurs in the microtiter plate wells during the course of the assay, the patterns of substrate utilization observed are shown to reflect the functional characteristics of only those microbes that are able to grow under the assay conditions. Despite limitations, this rapid method remains a valuable tool for microbial community analysis.

4.4 Immunochemical Methods

Immunochemical methods are based on the interaction between an antibody (detector) and an antigen (pollutant) (Van Emon and Gerlach 1998). The US EPA Pesticides and Groundwater Strategy advocates implementation of agricultural best management practices (BMPs) to groundwater contamination by pesticides. Direct immunofluorescence and ELISA can be used for near-real time quantification of hydrocarbon degrading organisms (Brigmon et al. 1998). Immunodetection was shown to be applicable to complex sample matrices for rapid field evaluation. Antibody mixtures could potentially be developed to target specific microbial groups although, in most situations, tracking the expression of specific genes, involved in hydrocarbon metabolism, would be of greater use. The immunoassay test kits are now commercially available and immunoassay methods are now used in a variety of applications. Some of the environmental applications include monitoring of underground storage tanks for leakage of volatile organic compounds, monitoring of agricultural runoff for pesticides, testing safety of dairy, meat and poultry products for chemical and biological residues, and monitoring characterization and bioremediation of hazardous waste sites.

5 Molecular Biology-Based Methods

5.1 Molecular Techniques

Molecular techniques to evaluate microbial community profiles include the polymerase chain reaction (PCR) in combination with denaturing and temperature gradient gel electrophoresis (DGGE and TGGE), reverse sample genome probing (RSGP), ribosomal intergenic spacer analysis (RISA), single-strand-conformation polymorphism (SSCP), ITS-restriction fragment length polymorphism (ITS-RFLP), random amplified polymorphic DNA (RAPD), and amplified ribosomal DNA restriction analysis (ARDRA) (Kent and Triplett 2002). Recently, developments in the use of DNA microarrays have attracted the attention of environmental microbiologists for more rapid throughput, to allow for tracking of thousands of catabolic genes at one time (Deniss et al. 2003). A few examples of community studies using various methods are discussed here.

PCR is now commonly used for specific detection of DNA in environmental samples and combining PCR with DNA probes, using nested primers or real-time detection systems (RT-PCR), can further enhance sensitivity (Power et al. 1998; Widada et al. 2002b). However, it is difficult to quantitatively measure the number of organisms or gene copies present in an environmental sample. For quantification of DNA by PCR techniques, three methods have been developed: MPN-PCR, replicative limiting dilution-PCR (RLD-PCR), and competitive-PCR (cPCR). Quantitative RT-PCR has been used to monitor bioremediation processes and to determine catabolic genes of bioaugmented bacteria in samples.

Domain probe analysis has been used to examine community structure in pristine and fuel-contaminated aquifers (Shi et al. 1999). Øvreås (2000) used DGGE, sequencing and DNA reassociation plots, in combination with methane and methanol-oxidation measurements, to show a decrease in diversity with a concomitant increase in known methanotrophs upon methane perturbation of agricultural soils. MacNaughton et al. (1999) used 16S rRNA PCR-DGGE and PFLA to identify populations responsible for decontamination, while evaluating oil spill bioremediation techniques. This method also helped to define an endpoint for substrate removal. Roony-Varga et al. (1999) also used a mixed approach including PFLA, MPN-PCR and DGGE of 16S rDNA, along with selective enrichment and biodegradation studies, to evaluate anaerobic benzene degradation in a petroleum-contaminated aquifer. Colores et al. (2000) studied surfactant effects on C_{16} and phenanthrene degradation, by a mixed culture in laboratory microcosms using respirometry, 16S rRNA DGGE and culture techniques.

Microbial community pattern analyses, using 16S rRNA PCR-DGGE, has certain limitations. Since one organism may produce more than one DGGE band, due to heterogeneous multiple rRNA operons, the number of bands in a DGGE gel does not always accurately reflect the number of corresponding species within the microbial community (Widada et al. 2002a; Siciliano et al. 2003). The vast majority of microbes in the environment have yet to be cultured; therefore, a major proportion of genetic diversity resides in uncultured organisms. Isolation of these genes is limited by lack of sequence information and PCR amplification techniques can be employed for the amplification of only partial genes. More information is often available when gene probes to specific isolates, genotypes or metabolic activities are used, and approaches to achieve this are being applied in both aerobic and anaerobic systems (Stapleton et al. 1998, 2000; Ringelberg et al. 2001; Thomassin-Lacroix et al. 2001).

DNA microarray technology is a powerful tool for analyzing microbes and their activities in environmental samples because a large number of hybridizations (over 100,000 spots cm^2 on glass slides) can be performed simultaneously (Cho and Tiedje 2002). Sample nucleic acids can be spotted onto carrier material and reverse hybridization can be performed using immobilization probes. When using PCR, partial or whole rRNA genes of the microorganisms of interest can be amplified with specific primers. For monitoring bioremediation processes, DNA microarrays can potentially be used to assess the functional diversity and distribution of selected genes.

Specific microorganisms may be tracked in a quantitative manner using reverse sample genome probing (RSGP) (Voordouw et al. 1993, 1996; Hubert et al. 1999). Sulfate-reducing bacterial (SRB) communities in oil fields play a key role in anaerobic corrosion in oil and gas fields, and elucidating modes of their action is important. The RSGP technique allows the total DNA from a community to be quantitatively analyzed in a single step. Use of the RSGP method in the evaluation of the effects of nitrate injection (Telang et al. 1994), diamine biocides (Telang et al. 1998) and dicyclopentadiene (Shen et al. 1998) on microbial community composition and functional properties have also been described. However, the drawback of this method is that the presence of a specific microorganism does not indicate that it is active.

Early work with gene probes following the Exxon Valdez spill revealed that bacterial populations containing both the *xyl*E and *alk*B genes could be detected in environmental samples (Stotsky and Atlas 1994). In laboratory columns, proportions of *xyl*E and *ndo*B (polycyclic aromatic hydrocarbon degradation) populations from an aquifer community were monitored during the degradation of creosote related PAHs (Hosein et al. 1997). Langworthy et al. (1998) found *nah*A and *alk*B in higher frequencies at PAH-contaminated sites, although these genes, along with *nah*H and *tod*C1/C2, were detected at pristine sites as well. Laurie and Lloyd-Jones (2000) used competitive PCR to illustrate that the newly described *phn* genes of *Burkholderia* sp. strain RP007 may have greater ecological significance than *nah*-like genes for PAH degradation.

Gene probes may be used for accurately evaluating bacterial degradative potential (Hamann et al. 1999; Mesarch et al. 2000) although the application of a few probes can be effective if meaningful hypotheses are tested (Siciliano et al. 2003). Recent advances in characterizing alkane metabolism in a number of organisms have allowed the production of a variety of primers to detect, for example, the *alkB* gene from *P. putida* GPo1 (Smits et al. 1999). As more strains and probes are tested, it is becoming clear that, while different alkane hydroxylases can be found in phylogenetically distant microorganisms (Andreoni et al. 2000), many probes will only provide information on the presence of a similar gene in closely related strains. Such gene probes will be more useful as the diversity of genes responsible for hydrocarbon metabolism is better appreciated (Power et al. 1998; Widada et al. 2002a).

Active biological containment strategies have also been developed to prevent the undesirable spread of genetically modified microorganisms in the environment after they have completed their intended tasks (Molin et al. 1993; Ronchel et al. 2000; Ronchel and Ramos 2001). One of the strategies is to construct a suicide system, where a suicide gene is induced and the introduced microorganism is eliminated from the population. An elimination of 100% is never achieved. The system is based on the expression of a gene encoding streptavidin under the control of a cascade system, which is ultimately controlled by the presence or absence of the growth substrate. Another strategy is to reduce the spread of genes by incorporation of a gene whose product kills non-immune recipients upon gene transfer.

5.2 Bacterial Biosensors

Many current analytical techniques used for monitoring pollutants require expensive equipment and extensive pretreatment of the environmental samples. Classical analytical methods cannot distinguish between unavailable and bioavailable compounds. Conventional analytical methods only provide information about concentrations in the contaminated phases. The assessment of bioavailability of a contaminant is also an important consideration of site remediation (Alexander 2000). The drawbacks of classical analytical methods have created an interest in the development of alternative methods including novel bacterial biosensors, which uniquely measure the interaction of specific compounds through highly sensitive biorecognition processes (Theron and Cloete 2000; Keane et al. 2002). The bacterial biosensor measurements have also been shown to be within a 3% range of those measured by standard GC-MS techniques in the case of toluene (Willardson 1998).

Whole cell biosensors, constructed by fusing a reporter gene to a promoter element induced by the target compound, offer the possibility to characterize, identify, quantify and determine the biodegradability of specific contaminants present in a complex mixture, without pretreatment of the environmental samples (Ramanathan et al. 1997; Errampalli et al. 1998; Daunert et al. 2000). The genetic information, located on a plasmid vector, is inserted into a bacterial strain so that the engineered fusion replicates along with the cell's normal DNA. The biosensor systems include a wide range of integrated devices that employ enzymes, antibodies, tissues or living microbes as the biological recognition element. Bacterial biosensors, developed for monitoring various soil contaminants, are shown in Table 4.

Table 4. Bacterial biosensors for monitoring contaminants

Bacterial biosensor	Contaminant	Reporter gene fusion	References
Pseudomonas fluorescens HK44	Naphthalene	*nahG-luxCDABE*	Heitzer et al. (1998)
Pseudomonas putida RB1401	Toluene, xylene	*xylR-luxCDABE*	Burlage (1997)
Pseudomonas putida B2	BTEX	*tod-luxCDABE*	Applegate et al. (1997)
Escherichia coli DH5α	Alkanes	*alkB-luxAB*	Sticher et al. (1997)
Escherichia coli DH5α	BTEX	*xylR-luc*	Willardson et al. (1998)
Escherichia coli	Benzene derivatives	*xylS-luc*	Ikariyama et al. (1997)
Rhodococcus eutropha ENV307	PCBs	*bph-luxCDABE*	Layton et al. (1998)
Rhodococcus eutropha JMP143-32	2,4-D	*tfd-luxCDABE*	Hay et al. (2000)

Broad specificity biosensors are used for toxicity testing and respond to a wide range of compounds. A good example of the latter is the commercially available Microtox assay used for measuring the toxicity of environmental samples, by monitoring the light production of naturally bioluminescent marine bacteria *Photobacterium phosphoreum* (Burlage 1997). Since bacterial bioluminescence is tied directly to cellular respiration, any inhibition of cellular metabolism due to toxicity results in a decrease in the light emission of the affected cells. In a non-specific bacterial biosensor, *lux* genes are fused to heat shock promoters, so that exposure of the cells to toxic organic compounds or metals rapidly induces light production (Daunert et al. 2000).

Several bioluminescent biosensors have been developed for the detection of benzene, toluene, ethylbenzene and xylene isomers (Burlage 1997; Ikariyama et al. 1997; Lau et al. 1997; Willardson et al. 1998), and quantitative assessment of PAH biodegradation in aqueous, soil and slurry systems (Heitzer et al. 1998; Gu and Chang 2000) and for measuring the bioavailable middle chain-length alkanes (Sticher et al. 1997). An advanced system consisting of biosensor cells interfaced with an integrated circuit called Bioluminescent Bioreporter Integrated Circuit (BBIC) has also been developed, which can detect the optical signal, distinguish it from the noise, perform signal processing, communicate the results and also carry out position sensing (Simpson et al. 1998). The toluene detection range of *E. coli* cells, carrying pGLTUR plasmid (fusion of firefly *luc* genes to transcriptional activator *xylR* gene), was between 10 and 20 μM (Willardson et al. 1998). The calculated toluene concentrations were within 3% of those measured by GC-MS techniques.

Molecular diagnostics in bioremediation have been advanced during the last decade. Qualitative detection methods have been replaced with quantitative measurements of the specific microbial population present in the contaminated soils. To assess the microbial treatment of contaminated sites, bioavailable concentrations of pollutants could be measured by using bacterial sensors and the overall genetic potential of the degradative pathways determined by DNA tests (Rogers and Gerlach 1999). Whether the pollutant concentrations are sufficiently high to induce the particular degradation can also be verified. However, the validity of these methods needs to be tested in the field, to assess the practicability and usefulness of these techniques in bioremediation. There are certain limitations with the bioluminescent bacterial biosensors. Light output of bioluminescent biosensors depends not only on the chemical complexity of the sample, but also on variations of the physiological state of the cells, such as protein synthesis, membrane permeability and metabolism.

With the rapid advances in molecular methods we are beginning to understand the astonishing diversity of microbial populations and communities in the environment. However, the inherent variability in microbial communities over time remains a major challenge. Rapid automated systems will be required to process and evaluate vast quantities of data in order to subtract background variability. Even then, care must be taken to realize that, while molecular methods are powerful and attractive, the genetic composition of a community cannot be used to precisely extrapolate ecosystem function. However, the advances in nanotechnology will continue to result in higher sensitivity and more versatile operational characteristics.

6 Toxicological Risk Assessments

Toxicity tests are useful tools in assessing the risks associated with contaminated soil and in monitoring bioremediation processes. Toxicity tests using biomonitors are used to assess the cumulative effects of pollutants on an organism or system, although the methods are time-consuming. Currently, bacteria, fungi, algae and plants are being applied as biomonitors in soil, and carnivorous fish or benthic macro-invertebrates are used to monitor bioaccumulative pollutants in aquatic systems. Toxicity tests involve the exposure of organisms or cultured cell lines to toxic compounds under defined experimental conditions followed by monitoring biological endpoints such as mortality, reproduction, growth and behavioral changes. Toxicity tests using biomonitors have been implemented in the remediation of PCP-contaminated (Middaugh et al. 1993), PAH-contaminated (Hund and Traunspurger 1994) and lead-contaminated sites (Chang et al. 1997). The toxicity evaluation of PAH-contaminated sites have shown a correlation between decreased toxicity and PAH degradation. Mostly these tests are used to assess the relative toxicity of pollutants on a specific organism, which sometimes makes extrapolation of laboratory tests to the field difficult.

A variety of biomarkers, which are indicators of cellular or biochemical responses to a pollutant, are used for toxicity assessment (Ahmad 1995). There are generally three classes of biomarkers. Some biomarkers are developed in response to exposure of organic and inorganic pollutants such as heat shock proteins, metallothiones and antioxidant enzymes. A second category of biomarkers can detect biological effects and disturbed structure and functions such as population size, DNA mutations, chromosomal aberrations and inhibition of the enzyme, acetylcholinesterase. The third category measures activities of detoxifying enzymes, cytochrome P450 monooxygenase and glutathione S transferase and DNA repair enzymes.

Toxicity varies with the hydrocarbon types and concentration, soil type and properties, microbial community composition and plant species. Kirk et al. (2002) have developed a phytotoxicity bioassay to select plant species for phytoremediation that were able to germinate and grow in petroleum-contaminated soil. Perennial ryegrass and alfalfa were more successful at germination and root growth than were little bluestem and crown vetch. The phytotoxicity assay provides a rapid method of prescreening potential plant species and eliminates those that are not able to germinate.

7 Conclusions

Challenges in monitoring, assessing and directing microbial processes during bioremediation of contaminated sites have represented a major research pursuit for developing more rapid and accurate methods over the past two decades. The presence of the toxic compounds and the ecological risk can be determined by using various advanced molecular and biochemical methods and toxicity tests. However, data on these advanced techniques in real field applications of bioremediation processes and environmental quality assessment are still limited. There are still outstanding questions. For example, are microbes capable of degrading the particular pollutant present in the contaminated site, and what biological treatment method will effectively remove the contaminants? What happens if the concentration of the contaminant is low compared to other biodegradable or metabolizable substrates? Although none of these questions may be answered, analytical tools available today will help to provide some of the answers in the coming years.

Although molecular techniques are at an early stage of investigation in field applications, they represent an important interdisciplinary pursuit encompassing biological and electronic expertise. Studies of community dynamics, related to microbes with bioremediation potential, will provide insights into the microbial diversity and, accelerated, accurate and sensitive molecular, immunochemical and automated methodologies will undoubtedly lead to the characterization of new microbial strains and biocatalytic activities.

References

Ahmad S (1995) Oxidative stress from enviromental samples. Arch Insect Biochem Physiol 29:135–157

Aichinger G, Leslie Grady Jr CP, Tabak HH (1992) Application of respirometric biodegradability testing protocol to slightly soluble organic compounds. Water Environ Res 64:890–900

Alexander M (2000) Aging, bioavailability, and overestimation of risk from environmental pollutants. Environ Sci Technol 34:4259–4265

Andreoni V, Bernasconi S, Colombo M, van Beilen JB, Cavalca L (2000) Detection of genes for alkane and naphthalene catabolism in *Rhodococcus* sp. strain 1BN. Environ Microbiol 2:572–577

Applegate BM, Kelly C, Lackey L, McPherson J, Kehrmeyer S, Menn F-M, Bienkowski P, Saylor G (1997) *Pseudomonas putida* B2: a tod-lux bioluminescent reporter for toluene and trichloroethylene cometabolism. J Ind Microbiol Biotechnol 18:4–9

Atlas RM (1981) Microbial degradation of petroleum hydrocarbons: an environmental perspective. Microbiol Rev 45:180–209

Atlas RM, Cerniglia CE (1995) Bioremediation of petroleum pollutants: diversity and environmental aspects of hydrocarbon biodegradation. BioScience 45:332–338

Balba MT, Al-Awandhi N, Al-Daher R (1998) Bioremediation of oil-contaminated soil: microbiological methods for feasibility assessment and field evaluation. J Microbiol Meth 32:155–164

Barajas-Aceves M, Hassan M, Tinoco R, Vazquez-Duhalt R (2002) Effect of pollutants on the ergosterol content as indicator of fungal biomass. J Microbiol Meth 50: 227–236

Bekins BA, Godsy EM, Warren E (1999) Distribution of microbial physiologic types in an aquifer contaminated by crude oil. Microb Ecol 37:263–275

Berthe-Corti L, Bruns A (1999) The impact of oxygen tension on cell density and metabolic diversity of microbial communities in alkane degrading continuous-flow cultures. Microb Ecol 37:70–77

Bogardt AH, Hemmington BB (1992) Enumeration of phenanthrene-degrading bacteria by an overlayer technique and its use in evaluation of petroleum-contaminated sites. Appl Environ Microbiol 58:2579–2582

Bouchez-Naïtali M, Rakatozafy H, Marchal R, Leveau J-Y, Vandecasteele JP (1999) Diversity of bacterial strains degrading hexadecane in relation to the mode of substrate uptake. J Appl Microbiol 86:421–428

Brigmon RL, Franck MM, Bray JS, Scott DF, Lanclos KD, Fliermans CB (1998) Direct immunofluorescence and enzyme-linked immunosorbent assays for evaluating organic contaminant degrading bacteria. J Microbiol Methods 32:1–10

Brohon B, Dlolme C, Gourdon R (2001) Complementarity of bioassays and microbial activity measurements for the evaluation of hydrocarbon-contaminated soils quality. Soil Biol Biochem 33:883–891

Brown EJ, Braddock JF (1990) Sheen screen, a miniaturized most-probable-number method for enumeration of oil-degrading microorganisms. Appl Environ Microbiol 56:3895–3896

Burlage RS (1997) Emerging technologies: bioreporters, biosensors and microprobes. In: Hurst CJ (ed) Manual of environmental microbiology. ASM, Washington, DC, pp 115–123

Chaîneau CH, Morel J, Dupont J, Bury E, Oudot J (1999) Comparison of the fuel oil biodegradation potential of hydrocarbon-assimilating microorganisms isolated from a temperate agricultural soil. Sci Total Environ 227:237–247

Chang LW, Meier JR, Smith MK (1997) Application of plant and earthworm bioassays to evaluate remediation of a lead-contaminated soil. Arch Environ Contam Toxicol 55:2924–2931
Cho J-C, Tiedje JM (2002) Quantitative detection of microbial genes by using DNA microarrays. Appl Environ Microbiol 68:1425–1430
Colores GM, Macur RE, Ward DM, Inskeep WP (2000) Molecular analysis of surfactant-driven microbial population shifts in hydrocarbon-contaminated soil. Appl Environ Microbiol 66:2959–2964
Daunert S, Barrett G, Feliciano JS, Shetty RS, Shrestha S, Smith-Spencer W (2000) Genetically engineered whole cell sensing systems: coupling biological recognition with reporter genes. Chem Rev 100:2705–2738
Dua M, Singh A, Sethunathan N, Johri AK (2002) Biotechnology and bioremediation: successes and limitations. Appl Microbiol Biotechnol 59:143–152
Delille D, Bassères A, Dessommes A (1997) Seasonal variation of bacteria in sea ice contaminated by diesel fuel and dispersed crude oil. Microb Ecol 33:97–105
Dennis P, Edwards EA, Liss SN, Fulthorpe R (2003) Monitoring gene expression in mixed microbial communities by using DNA microarrays. Appl Environ Microbiol 69:769–778
Donaldson EC, Chilingarian GV, Yen TF (1989) Introduction. In: Donaldson EC, Chilingarian GV, Yen TF (eds) Microbial enhanced oil recovery. Elsevier, New York, pp 1–15
Drobnik J (1999) Genetically modified organisms (GMO) in bioremediation and legislation. Int Biodet Biodeg 44:3–6
Eckford RE, Fedorak PM (2002a) Planktonic nitrate-reducing bacteria and sulfate-reducing bacteria in some western Canadian oil field waters. J Ind Microbiol Biotechnol 29:83–92
Eckford RE, Fedorak PM (2002b) Chemical and microbiological changes in laboratory incubations of nitrate amendment "sour" produced waters from three western Canadian oil fields. J Ind Microbiol Biotechnol 29:243–254
Ensley BD (1991) Biochemical diversity of trichloroethylene metabolism. Annu Rev Microbiol 45:283–299
Eriksson M, Swartling A, Dalhammer G (1998) Biological degradation of diesel fuel in water and soil monitored with solid-phase micro-extraction and GC-MS. Appl Microbiol Biotechnol 50:129–134
Errampalli D, Okamura H, Lee H, Trevors JT, van Elsas JD (1998) Green fluorescent protein as a marker to monitor survival of phenanthrene-mineralizing *Pseudomonas* sp. UG14Gr in creosote-contaminated soil. FEMS Microbiol Ecol 26:181–191
Foght J, Semple K, Westlake DWS, Blenkinsopp S, Sergy G, Wang Z, Fingas M (1998) Development of a standard bacterial consortium for laboratory efficacy testing of commercial freshwater oil spill bioremediation agents. J Ind Microbiol Biotechnol 21:322–330
Foght JM, Semple K, Gauthier C, Westlake DWS, Blenkinsopp S, Sergy G, Wang Z, Fingas M (1999) Effect of nitrogen source on biodegradation of crude oil by a defined bacterial consortium incubated under cold, marine conditions. Environ Technol 20:839–849
Geesey GG, Neal AL, Suci PA, Peyton BM (2002) A review of spectroscopic methods for characterizing microbial transformations of minerals. J Microbiol Meth 51:125–139
Graves DA, Lang CA, Leavitt ME (1991) Respirometric analysis of the biodegradation of organic contaminates in soil and water. Appl Biochem Biotechnol 28/29:813–826

Gu MB, Chang ST (2000) Soil biosensor for the detection of PAH toxicity using an immobilized recombinant bacterium and a biosurfactant. Biosens Bioelectron 16: 667–674

Haines JR, Koran KM, Holder EL, Venosa AD (2003) Protocol for laboratory testing of crude-oil bioremediation products in freshwater conditions. J Ind Microbiol Biotechnol 30:107–113

Hamann C, Hegemann J, Hildebrandt A (1999) Detection of polycyclic aromatic hydrocarbon degradation genes in different soil bacteria by polymerase chain reaction and DNA hybridization. FEMS Microbiol Lett 173:255–263

Hanson JG, Desai JD, Desai AJ (1993) A rapid and simple screening technique for potential crude oil degrading microorganisms. Biotechnol Tech 7:745–748

Harju JH, Nakles DV, DeVaull G, Hopkins H (1999) Application of risk-based approaches for the management of E & P sites. SPE/EPA Exploration and Production Environmental Conference, Society of Petroleum Engineers, Feb/Mar 1999

Hay AG, Rice JF, Applegate BM, Bright NG, Sayler GS (2000) A bioluminescent whole cell reporter for detection of 2,4-dichlorophenoxyacetic acid and 2,4-dichlorophenol in soil. Appl Environ Microbiol 66:4589–4594

Heitzer A, Applegate B, Kehrmeyer S, Pinkart H, Webb OF, Phelps TJ, White D, Sayler GS (1998) Physiological considerations of environmental applications of *lux* reporter fusions. J Microbiol Meth 33:45–57

Höhener P, Hunkeler D, Hess A, Bregnard T, Zeyer J (1998) Methodology for evaluation of engineered in situ bioremediation: lessons from a case study. J Microbiol Meth 32:179–192

Hosein SG, Millette D, Butler BJ, Greer CW (1997) Catabolic gene probe analysis of an aquifer microbial community degrading creosote-related polycyclic aromatic and heterocyclic compounds. Microb Ecol 34:81–89

Hubert C, Shen Y, Voordouw G (1999) Composition of toluene-degrading microbial communities from soil at different concentrations of toluene. Appl Environ Microbiol 65:3064–3070

Huesemann MH (1995) Predictive model for estimating the extent of petroleum hydrocarbon biodegradation in contaminated soils. Environ Sci Technol 29:7–18

Hund K, Traunspurger W (1994) Ecotox-evaluation strategy for soil bioremediation exemplified for a PAH-contaminated site. Chemosphere 29:371–390

Ijah UJJ (1998) Studies on relative capabilities of bacterial and yeast isolates from tropical soils in degrading crude oil. Waste Manage 18:293–299

Ikariyama Y, Nishiguchi S, Koyama T, Kobatake E, Aizawa M (1997) Fiber-optic biomonitoring of benzene derivatives by recombinant *E. coli* bearing luciferase gene fused TOL-plasmid immobilized on the fiber-optic end. Anal Chem 69:2600–2605

Kämpfer P, Steiof M, Becker PM, Dott W (1993) Characterization of chemoheterotrophic bacteria associated with the in situ bioremediation of a waste-oil contaminated site. Microb Ecol 26:161–188

Keane A, Phoenix P, Ghoshal S, Lau PCK (2002) Exposing culprit organic pollutants: a review. J Microbiol Meth 49:103–119

Kent AD, Triplett EW (2002) Microbial communities and their interactions in soil and rhizosphere ecosystems. Annu Rev Microbiol 56:211–236

Kirk JL, Klironomos JN, Le H, Trevors JT (2002) Phytotoxicity assay to assess plant species for phytoremediation of petroleum-contaminated soil. Biorem J 6: 57–63

Korda A, Sanatas P, Tenente A, Santas R (1997) Petroleum hydrocarbon bioremediation: sampling and analytical techniques, in situ treatment and commercial microorganisms currently used. Appl Microbiol Biotechnol 48:677–686

Kozdroj J, van Elsas JD (2001) Structural diversity of microorganisms in chemically perturbed soil assessed by molecular and cytochemical approaches. J Microbiol Meth 43:197–212

Langworthy DE, Stapleton RD, Sayler GS, Findlay RH (1998) Genotypic and phenotypic responses of a riverine microbial community to polycyclic aromatic hydrocarbon contamination. Appl Environ Microbiol 64:3422–3428

Lau PCK, Wang Y, Patel A, Labbe D, Bergeron H, Brousseau R, Konishi Y, Rawlings M (1997) A bacterial basic region leucine zipper histidine kinase regulating toluene degradation. Proc Natl Acad Sci USA 95:1453–1458

Layton AC, Muccini M, Ghosh MM, Sayler GS (1998) Construction of a bioluminescent reporter strain to detect polychlorinated biphenyls. Appl Environ Microbiol 64: 5023–5026

Laurie AD, Lloyd-Jones G (2000) Quantification of *phnAc* and *nahAc* in contaminated New Zealand soils by competitive PCR. Appl Environ Microbiol 66: 1814–1817

Leahy JG, Colwell RR (1990) Microbial degradation of hydrocarbons in the environment. Microbiol Rev 54:305–315

Lee SM, Jung JY, Chung YC (2000) Measurement of ammonia inibition of microbial activity in biological wastewater treatment process using dehydrogenase assay. Biotechnol Lett 22:991–994

MacNaughton SJ, Stephen JR, Venosa AD, Davis GA, Chang Y-J, White DC (1999) Microbial population changes during bioremediation of an experiment oil spill. Appl Environ Microbial 65:3566–3577

Magot M, Ollivier B, Patel BKC (2000) Microbiology of petroleum reservoirs. Anton van Leeuwen 77:103–116

Margesin R, Zimmerbauer A, Schinner F (1999) Soil lipase activity – a useful indicator of oil biodegradation. Biotechnol Tech 13:859–863

Margesin R, Zimmerbauer A, Schinner F (2000) Monitoring of bioremediation by soil biological activities. Chemosphere 40:339–346

Mathew M, Obbard JP (2001) Optimisation of the dehydrogenase assay for measurement of indigenous microbial activity in beach sediments contaminated with petroleum. Biotechnol Lett 23:227–230

Mathew M, Obbard JP, Ting YP, Gin YH, Tan HM (1999) Bioremediation of oil contaminated beach sediments using indigenous microorganisms in Singapore. Acta Biotechnol 19:225–233

Mesarch MB, Nakatsu CH, Nies L (2000) Development of catechol 2,3-dioxygenase-specific primers for monitoring bioremediation by competitive quantitative PCR. Appl Environ Microbiol 66:678–683

Middaugh DP, Resnick SM, Lantz SE, Heard CS, Mueller JG (1993) Toxicological assessment of biodegraded pentachlorophenol: microtox and fish embryos. Arch Environ Contam Toxicol 24:165–172

Molin S, Boe L, Jensen LB, Kristensen CS, Givskov M, Ramos JL, Bej AK (1993) Suicidal genetic elements and their use in biological containment of bacteria. Annu Rev Microbiol 47:139–166

Naseby DC, Lynch JM (1997) Functional impact of genetically modified microorganisms on the soil ecosystem. In: Zelikoff JT (ed) Ecotoxicology: responses, biomarkers and risk assessment. SOS, Fair Haven, pp 419–442

Øvreås L (2000) Population and community level approaches for analyzing microbial diversity in natural environments. Ecol Lett 3:236–251

Petrikevich SB, Kobzev EN, Shkidchenko AN (2003) Estimation of hydrocarbon oxidizing activity of microorganisms. Appl Biochem Microbiol 39:19–23

Power M, van der Meer JR, Tchelet R, Egli T, Eggen R (1998) Molecular-based methods can contribute to assessments of toxicological risks and bioremediation strategies. J Microbiol Meth 32:107–119

Ramanathan S, Ensor M, Daunert S (1997) Bacterial biosensors for monitoring toxic metals. Trends Biotechnol 15:500–506

Randall JD, Hemmington BB (1994) Evaluation of mineral agar plates for the enumeration of hydrocarbon-degrading bacteria. J Microbiol Meth 20:103–113

Raymond JW, Rogers TN, Shonnard DR (2001) A review of structure-based biodegradation estimation methods. J Hazard Mater B84:189–215

Ringelberg DB, Talley JW, Perkins EJ, Tucker SG, Luthy RG, Bouwer EJ, Fredrickson HL (2001) Succession of phenotypic, genotypic, and metabolic community characteristics during in vitro bioslurry treatment of polycyclic aromatic hydrocarbon-contaminated sediments. Appl Environ Microbiol 67:1542–1550

Rogers KR, Gerlach CL (1999) Update on environmental biosensors. Environ Sci Technol 33:500A–506A

Ronchel MC, Ramos JL (2001) Dual system to reinforce biological containment recombinant bacteria designed for rhizoremediation. Appl Environ Microbiol 67:2649–2656

Ronchel MC, Ramos-Diaz MA, Ramos JL (2000) Retrotransfer of DNA in the rhizosphere. Environ Microbiol 2:319–323

Rooney-Varga JN, Anderson RT, Fraga JL, Ringelberg D, Lovley DR (1999) Microbial communities associated with anaerobic benzene degradation in a petroleum-contaminated aquifer. Appl Environ Microbiol 65:3056–3063

Rosenberg E (1992) The hydrocarbon-oxidizing bacteria. In: Balows A (ed) The prokaryotes: a handbook on the biology of bacteria: ecophysiology, isolation, identification, applications. Springer, Berlin Heidelberg New York, pp 446–459

Rossel D, Tarradellas J, Bitton G, Morel JL (1997) Use of enzymes in ectotoxicology: a case for dehydrogenase and hydrolytic enzymes. In: Tarradellas J, Bitton G, Rossel D (eds) Soil ecotoxicology. CRC Lewis, Boca Raton, pp 179–192

Shen Y, Stehmeier LG, G Voordouw G (1998) Identification of hydrocarbon-degrading bacteria in soil by reverse sample genome probing. Appl Environ Microbiol 64: 637–645

Shi Y, Zwolinski MD, Schreiber ME, Bahr JM, Sewell GW, Hickey WJ (1999) Molecular analysis of microbial community structures in pristine and contaminated aquifers: field and laboratory microcosm experiments. Appl Environ Microbiol 65:2143–2150

Siciliano SD, Germida JJ, Banks K, Greer CW (2003) Changes in microbial community composition and function during a polyaromatic hydrocarbon phytoremediation field trial. Appl Environ Microbiol 69:483–489

Simpson ML, Sayler GS, Applegate BM, Ripp S, Nivens DE, Pauls MJ, Jellison GE Jr (1998) Bioluminescent-bioreporter integrated circuits from novel whole cell biosensors. Trends Biotechnol 16:332–338

Singh A, Mullin B, Ward OP (2001) Reactor-based process for the biological treatment of petroleum wastes, PN # 200. Proceedings, Middle East Petrotech 2001 Conference, Bahrain, pp 1–13

Smits THM, Röthlisberger M, Witholt B, van Beilen JB (1999) Molecular screening for alkane hydroxylase genes in Gram-negative and Gram-positive strains. Environ Microbiol 1:307–317

Stapleton RD, Ripp S, Jimenez L, Cheol-Koh S, Fleming JT, Gregory IR, Sayler GS (1998) Nucleic acid analytical approaches in bioremediation: site assessment and characterization. J Microbiol Meth 32:165–178

Stapleton RD, Bright NG, Sayler GS (2000) Catabolic and genetic diversity of degradative bacteria from fuel-hydrocarbon contaminated aquifers. Microb Ecol 39:211–221

Sticher P, Jasper MCM, Stemmler K, Harms H, Zehnder AJB, van der Meer JR (1997) Development and characterization of a whole cell bioluminescent sensor for bioavailable middle-chain alkanes in contaminated groundwater samples. Appl Environ Microbiol 63:4053–4060
Stotsky JB, Atlas RM (1994) Frequency of genes in aromatic and aliphatic hydrocarbon biodegradation pathways within bacterial populations from Alaskan sediments. Can J Microbiol 40:981–985
Telang AJ, Voordouw G, Ebert S, Sifeldeen N, Foght JM, Fedorak PM, Westlake DWS (1994) Characterization of the diversity of sulfate-reducing bacteria in soil and mining wastewater environments by nucleic acid hybridization techniques. Can J Microbiol 40:955–964
Telang AJ, Ebert S, Foght JM, Westlake DWS, Jenneman GE, Gevertz D, Voordouw G (1997) Effect of nitrate injection on the microbial community in an oil field as monitored by reverse sample genome probing. Appl Environ Microbiol 63:1785–1793
Telang AJ, Ebert S, Foght JM, Westlake DWS, Voordouw G (1998) Effects of two diamine biocides on the microbial community of an oil field. Can J Microbiol 44:1060–1065
Theron J, Cloete TE (2000) Molecular techniques for determining microbial diversity and community structure in natural environments. Crit Rev Microbiol 26:37–57
Thomassin-Lacroix EJM, Yu Z, Eriksson M, Reimer KJ, Mohn WW (2001) DNA-based and culture-based characterization of a hydrocarbon-degrading consortium enriched from Arctic soil. Can J Microbiol 47:1107–1115
Thouand G, Bauda P, Oudot G, Kirsch G, Sutton C, Vidalie JF (1999) Laboratory evaluation of crude oil biodegradation with commercial or natural microbial inocula. Can J Microbiol 45:106–115
Thurmann U, Zanto C, Schmitz C, Vomberg A, Püttman W, Klinner U (1999) Correlation between microbial ex situ activities of two neighboring uncontaminated and fuel oil contaminated subsurface sites. Biotechnol Tech 13:271–275
Torstensson L (1997) Microbial assays in soils. In: Taradellas J, Bitton G, Rossel D (eds) Soil ecotoxicology. CRC Lewis, Boca Raton, pp 207–233
Van Elsas JD, Duarte GF, Rosado AS, Smalla K (1998) Microbiological and molecular biological methods for monitoring microbial inoculants and their effects in the soil environment. J Microbiol Meth 32:133–154
Van Emon JM, Gerlach CL (1998) Environmental monitoring and human exposure assessment using immunochemical techniques. J Microbiol Meth 32:121–131
Van Hamme JD, Ward OP (2000) Development of a method for the application of solid-phase microoextraction to monitor biodegradation of volatile hydrocarbons during bacterial growth on crude oil. J Ind Microbiol Biotechnol 25:155–162
Van Hamme JD, Odumeru JA, Ward OP (2000) Community dynamics of a mixed-bacterial culture growing on petroleum hydrocarbons in batch culture. Can J Microbiol 46:411–450
Van Straalen NM (2002) Assessment of soil contamination-a functional perspective. Biodegradation 13:41–52
Venosa AD, Haines JR, Nisamaneepong W, Govind R, Pradhan S, Siddique B (1992) Efficacy of commercial products in enhancing oil biodegradation in closed laboratory reactors. J Ind Microbiol 10:13–23
Vomberg A, Klinner U (2000) Distribution of *alkB* genes within n-alkane degrading bacteria. J Appl Microbiol 89:339–348
Voordouw G, Shen Y, Harrington CS, Telang AJ, Jack TR, Westlake DWS (1993) Quantitative reverse sample genome probing of microbial communities and its application to oil-field production waters. Appl Environ Microbiol 59:4101–4114
Voordouw G, Armstrong SM, Reimer MF, Fouts B, Telang AJ, Shen Y, Gevertz D (1996) Characterization of 16S rRNA genes from oil field microbial communities indicates

the presence of a variety of sulfate-reducing, fermentative, and sulfide-oxidizing bacteria. Appl Environ Microbiol 62:1623–1629

US EPA (1999) Guide for Industrial waste management. United States Environmental Protection Agency, EPA 530-R-99-001

Ward OP, Singh A (2000) Biodegradation of oil sludge. Canadian Patent #2,229,761

Ward OP, Singh A, Van Hamme J (2003) Accelerated biodegradation of petroleum hydrocarbon waste. J Ind Microbiol Biotechnol 30:260–270

White DC, Flemming CA, Leung KT, Macnaughton SJ (1998) In situ microbial ecology for quantitative appraisal, monitoring, and risk assessment of pollution remediation in soils, the subsurface, the rhizosphere and in biofilms. J Microbiol Meth 32: 93–105

Widada J, Nojiri H, Omori T (2002a) Recent developments in molecular techniques for identification and monitoring of xenobiotic-degrading bacteria and their catabolic genes in bioremediation. Appl Microbiol Biotechnol 60:45–59

Widada J, Nojiri H, Kasuga K, Yoshida T, Habe H, Omori T (2002b) Molecular detection and diversity of polycyclic aromatic hydrocarbon-degrading bacteria isolated from geographically diverse sites. Appl Microbiol Biotechnol 58:202–209

Willardson BM, Wilkins JF, Rand TA, Schupp JM, Hill KK, Keim P, Jackson PJ (1998) Development and testing of a bacterial biosensor for toluene based environmental contaminants. Appl Environ Microbiol 64:16–1012

Zelles L (1997) Phospholipid fatty acid profiles in selected members of soil microbial communities. Chemosphere 35:275–294

Zelles L (1999) Fatty acid patterns of phospholipids and lipopolysaccharides in the characterization of microbial communities in soil: a review. Biol Fertil Soils 29: 111–129

Subject Index

Printing: Mercedes-Druck, Berlin
Binding: Stein+Lehmann, Berlin